WHALES, DOLPHINS & PORPOISES

EDITED BY ANNALISA BERTA

WHALES, DOLPHINS & PORPOISES

A NATURAL HISTORY AND SPECIES GUIDE

THE UNIVERSITY OF CHICAGO PRESS

Chicago

Annalisa Berta has been Professor of Biology at San Diego State University, California, for more than 30 years specializing in the anatomy and evolutionary biology of marine mammals. Past President of the Society of Vertebrate Paleontology and co-Senior Editor of the *Journal of Vertebrate Paleontology*, Berta has authored and co-authored numerous scientific articles and several books.

The University of Chicago Press, Chicago 60637
The University of Chicago Press, Ltd., London
© 2015 by Ivy Press Limited
All rights reserved. Published 2015.
Printed in China

24 23 22 21 20 19 18 17 16 15   1 2 3 4 5

ISBN-13: 978-0-226-18319-0 (cloth)
ISBN-13: 978-0-226-18322-0 (e-book)
DOI: 10.7208/chicago/978-0226183220.001.0001

Library of Congress Cataloging-in-Publication Data

Berta, Annalisa, author.
  Whales, dolphins, and porpoises : a natural history and species guide / Annalisa Berta.
      pages cm
  Includes bibliographical references and index.
  ISBN 978-0-226-18319-0 (cloth : alk. paper) — ISBN 978-0-226-18322-0 (ebook)
1. Whales—Anatomy. 2. Dolphins—Anatomy. 3. Porpoises—Anatomy. 4. Whales—Behavior. 5. Dolphins—Behavior. 6. Porpoises—Behavior. I. Title.
  QL737.C4B6515 2015
  599.5—dc23
                                                    2015008715

Color origination by Ivy Press Reprographics

JACKET AND LITHOCASE IMAGES
Nature Picture Library/Martin Camm (WAC): Andrews' beaked whale, Arnoux's beaked whale, Atlantic humpback dolphin, beluga, blue whale, Ganges River dolphin, gray whale, Irrawaddy dolphin, narwhal, pantropical spotted dolphin, Peale's dolphin, short-beaked common dolphin, southern bottlenose whale, southern right whale, spectacled porpoise, sperm whale, striped dolphin.
Nature Picture Library/Rebecca Robinson: Australian snubfin dolphin, Guiana dolphin.
Sandra Pond: Commerson's dolphin, Indo-Pacific finless porpoise.

This book was conceived, designed, and produced by
Ivy Press
210 High Street, Lewes
East Sussex BN7 2NS
United Kingdom
www.ivypress.co.uk

Publisher SUSAN KELLY
Creative Director MICHAEL WHITEHEAD
Editorial Director TOM KITCH
Senior Project Editor CAROLINE EARLE
Commissioning Editor KATE SHANAHAN
Design J C LANAWAY

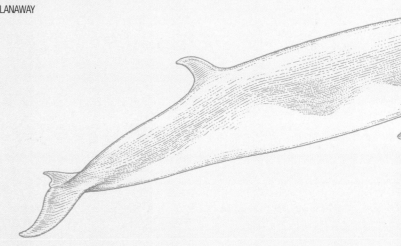

CONTENTS

Introduction

Whales, dolphins, and porpoises, also known as cetaceans, include 90 currently recognized living species. Although some cetacean species are on the brink of extinction, there are also exciting discoveries of new species. This guide is intended to introduce the reader to the identification and biology of these magnificent and charismatic mammals of the sea.

Part One of this guide includes information about cetacean biology. The Phylogeny & Evolution section highlights where whales originated and how they evolved and diversified from the tropics to polar waters. The Anatomy & Physiology section includes key features of the head, body, and appendages (fins, flippers, and flukes) that enable a fully aquatic life—emphasizing a few novel adaptations, such as high-frequency sound production and reception in some whales. These adaptations provide a historical framework for understanding how these mammals make a living today and guide our efforts in their conservation. The Behavior section highlights the social organization of cetaceans ranging from solitary species to the highly complex societies of some toothed whales. Cetaceans have evolved to feed on a diverse prey. Whales feed on aggregations of zooplankton averaging less than an inch (1-2 mm) in length to large squid 10 ft (3 m) or more in length.

Gregariousness
Groups of common bottlenose dolphins often travel together exhibiting playful behavior.

The section on Food & Foraging identifies how whales locate and catch their prey using techniques that range from the pursuit of individual fish to bulk feeding of large aggregations of zooplankton. The Life History section highlights the growth, reproduction, and survival of cetaceans including techniques for determining the age of whales. The reproductive biology of cetaceans reveals that many species do not reproduce annually, which is a key factor guiding our conservation efforts. The Range & Habitat sections reveal how new techniques such as digital devices and satellite telemetry track the location, movement patterns, and ranges of cetaceans. The Conservation & Management section discusses the status of some endangered species, major threats, and notable conservation actions designed to protect cetacean species.

Part Two of this guide includes Identification Tools & Maps, which provides keys to the identification of whales, dolphins, and porpoises using distinctive body features—such as size, color, and markings, and fluke and flipper shapes. There are many ways to watch cetaceans—from the air, on land, and at sea. Many display distinctive surface behaviors described in this section such as leaping out of the water, which aids in their identification. Another section describes whale watching, which brings people in close contact with whales, covering the gear involved as well as some top viewing locations around the world. Checklists provide species assemblages encountered in different regions of the world.

The largest section of this guide, **Part Three**, is the Species Directory (see pages 62–275). This is followed by several appendices including a classification of cetaceans list, glossary of commonly used terms, and an index. We hope that you are inspired to find, recognize, watch, and appreciate whales, dolphins, and porpoises. Their future and ultimately our own depends on our abilities and efforts to conserve and protect the world's oceans and its inhabitants.

Breaching
This image shows a humpback whale displaying a typical "breaching" behavior in which whales, dolphins, and some porpoises leap out of the water. There are a number of possible explanations for this behavior including signaling, dominance, or warning other whales of danger.

THE BIOLOGY

Phylogeny & Evolution

The majority of marine mammals belong to the order Cetacea, which includes whales, dolphins, and porpoises. The name Cetacea comes from the Greek *ketos* meaning whale. Two major groups of extant whales are recognized—toothed whales (Odontoceti) and baleen whales (Mysticeti). Toothed whales are much more diverse with 10 families, 34 genera, and 76 extant species (one of which is likely extinct) compared to mysticetes that include 4 families, 6 genera, and 14 extant species. Toothed cetaceans include sperm whales, oceanic whales, river dolphins, monodontids (beluga and narwhal), ocean dolphins, and porpoises. Baleen whales include the right whales, pygmy right whale, gray whale, and blue, fin, sei, Bryde's, humpback, minke, Antarctic minke, and the recently described Omura's whale.

Evolutionary relationships of whales

Cetaceans originated from land mammals and there is strong support for whales being most closely related to artiodactyls (even-toed ungulates), which include cows, goats, camels, and hippos. Because cetaceans and artiodactyls are linked, they are grouped together in the clade Cetartiodactyla. Within cetaceans, relationships among families are still debated. There is general agreement from both molecular and anatomical data for the family level evolutionary history among odontocetes. Basal odontocetes include sperm whales (Physeteridae and Kogiidae). Asiatic river dolphins, Platanistidae, are the next diverging lineage followed by beaked whales (Ziphiidae), Chinese river dolphin (Lipotidae), South American river dolphins (Iniidae and Pontoporiidae),

and the latest most recent divergent lineage the Monodontoidea—beluga and narwhal (Monodontidae) and porpoises (Phocoenidae), and oceanic dolphins (Delphinidae).

Unlike odontocetes, higher-level relationships among mysticetes conflict based on molecular (i.e. DNA sequences) versus anatomical data. Using molecular data right whales and the bowhead (Balaenidae) are recognized as basal mysticetes whereas anatomical data positions pygmy right whales (Neobalaenidae) as sister to Balaenidae followed by an alliance of the remaining baleen whales: rorquals (Balaenopteridae) and gray whales (Eschrichtiidae). The position of the gray whale is also debated. Anatomical data places Eschrichtiidae and Balaenopteridae as close relatives whereas molecular data nests gray whales within balaenopterids.

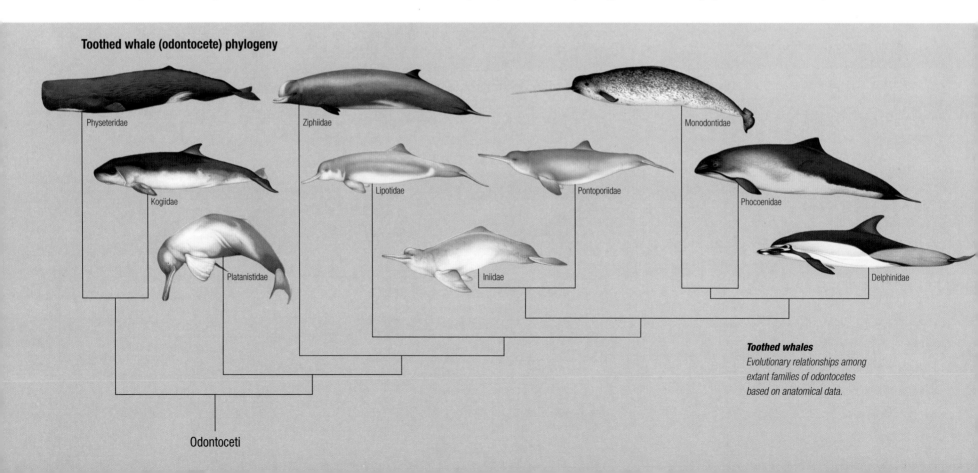

Toothed whale (odontocete) phylogeny

Physeteridae

Ziphiidae

Monodontidae

Kogiidae

Lipotidae

Pontoporiidae

Phocoenidae

Platanistidae

Iniidae

Delphinidae

Toothed whales
Evolutionary relationships among extant families of odontocetes based on anatomical data.

Odontoceti

Whale origins

Cetaceans are first found in the fossil record approximately 52.5 million years ago (MYA) during the early Eocene period in current-day India and Pakistan. Recent discoveries in Pakistan and southern India have suggested that extinct artiodactyls, the raoellids, such as *Indohyus*, are the closet extinct relatives of whales. *Indohyus* was a cat-sized animal with a long nose, tail, and slender limbs. At the end of each limb were four to five toes that ended in hooves, similar to those of deer. Raoellids also had very thick dense limb bones, an adaptation for buoyancy control. Since raoellids were largely aquatic, this indicates that an aquatic lifestyle arose before whales evolved.

Whale relatives
Life restoration of the closest whale relative, the aquatic, deer-like raoellid Indohyus.

Baleen whale (mysticete) phylogeny

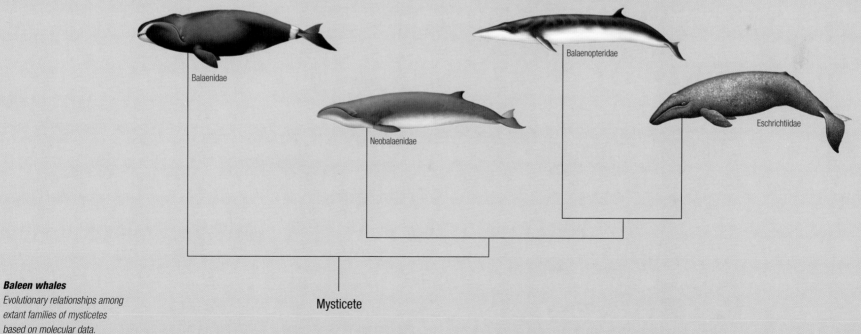

Balaenidae

Balaenopteridae

Neobalaenidae

Eschrichtiidae

Mysticete

Baleen whales
Evolutionary relationships among extant families of mysticetes based on molecular data.

Phylogeny & Evolution

Early whales had legs

The earliest stem cetaceans—such as Pakicetidae (e.g. *Pakicetus*), Ambulocetidae (e.g. *Ambulocetus*), and Remingtonocetidae (e.g. *Kutchicetus*)—are all known from the early and middle Eocene (50 MYA) of current-day India and Pakistan. They are all thought to have been semiaquatic, able to move on land as well as in the water. These stem whales had well-developed forelimbs and hind limbs. Wear on the teeth is consistent with a fish-eating habit. The occurrence of later diverging whales (such as Protocetidae, e.g. *Rodhocetus*), in Asia, Africa, Europe, and North America) indicates that cetaceans had spread across the globe between 49–42 MYA. They differed from other early cetaceans in having large eyes with the nasal opening that had migrated further back on the skull. Basilosaurids (such as *Dorudon*)—the closest relatives of modern cetaceans—were widely distributed and lived between 41–35 MYA. Best known is *Basilosaurus isis*, which had a snake-like body with a maximum length of 56 ft (17 m), with several hundred skeletons reported from the middle-Eocene Valley of Whales in north-central Egypt.

Fossil whales

This diagram shows the time ranges of some fossil and extant lineages of whales.

Fossil whales and their closest relative

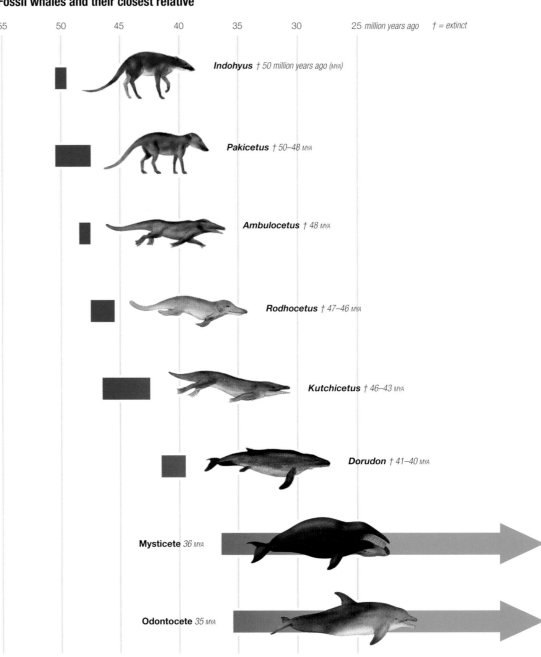

55   50   45   40   35   30   25 *million years ago*   † = *extinct*

Indohyus † *50 million years ago* (MYA)

Pakicetus † *50–48* MYA

Ambulocetus † *48* MYA

Rodhocetus † *47–46* MYA

Kutchicetus † *46–43* MYA

Dorudon † *41–40* MYA

Mysticete *36* MYA

Odontocete *35* MYA

Modern whales

Modern (crown) cetaceans originated from archaic (stem) cetaceans, such as *Basilosaurus*, approximately 33.7 MYA during the Oligocene period. The diversification of modern cetaceans (Neoceti) has been associated with the breakup of the southern continents and restructuring of ocean circulation patterns—signaled by higher oxygen isotope levels—which resulted in increased food production (indicated by diatoms, a type of tiny algae) and upwelling of nutrient-rich water.

Crown cetaceans differ from stem cetaceans in having a telescoped skull. In "telescoping" the bones of the rostrum are extended and displaced posteriorly and the nostrils have moved to the top of the head where they form the blowholes (see page 16).

Odontocetes differ from mysticetes by the presence of teeth. Odontocetes acquired echolocation, which enabled them to produce high-frequency sounds that are reflected from objects that surround them—these reflections allow them to pursue individual prey items. Mysticetes acquired a novel feeding mechanism, bulk filter-feeding using baleen plates, strainers in the mouth.

Although most whales today are entirely marine, early fossil members of this lineage, such as pakicetids, likely foraged exclusively in freshwater based on analysis of the carbon and oxygen isotope levels of their teeth and bones.

Whale diversification

Mysticete diversity
(number of species)

Odontocete diversity
(number of species)

Diatom (algae) diversity
(number of species)

Ocean temperature $\delta^{18}O$ (‰)
(oxygen isotope values)

Time (MYA)   25   20   15   10   5   0

Diversity, food, and ocean temperature
Whale diversity is linked to an increase in food production driven by climatic changes (such as ocean temperature). Differences in oxygen isotope values reveal temperature changes in the geologic past.

Phylogeny & Evolution

The oldest named odontocetes are from the North Atlantic (North America). A recently discovered stem odontocete, *Cotylocara*, has dense bones and air sinuses, features that support the theory that echolocation originated early— between 32–35 MYA. Crown odontocetes, or modern families, diversified in the Miocene, approximately 23–26 MYA. Extant genera of both mysticetes and odontocetes appeared during the Pleistocene, approximately 1.6 MYA. Analysis of the morphology and evolutionary relationships of river dolphins supports the hypothesis that marine odontocetes invaded river systems on multiple occasions. The range and habitat of some whales is much different today than in the past. For example, distant relatives of the South American La Plata River dolphin had a broader range in the past that included southern California. A similar range expansion is indicated for fossil relatives of the beluga that today occupies Arctic waters but inhabited temperate waters as far south as Baja California in the Miocene.

Several fossil odontocetes exhibit unique feeding adaptations. An extinct relative of monodontids (narwhal and beluga), *Odobenocetops* lived in Peru during the early Pliocene (3–4 MYA). The presence of tusks and a presumed mollusk-eating suction feeding habit are convergences (similarities based on ecology rather than relationship) with the walrus. A recently described fossil porpoise, *Semirostrum ceruttii* from the Pliocene of California is reconstructed to have employed a form of benthic skim-feeding by using its lower jaw, which extended further beyond the rostrum than in any other known mammal, to probe for and obtain prey.

Odobenocetops

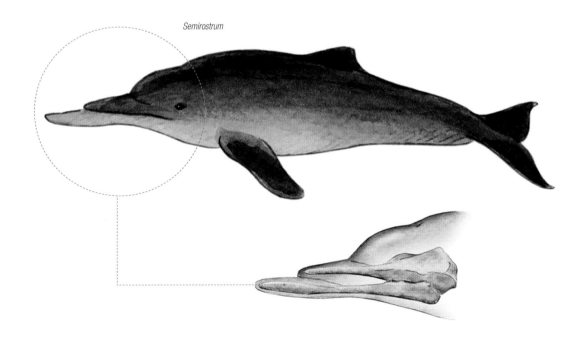

Semirostrum

Feeding specializations

An extinct fossil relative of the beluga and narwhal, Odobenocetops *spp. (top) had large, down-turned tusks and a blunt snout. Strong muscle scars at the front of the rostrum, a vaulted palate, and absence of teeth suggests that they fed on benthic mollusks using suction. The extinct skimmer porpoise,* Semirostrum ceruttii *(bottom), had an elongated lower jaw extending well beyond the rostrum that it may have used to probe or skim along the seafloor for prey.*

Aetiocetus

Fossil baleen whales

The stem whale, Aetiocetus weltoni, *that lived 24–28 MYA may have been the earliest bulk filter-feeder employing both teeth and baleen in prey capture. The fossil whale species,* Herpetocetus morrowi, *described from southern California is one of the smallest baleen whales with a length of 14¾ ft (4.75 m).*

Herpetocetus

The earliest named mysticetes are from the South Pacific (Australia and New Zealand). These stem mysticetes, some of which were of large size ranging from 16–40 ft (5–12 m) such as *Llanocetus*, possessed well-developed teeth with multiple accessory cusps and likely hunted individual prey. Other fossil taxa, such as the *Aetiocetus weltoni*, were smaller-bodied and may have had both teeth and baleen employed in batch filter-feeding as seen in modern baleen whales.

The earliest known and earliest diverging toothless fossil mysticetes, the eomysticetids, were relatively large bodied at around 33 ft (10 m) in length, with long skulls. They appeared in the Oligocene in both the North and South Pacific and were contemporaneous with some stem-toothed mysticetes. Although the ancestral feeding strategy among crown mysticetes is debated, functional analysis of the late Pliocene (2.5–3.5 MYA) mysticete, *Herpetocetus morrowi*, suggests a lateral suction-feeding strategy similar to but evolved independently from feeding in living gray whales. As was the case for odontocetes, crown mysticetes underwent an explosive radiation in the Miocene. Whale diversity peaked in the late middle Miocene (14 MYA) and fell thereafter, yielding a modern fauna that is much less diverse today than in the past.

Anatomy & Physiology

Cetaceans display considerable diversity in size. Included among the mysticetes or baleen whales are some of the largest species such as the blue whale, the largest animal on earth at 110 ft (33 m) long and weighing 330,000 lb (150,000 kg). Odontocetes show a wider range of sizes, from the sperm whale that is as large as some baleen whales to the vaquita that is about 4¾ ft (1.4 m) in length, weighing 92 lb (42 kg). In odontocetes or toothed whales, males are typically larger, whereas in mysticetes, females are generally larger than males. Since most mysticetes depend upon stored body fat to support their metabolic requirements, particularly during the winter months far from feeding grounds, the extra weight is necessary for their survival, promoting greater reproductive success and aiding females in the nursing of their offspring.

Adaptations

Cetaceans exhibit numerous adaptations for a fully aquatic life. Breathing occurs through blowholes that have migrated to the top of the head. Odontocetes have only a single blowhole instead of the two blowholes of mysticetes. The heads of mysticetes are very large, up to one-third of the body length.

The vertebral region does not contain a sacral region in whales because the pelvic girdle is absent. External hind limbs are very reduced or absent in cetaceans and the forelimbs have been modified into flippers or pectoral fins with an inflexible elbow that functions in steering. The broad flippers of some mysticetes, such as right and bowhead whales, aid in slow turns. The flippers of the humpback are exceptionally long and maintain hydrodynamic efficiency; they are also "waved" during feeding and social displays. The flippers of most odontocetes assist in turning during high-speed maneuvers while chasing prey. In odontocetes that occupy pack ice or rivers, such as the beluga or river dolphins, flipper shapes allow for angled maneuvers in those environments.

Size comparison among whales

Blue whale

Sperm whale

Body sizes

Cetaceans differ considerably in their size. Their size categories compared to a human for scale include: small, up to 10 ft (3 m); intermediate, 0–33 ft (3–10 m); or large, more than 33 ft (10 m). The majority of species (47) are small, 31 species are intermediate, and only 11 species are large (see pages 50–55).

Diver

Vaquita

Humpback whale

50 ft (15.25 m)

Anatomy & Physiology

Anatomy of a mysticete and odontocete

The whale's skeleton shows numerous adaptations for life in the water. The forelimbs are reduced and flattened into paddles. The elbow joint is immobile and since it is enclosed in the flipper the forelimb is used mostly for steering. The finger bones are lengthened by additional bony elements that serve to increase the surface area of the flipper. The hind limbs are reduced to a few vestigial bones embedded in muscle. The vertebral column has large spines to anchor the powerful fluke muscles that provide propulsion. Some or all of the neck vertebrae (seen in the bowhead, right) are fused; this inhibits neck mobility, which is important in maintaining hydrodynamic efficiency. The dorsal fin (seen in the dolphin below) is similar to the fluke in its lack of bony support and connective tissue composition.

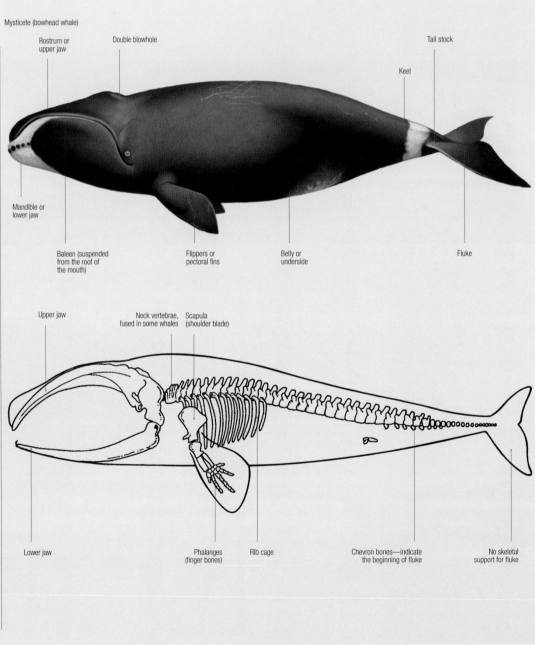

Mysticete (bowhead whale)

Rostrum or upper jaw

Double blowhole

Tail stock

Keel

Mandible or lower jaw

Baleen (suspended from the roof of the mouth)

Flippers or pectoral fins

Belly or underside

Fluke

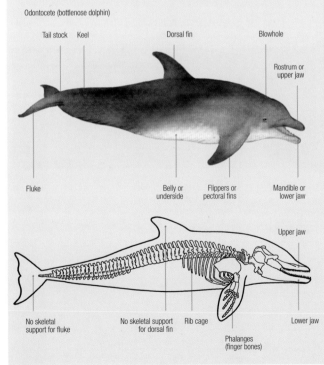

Odontocete (bottlenose dolphin)

Tail stock   Keel

Dorsal fin

Blowhole

Rostrum or upper jaw

Fluke

Belly or underside

Flippers or pectoral fins

Mandible or lower jaw

Upper jaw

No skeletal support for fluke

No skeletal support for dorsal fin

Rib cage

Phalanges (finger bones)

Lower jaw

Upper jaw

Neck vertebrae, fused in some whales

Scapula (shoulder blade)

Lower jaw

Phalanges (finger bones)

Rib cage

Chevron bones—indicate the beginning of fluke

No skeletal support for fluke

Generalized dolphin's head

Sound production and reception

The dolphin's sound production and hearing systems have undergone extensive modification to enable them to perceive and interpret underwater sound. Sounds are produced by the movement of air between the phonic lips. The opening and closing of the phonic lips breaks up the air flow and produces pulsed sounds or clicks. The melon acts as an acoustic lens to focus sounds into the water. The external ears have disappeared and new pathways of sound reception to the inner ear have evolved, including the fat-filled channels of the lower jaw.

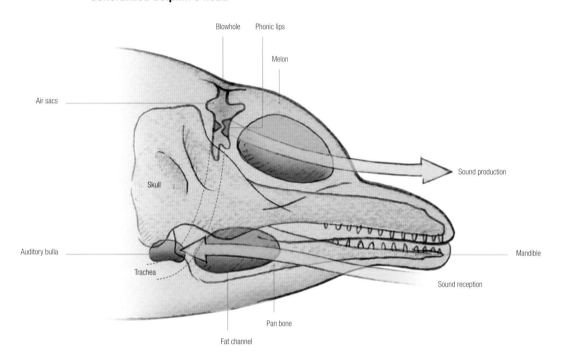

The muscular horizontal tail or fluke lacks bony support and is composed of tough, fibrous connective tissue. The tail provides propulsion by vertical movement. Tail shape differs among cetaceans and most provide increased efficiency at high speeds (see Identification). To help minimize drag in the water many smaller odontocetes, for example delphinids, move at high speed and leap and glide (porpoising) above the water's surface. Most whales have a dorsal fin (see also Identification) that provides stability and balance.

The fins, flippers, and flukes of cetaceans have arteries and veins that pass close to one another in opposite directions and function as radiators (counter current exchangers) to control heat balance. The skin of cetaceans is generally smooth and rubbery to the touch. Hair is absent except for sparse bristles (vibrissae) found on the head of some species. The blubber layer, thickest in large baleen whales, enhances streamlining and provides insulation and energy storage.

Mysticetes don't have teeth as adults and have evolved novel feeding structures—baleen plates composed of keratin (the material that makes up the hair, claws, and fingernails of mammals)—that hang down from the upper jaw and strain bulk prey, for example, krill. Rorquals, such as fin whales, can engulf a volume of water that is greater than their body mass. For example, fin whales have been reported to engulf 18,000 gallons (70,000 liters) of water in each gulp, containing 22 lb (10 kg) of krill. Expansion of the mouth and throat in mysticetes is facilitated by external throat grooves or pleats below the mouth and throat. Most odontocetes, especially those with diets of schooling fish, employ their many teeth to obtain prey. Odontocetes are active hunters and pursue prey using echolocation, in which high-frequency sounds are emitted from "phonic lips" near the blowhole. Sounds are focused by the melon, a fatty structure on the top of the head, and returning echoes pass through and under fat bodies on the lower jaw before being transmitted to the ears.

Beaked and sperm whales have fewer or no teeth and are deep divers, feeding primarily on squid. Cuvier's beaked whales hold the diving record and are the longest- and deepest-diving vertebrates, with dives lasting 137 minutes to depths of more than 1.86 miles (2,992 m) on a single breath. Deep-diving cetaceans exhibit a variety of circulatory and respiratory modifications including high blood volume, flexible ribs, and tolerance of complete lung collapse. By contrast with human lungs, large oxygen stores are located in the muscles and blood.

A feature of some whale brains is their large size, especially the cerebrum, the front portion of the brain responsible for movement and mental functions. Brain size relative to body size is large in odontocetes. In comparison to other similar-sized animals most odontocetes have brains that are four to five times larger. Only the human brain is proportionally larger. The high brain:body size ratio of odontocetes, such as dolphins and killer whales, are partly explained by their complex social structure and behavior.

Behavior

"Behavior" refers to the ways that organisms respond to each other and to environmental cues. Like other mammals, cetacean behavior is driven by the need to obtain food, remain safe from predators, find mates, and rear offspring. However, what makes cetaceans unique is that all of these activities are conducted underwater, yet they are tied to the surface to breathe. This has led to some unique adaptations, particularly in foraging behavior (see pages 26–31). Unraveling behavior in animals that spend the majority of their lives underwater can be a challenging task, but as long-term studies continue and technology advances, we are gaining a greater understanding of the intricacies of cetacean behavior.

Grouping behavior

Cetaceans are social animals and so it is important to understand how their behaviors occur within the context of associating with other individuals. The gregariousness of cetaceans varies along a spectrum of being relatively solitary to highly social. In the case of some pelagic dolphins, such as striped and common dolphins, groups may contain hundreds or even thousands of individuals. In general, odontocetes tend to be more gregarious than mysticetes (baleen whales). This is due in part to the larger size of mysticetes, decreased predation risk, and the need to reduce competition for food. However, there are exceptions, as humpback whales can be quite gregarious and river dolphins may be relatively asocial. It is also possible that species typically viewed as being solitary may in fact be socializing acoustically across large distances via low-frequency vocalizations, such as in blue whales.

Anti-predation behavior

The group is the primary line of defense against predators. In the ocean, there is no place to hide but individuals can gain safety in numbers—by forming groups. Group members may also actively deter predators, for example by mobbing a predator or arranging themselves in defensive formations.

Marguerite formation

Adult sperm whales use the marguerite formation to protect vulnerable calves. Calves are positioned in the center with the females' flukes radiating outward. If needed, the flukes will be slapped against the surface of the water to deter predators such as killer whales.

Costs and benefits of group-living

Many delphinid species solve the problem of balancing the costs and benefits of living together by living in ever-changing groups of varying size and composition. In societies that exhibit these fission–fusion dynamics, individuals join and split from groups according to changing situations. For example, when large aggregations of schooling prey are present, individuals may work together to corral the prey and herd it toward the surface (see pages 28–31). When predation risk is high, group size may increase. Alternatively, when prey are less abundant or predation risk is low, group size may decrease. Group living may also be affected by habitat, with the largest groups generally found in open pelagic waters and smaller groups generally found in more enclosed bays and rivers.

Costs
- Increased competition for food
- Increased competition for mates
- Increased aggression
- Increased spread of disease and parasites
- Increased detectability by predators

Benefits
- Protection from predators ("safety in numbers")
- Cooperation to find and secure prey
- Assistance with calf-rearing
- Offspring socialization
- Increased access to mates
- Kin selection (see page 23)

Grouping patterns
Some dolphins have different grouping patterns during the day and night. Hawaiian spinner dolphins off the Kona Coast form groups of 200–400 individuals at night when feeding offshore. During the day, they split into smaller groups of 20–100 individuals when resting in nearshore bays.

Behavior

Mating behavior

Most cetaceans have a multi-mate mating strategy, where males and females both have several mating partners during the course of one breeding season. Females typically do not exhibit aggressive or competitive behaviors but they often exhibit mate choice by enticing males into a chase or competition. This may provide the female with the opportunity to compare males and judge the fitness of each. In contrast, males often compete with one another, either directly or indirectly. Humpback whales have a lekking mating system where several males jockey for the position closest to a receptive female. Males may attempt to deter one another through astonishing displays of breaching, flipper-slapping, and lobtailing (see pages 58–59). Sperm competition occurs in some cetacean species, as indicated by males' disproportionately large testes as compared to their body weight, such as in right whales. This enables a male to "swamp" females with his genetic material so as to improve his chances of siring offspring. Although many multi-mate strategies involve male competition, males in some species cooperate with one another. For example, bottlenose dolphin males work together by forming strong alliances in order to gain access to receptive females. In species where female groups are more widely dispersed, such as in sperm whales, males mate by roving between groups of females.

Humpback whale heat run

During a "heat run" several humpback whale males compete with each other to mate with the female. The heat run becomes increasingly aggressive as they compete for the attention of the female.

Heat run

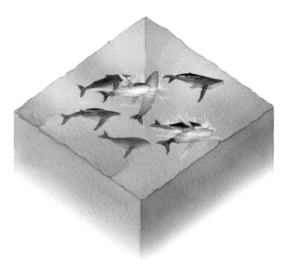

1. Competitive group forms

A heat run occurs when male humpback whales compete for mating access to a receptive female. They vie for position close to the female.

2. Activity level intensifies

The female swims in front of the males. Some males attempt to stop others approaching the female by performing rolls and breaches.

3. Bubble streams

The entire group dives underwater. The males emit bubble streams from their blowholes as a form of aggression.

4. Ramming

Males ram and head butt each other. At the end of the heat run, one male often swims away with the female. Copulation is rarely seen in humpback whales, but it is thought mating occurs at this time.

Parental behavior

Cetacean calves are precocial, or well developed at birth. They are able to breathe and move freely on their own straight from birth, but nurse, and are dependent on their mothers for periods ranging from less than a year (as in most mysticetes) to 3 years (as in pilot whales) or even 13 years (as in sperm whales). Typically, females are the sole care-givers for the young. After weaning, most calves leave their natal group and birth area. However, pilot whales and resident killer whales have natal philopatry whereby neither sex disperses from the natal group. Individuals in these matrilineal societies are closely related and inbreeding is avoided through mating interactions with other groups. Importantly, kin selection may influence how

individuals behave in these societies. For example, a female killer whale may assist her son in foraging to help ensure his survival and therefore increase the probability that her genes will perpetuate into future generations.

Females of some species may form nursery groups and help in rearing one another's offspring. This is known as allomaternal care and it is well-documented in bottlenose dolphins, sperm whales, and pilot whales. If the females are related, kin selection is again at play. A female that has a "babysitter" may have an energetic advantage as she is able to dive longer and deeper in search of prey while leaving her calf—with its less-developed diving abilities— at the surface with other group members. Offspring

socialization is also an important benefit of these nursery groups as it allows youngsters to learn the skills necessary to integrate into the social fabric of the pod in order to become a successful hunter and breeder as an adult.

Mom-calf pairs

Two presumed mom-calf striped dolphin pairs surface in the Gulf of Corinth, Greece. Each calf swims in infant position, alongside the mother and behind her dorsal fin. In this region, mom–calf pairs may form subgroups within larger mixed sex groups.

Behavior

Foraging behavior

Cetacean foraging behavior has evolved to overcome the challenge of diving to depth to forage while being tied to the surface to breathe. In addition to a number of anatomical and physiological adaptations that have evolved to solve this problem (see pages 16–19), some cetaceans work with their group members to find and secure prey. Coordinated or cooperative foraging has been well-documented in several species, including killer whales, humpback whales (see pages 26–29), dusky dolphins, and bottlenose dolphins. Individuals may cooperate to find prey by spreading out in broad ranks to search for prey and performing acoustic or visual displays to alert other group members when prey are found. Cooperation may also occur to secure prey, for example by swimming around schooling fish to corral them into prey balls—for example in dusky dolphins and common dolphins—or driving prey against barriers—such as bubble curtains, mud plumes, and mud banks.

Studying behavior

Ethology is the study of behavior, and in cetaceans this can be both rewarding and challenging. It requires patience and long hours at sea, and often involves the use of sophisticated instruments. The method or technique used for collecting behavioral data depends on the question being asked (see box opposite). Regardless of the technique, a fundamental component of all behavioral studies is the ethogram (see opposite).

Corralling prey

Long-beaked common dolphins off Port St. Johns, KwaZulu Natal, South Africa work together to corral sardines into tight prey balls and herd them toward the surface. Individual dolphins take turns feeding on the prey ball, corralling it, and surfacing to breathe. This behavioral strategy helps to solve the challenge of the air supply being physically separated from the food supply. Other predators such as cooper sharks, dusky sharks, and Cape gannets are also attracted to these large prey balls.

Collecting behavioral data

Examples of questions that could be asked about cetacean behavior and a corresponding technique that could be used to answer that question.

Question	Method
Do individuals preferentially associate with other individuals in the population?	Boat-based behavioral observation with paper-and-pencil or digital datasheets and photo-identification (see below)
How does mother-calf distance change with calf age?	Land-based behavioral observation with binoculars, spotting scope, and theodolite (a surveying instrument)
How do females exert mate choice?	Underwater videography while snorkeling
How do diving behaviors vary according to age and sex class?	Attachment of short-term (less than 24 hours) suction cup tags with a time-depth recorder

Ethogram: dusky dolphin behavioral states

Ethograms provide a systematic way to identify and record behavior. They may contain behavioral states (broad-level behaviors of long duration) and events (fine-level behaviors of short duration).

Foraging	Searching for or consuming prey, as indicated by long, deep dives followed by loud forceful exhalations ("chuffs"), and directionless movement—may include coordinated "burst swims" (rapid bursts of speed), "clean" noiseless headfirst re-entry leaps, coordinated clean leaps, and tail slaps
Resting	Little motion and travel speed typically less than 3 knots
Socializing	Interacting with each other or inanimate objects—usually directionless movement and may include body and pectoral fin-rubbing, rolling, belly-up swimming, spyhops (projection of the head above water), splashing at the surface, chasing, leaping, mating, and playing with seaweed
Traveling	Slow directionless movement at speeds of less than 3 knots close to the surface with low activity level—often includes slow surfacings and floating near the surface

Bottlenose dolphin

Humpback whale

Killer whale

Sperm whale

Photo-identification technique

Photo-identification is a technique used to capture and recapture photographically individuals based on natural, distinctive markings. Bottlenose dolphins are identified by the pattern of nicks and notches on the trailing edge of the dorsal fin, killer whales are identified by nicks and notches on the dorsal fin and saddle patch markings, sperm whales are identified by the outer contour of their flukes, and humpback whales are identified by the distinct black-and-white pattern on the underside of their flukes.

Food & Foraging

Cetaceans have developed unique mechanisms to feed on a wide variety of prey from small zooplankton to the largest squid. As warm-blooded mammals and among the largest ocean predators, cetaceans have enormous energetic demands that they must satisfy through feeding frequently and on dense aggregations of prey. Cetaceans have also evolved a wide variety of behaviors to both locate and acquire prey in a dynamic ocean environment. Whereas the toothed whales and dolphins feed on individual prey items, the largest baleen whales generally feed on batches of small zooplankton.

Catching and detecting prey

One of the most interesting feeding adaptations among mysticete cetaceans is baleen. This keratin-based structure acts like a comb or sieve to strain out vast quantities of small prey. Combined with a massive tongue and expandable throat pouch—made of elastic blubber and muscle—baleen whales feed efficiently in bulk on small schooling fish and zooplankton. The mouth of a baleen whale is massive. For example, a blue whale can fit up to 150 percent of its body mass in water into its mouth in a single gulp. In order to capture high-density prey patches, the rorqual whales (blue whale, fin whale, sei whale, Bryde's whale, Omura's whale, humpback whale and minke whale) must perform an energetically costly lunge. In this maneuver, the whale accelerates quickly and then opens its huge mouth. When it does, the motion is similar to that of opening a parachute: the whale's mouth fills with water and the animal comes to a near-complete stop. The whale must then close its mouth, push the prey-laden water through the plates of baleen with its enormous tongue, and then swallow its catch. While this bulk feeding strategy is extremely costly for the whales to perform, it is clearly profitable as it has allowed baleen whales to become the largest animals that have ever lived on Earth.

Lunge-feeding (left)
A humpback whale surfaces with its mouth agape after lunge-feeding at the surface while feeding on sand lance. The baleen plates hanging from the upper jaw are visible including the comb-like outer side and the more matted inner side. Lunge-feeding is an energetically costly, yet very efficient way for baleen whales to feed on schools of small-bodied prey.

Echolocation (opposite)
This image depicts how dolphins use echolocation to locate prey. Sound is emitted through the melon (yellow structure at the front of the head) and projects outward. Signals bounce off targets then return to the animal where the echoes are received at the lower jaw (orange spot) and are transmitted to the ear.

Of great interest to scientists is just how whales locate their prey. While there is no single way that all whales find prey, they likely use a combination of senses. Whales that feed in shallow waters have the opportunity to detect prey patches visually by either silhouetting them from below or swimming close enough to see them in the water. Whales may also use passive listening to pick up the sounds of schools of fish and krill. Some whales, including humpback whales, have a small number of hair follicles on their rostrum and it has been postulated that these act as detectors—prey items may hit these hairs indicating to the whale when it is in an area of high prey density. More recently, it was also discovered that some whales have a receptor organ located between the two bones that form the lower jaw. While the function of this organ is not fully understood, it may act as a receptor for information coming to the whale through the water similar to the way in which sharks detect movement.

Odontocete cetaceans find food largely through echolocation. Biosonar sounds produced in the nasal passages are projected through the melon on the front of their heads (see page 19) and allows the animal to sense their environment in a precise and effective way similar to active sonar used by many ships and other animals, such as bats. This remarkable adaptation is so precise that dolphins can distinguish very small differences in the size and type of fish they are targeting and can also penetrate into sand or mud to locate prey that has burrowed. Odontocetes feed mostly on fish and cephalopods, however, some species are known to feed on larger prey items including other marine mammals. While the majority of odontocetes feed with the aid of echolocation, some have evolved strategies that are either visual or take advantage of tools. For example, bottlenose dolphins are known to drive schools of fish onto mud banks and then follow them onto the beach where they capture the fish before returning to the water. Bottlenose dolphins have also been known to feed with the aid of a sponge on their rostrums to help them locate and dislodge prey that may be hidden (see page 180).

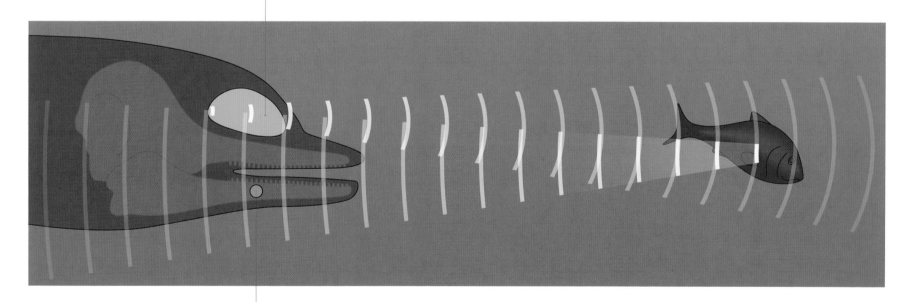

Biosonar sounds projected through the melon travel toward fish

Echoes bounce back through the dolphin's jaw

Food & Foraging

Feeding techniques

Baleen whales have a wide variety of feeding strategies that are species-specific. Gray whales lack an elastic throat so largely feed on the seafloor employing suction-feeding in which the tongue is used to draw water and sediment containing crustaceans and msyid shrimp into the mouth. These whales make feeding pits in the sand by rolling on their sides. Because of this strategy, the baleen of gray whales is short, coarse, and thick so that it can withstand the constant rubbing and sediment that must pass through it. Alternatively, the balaenid whales, including the right and bowhead whales, feed by skim-feeding. Rather than taking large gulps, these whales open their mouths wide and swim continuously through patches of small copepods and krill—forcing water through the baleen and leaving the small prey behind inside the mouth. This is similar to how basking sharks and whale sharks feed. The baleen of right and bowhead whales is the longest and finest among the cetaceans and it can reach lengths of 13 ft (4 m).

Bubble-net feeding

One of the most visible and striking feeding behaviors known in baleen whales is performed by humpback whales. Throughout their range, humpback whales are known to cooperatively feed in groups by producing nets and clouds of bubbles to corral prey. In Alaska, where humpback whales feed on schooling herring, groups of up to 15 whales have been seen feeding in concert using a bubble net.

Bubbles are released to concentrate and contain schools of fish such as herring, sand lance, and capelin. The avoidance responses of herring to bubbles suggests that whales may use the bubbles as tools to herd fish into tighter aggregations that are easier to exploit, or to force the school upward through the water column, thus trapping them against the surface. Humpback whales also deploy bubbles in a more elaborate manner, frequently in combination with other behaviors such as broadcasting of sounds, group hunting, and flashing their large pectoral flippers at prey. Bubble-net feeding may be conducted by individual whales or in large groups. The recent use of multi-sensor tags—such as DTAGs, (see pages 30–31) attached to humpbacks demonstrates that they employ several distinct techniques when feeding via bubble nets.

Two have been described: the "double loop" and the "upward spiral." The "double loop" consists of two individual dive loops separated by surfacing to make one or more fluke slaps, or lobtails (see Surface Behaviors, page 58). The "upward spiral" net is described to the right.

A feeding whale pod, Southeast Alaska
The bubble net curtain varies in size from 33–66 ft (10–20 m) across. Once the fish are at the surface, the whales lunge upward, huge mouths agape. The force of the lunge presses water against their throat grooves, they expand, and the whales engulf the trapped fish with their mouths.

Ram (skim) feeding and benthic feeding

Other modes that baleen whales use to feed include ram feeding in which whales move continuously through the water with open mouths to catch smaller prey, and benthic feeding in which whales take in mouthfuls of sediment from the bottom and sift it out through their baleen. Ram feeding is generally done when feeding on small copepods, and benthic feeding typically targets bivalves or crustaceans.

Ram feeding

Benthic feeding

The "upward spiral" technique

1. Positioning the bubbles

One or more humpbacks dive beneath the school of fish. Usually a single whale will release a ring of bubbles that spirals upward beneath the prey. The combination of an increasing turn-rate and changes in the body roll of the whale forms a constricting spiral that becomes narrower as it rises. This "upward spiral" net occurs in many parts of the water column, but bubbles are typically released less than 82 ft (25 m) deep, at the deepest portion of the whale's dive.

2. Confining the school of fish

Bubble net feeding involves a coordinated release of bubbles, generally other participating whales, and in some locations coordinating vocalizations. Recent studies have found evidence that humpbacks may manipulate their prey using vocalizations: the fish display defensive behavior at the sounds, retreating into ever-tighter bait balls. It is also thought that these vocalizations may serve to help coordinated movements among groups of whales feeding together. The bubble net absorbs the sound, shielding the fish, while at the same time increasingly constricting their movements. -The sounds discourage the fish from breaking out through the net.

3. Coordinated feeding

Whales feed at the center of the bubble net at or near the surface. In some bubble-netting groups, whales will surface in the same relative orientation as the bubble net. This coordinated behavior is highly organized and it appears that specific roles may be partitioned among individual whales, resulting in an impressive display of cooperative foraging.

Food & Foraging

Cooperative feeding

While most baleen whales—with the exception of humpback whales—feed individually, odontocetes often travel and feed in large groups. A well-known example of this can be seen among the orcas. Throughout their range, different populations of orcas specialize on different prey types. In the Antarctic, different ecotypes of orcas—distinguished by their size, coloration, and feeding preferences—are known to prey on fish, seals, penguins, and other whales. One of the most striking examples of cooperative feeding has been reported for seal-eating orcas. In order to capture seals, these orcas use a variety of methods to flush seals from the safety of an ice floe into the water where they can be captured. Initially, orcas will spyhop around ice floes until they locate both the species of seal they prefer—typically Weddell seals—and one that is in a position to be captured. Once this has been

determined, the orcas will coordinate themselves to tip the ice floes by pushing it from beneath, break the ice floe apart, or wash waves of water over the ice to push the seals off. When performing the latter, the orcas will swim in a tightly formed line just below the surface toward the ice floe and dive beneath at the last moment sending a large wave of water onto the ice that washes the seal into the water and toward the waiting jaws of other whales.

Mud-ring feeding

Another example of the extraordinary cooperative feeding abilities of cetaceans is mud-ring feeding. Bottlenose dolphins in muddy-bottomed estuaries and flats are known to locate schools of fish and then encircle them by swimming around the school in a tight circle and agitating the mud with the strokes of their flukes. This produces a circular curtain around the fish that spreads throughout the

shallow waters. When presented with a barrier like this, the fish will not swim through but rather will try to jump over. The dolphins spread themselves out around the mud ring oriented vertically in the water with their heads out of the water. As the fish begin to jump over the mud ring to try and escape, the dolphins see and catch the fish in mid air.

Teamwork
Bottlenose dolphins demonstrate their ingenious method of cooperative foraging. One dolphin corrals the fish by encircling them in a curtain of muddy water. The agitated fish jump out of the ring and are caught by the dolphins waiting outside the circle.

Mud-ring feeding

1. Mud ring formed
One dolphin initiates a mud ring around a school of fish and forms the ring by swimming close to the bottom and disturbing sediment with its flukes.

2. Fish encircled
The mud ring is completed and the school of fish is encircled while other dolphins remain on the outside of the ring.

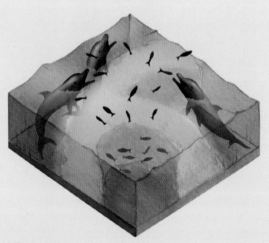

3. Prey caught
The dolphins position themselves around the mud ring with their heads above the water to catch the fish as they jump out trying to escape.

DTAG key functions are to:
• Provide information on body orientation (acceleration, pitch, role, and heading) and depth
• Record all sounds made and heard by the tagged whale(s)
• Record for up to 24 hours before becoming detached
• Record various other parameters including the environment (for example, water temperature and depth)

The examples below of cetacean species and their preferred prey provide a small sample of the extraordinary feeding behaviors and foraging strategies found in cetaceans. Given that these animals live most of their lives underwater, we now understand only a small fraction of the enormous repertoire of feeding behaviors that are generally performed at or near the surface. Recent advances in our ability to record underwater behavior, such as multi-sensor tags (as detailed above) now offer us the opportunity to see below the surface and explore the underwater world of cetaceans.

Selective diet
An adult male killer whale emerges from the water in front of an iceberg with a crabeater seal in its mouth. In Antarctica, there are several ecotypes of killer whales with distinct diets, including those that feed almost exclusively on seals.

A key research development

Developed in the late 1990s, the digital acoustic tag, or DTAG, has revolutionized research into a wide range of cetacean species, having been deployed on more than 20 species. These tags attach to the whale for up to 24 hours and after the tags detach they are retrieved, data is downloaded, and the tags are ready for redeployment on other animals. Attachment is by means of four suction cups placed on the whale using a 45-ft (13.7-m) cantilevered or 25-ft (7.7-m) handheld carbon-fiber pole positioned from an inflatable boat.

Examples of cetacean species and their preferred prey

Species	Feeding mode	Preferred prey
Blue whale	Baleen: lunge	Krill
Humpback whale	Baleen: lunge	Krill and fish up to 12 in (30 cm) long, such as herring
Right whale	Baleen: ram (skim)	Copepods
Gray whale	Baleen: suction	Mysid shrimp
Bottlenose dolphins	Teeth: mud ring, strand, sponge, echolocation	Numerous small fish and squid
Orcas	Teeth: strand, cooperative ice breaking, wave washing	Fish (including salmon), squid, seals, dolphins, porpoises, baleen whales
Beaked whales	Teeth: suction	Deep water squid

Life History

An organism's life-history strategy is defined by how individuals allocate resources to growth, reproduction, and survival. The characteristics that describe these strategies include how long individuals live, the ages at which they become sexually mature and first reproduce, the number of offspring females will produce in a lifetime, and when and where they travel to find sufficient food to survive.

Life-history studies

The life-history characteristics of cetaceans are well known for a few species and limited, or incomplete, for many. The common bottlenose dolphin, humpback whale, killer whale, and North Atlantic right whale are among the most well-known cetaceans. Each of these species have been studied for several decades, documenting the births and deaths of individual animals through time. These long-term studies, often referred to as longitudinal studies, provide unique insights about the species' life-history strategies, because they combine information about individual variability in life patterns with observations about social behavior and ecology. Additional insight about a species' strategy is gained by comparing characteristics of multiple discrete populations. These comparisons can reveal intra-specific variability in strategies and, in some cases, habitat-specific adaptations. Longitudinal studies are possible for species with populations that are relatively accessible—they live near the coast where boats are easily launched—and individual animals are readily distinguished by natural markings, such as color patterns or dorsal fin notches.

Natural markings

Numerous common bottlenose dolphin populations have been studied because they live close to shore. A coastal population of common bottlenose dolphins lives within ½ mile (800 m) of shore off southern California, USA (right), and many individuals in this small population can be recognized by naturally distinctive dorsal fins (above right).

Our knowledge of cetacean life-history strategies continues to grow from both longitudinal and cross-sectional studies. Several technological advances have provided researchers with new tools to augment the more traditional study techniques (for example, photo-identification in longitudinal and specimen-based in cross-sectional studies) used to study cetaceans. These tools include photogrammetry (photographs used to obtain accurate measurements of animals and estimate calf production), projectile biopsy techniques (gun- or crossbow-delivered dart with a coring tip to collect skin and blubber samples from free-swimming animals), and laboratory techniques to quantify molecular markers (for example, steroid hormones to identify pregnant females) from very small samples of tissue. In addition to contributing new information to ongoing studies, the new techniques provide researchers with tools for studying populations of some of the more remote, and largely unknown, cetacean species.

Most other cetacean species are more difficult to study, and for many species, knowledge of their life-history strategies is based on cross-sectional studies. Cross-sectional studies provide a snapshot of the life-history characteristic for species, or populations of species, using data collected from individual animals. For some species, data are collected from animals that died as a result of direct or indirect (for example, fisheries bycatch) exploitation, while for others, data are collected from individual animals that died on beaches (or stranded).

Another advantage of cross-sectional studies is that the length, age, and reproductive maturity of each individual is known. Age is determined by counting growth layer groups in ear plugs or ear bones for most species of baleen whales and teeth of toothed whales (see right). With length and age data for a representative sample of the population, growth rates from birth to full adult size can be estimated along with the size of individuals at different life stages (calf, juvenile, adult). Reproductive maturity is determined by examining reproductive organs. In females, the presence of at least one corpus—scar tissue remaining from an ovarian follicle after ovulation or pregnancy—indicates sexual maturity. In males, the presence of sperm and large seminiferous tubules in the testes indicates sexual maturity. These determinations give vital information for characterizing life-history strategies.

Age

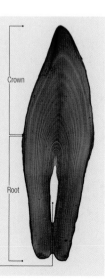

The light-dark patterns evident in this stained cross-section of a common bottlenose dolphin tooth are called "growth layer groups." They are similar to growth rings in trees and are counted to determine the dolphin's age. As dolphins age, the pulp cavity is filled in.

Crown

Root

Pulp cavity

Life History

Strategies

Cetacean life-history strategies are diverse and differ markedly between baleen and toothed whales. All baleen whales are large and long-lived, and many species make long-range annual migrations between wintering grounds in tropical waters and feeding grounds in temperate or polar waters. The characteristics of toothed whales are much more varied. They range in size from the small, relatively short-lived vaquita and harbor porpoise to the large, relatively long-lived sperm whale. They also live in diverse habitats ranging from pelagic and coastal ocean waters to estuarine and freshwater rivers. None of these species migrate but some make seasonal movements.

Despite their diversity, all cetacean species share several life-history characteristics. All species give birth to single, large, and precocial young—calves are well developed and move freely from birth. Multiple fetuses or births have been rarely documented, and there are no known cases of multiple offspring being successfully reared. Cetacean neonates range in size from about one-third of adult female size in baleen whales to nearly half the size of adult female toothed whales. Gestation is approximately one year for most species, but longer in several species, including the sperm whale (14–15 months) and killer whale (17 months). The lengthy gestation period helps balance the cost of producing large offspring.

Sexual dimorphism

Males and females differ in adult size (sexual dimorphism) in many cetacean species. Both sexes grow rapidly from birth to weaning, and then growth rates begin to slow until they stop when full adult size is reached. However, the sex that grows largest tends to grow faster for a longer period of time. For example, female killer whales are approximately 10 years old and male killer whales approximately 16 years old when they reach sexual maturity. Among baleen whales, females are generally five percent larger than males. Among toothed whales, females are slightly larger than males in porpoises and river dolphins, but males are larger than females in all other species. The size differences between the sexes range from males being around 2–10 percent larger than females in most of the small delphinids, such as the common bottlenose, pantropical spotted, and common dolphins, to an approximately 60 percent difference in sperm whales.

Body size

Body-size differences between the sexes (sexual dimorphism) vary among the cetaceans and are most variable among toothed whales. In toothed whales, females are larger than males in the porpoises and river dolphins, and males are larger than females in all other species with sperm whales being the most sexually dimorphic.

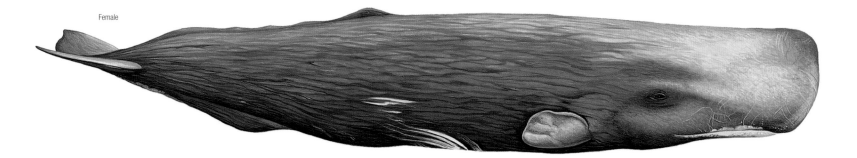

Female

Size is just one form of sexual dimorphism seen in cetaceans. Another is tooth morphology (see right). Whatever the form, sexual dimorphism can be a predictor of cetacean mating systems. For example, a polygynous—in which males mate with multiple females—mating system has been hypothesized for the sperm whale, short-finned pilot whale, and killer whale. All three species have marked sexual dimorphism—adult males are much larger than adult females—which has been interpreted as an indicator of male–male competition. In sperm whales, the presence of scars inflicted by other males provides supporting evidence of male–male competition. However, limited male dispersal from natal pods has been observed in short-finned pilot whales and killer whales, suggesting that their mating system is not polygyny and that sexual dimorphism likely evolved for other reasons. One hypothesis proposed for killer whales is that large males enhance their reproductive fitness by improving the foraging efficiency of their natal pod. Polygynous mating systems are the exception in cetaceans, and most species have a promiscuous—both males and females having multiple partners— mating system.

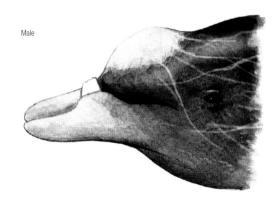

Male

Female

Male

Tooth morphology

Another form of sexual dimorphism in cetaceans is differential tooth morphology. The most dramatic examples are seen among the mesoplodont beaked whales. These whales possess only a single pair of teeth, which vary in shape, size, and position in the lower jaw. The teeth develop only in adult males. Scientists think the teeth play a role in reproductive success, with males using them in combat for access to mates and females using them as a cue for mate selection. The Hubbs' beaked whale, shown above, is one example.

Life History

Reproductive cycles

The reproductive cycle for all cetacean species has three parts: a gestation period, a lactation period, and a resting period. Females of most species produce calves throughout their lives. The minimum time between pregnancies is two years but is typically longer because the energetic demands of pregnancy and lactation are high for reproductive females. Among baleen whales, the cycle for blue, Bryde's, humpback, sei, and gray whales includes an 11-month gestation period, a 6–7-month lactation period, and a 6–7-month resting period with females becoming pregnant every 2–4 years. The reproductive cycle for the bowhead and right whales starts with a 12–13-month gestation and females become pregnant every 3–7 years (see right). Many species make long-distance annual migrations between high-latitude feeding grounds and low-latitude (typically tropical) wintering grounds. Although the reproductive cycle of baleen whales is synchronized with their migration cycle, the timing for individual whales depends on their reproductive condition. Newly pregnant females are the first to leave the wintering grounds and return to summer feeding grounds. This maximizes a pregnant female's time on the feeding grounds, allowing her to put on enough fat to sustain her and her growing fetus through the coming year. Near-term pregnant females are among the first to leave the feeding grounds and return to the wintering grounds. Females with new calves are the last to leave the wintering grounds. This gives calves more time to grow before making their first migration. Several hypotheses have been proposed for the adaptive significance of these large-scale migrations, including increased survival rates for neonates born in tropical waters where the thermoregulatory demands and risk of predation by killer whales are reduced.

Among toothed whales, the reproductive cycle (see right) is more variable than that of baleen whales. Toothed whales have multi-year lactation periods and no large-scale migrations. Many smaller dolphin species have a calving interval of around three years. The cycle starts with an 11–12-month gestation, a 1–2-year lactation period, and several months of rest before breeding again. Several of the other, larger, toothed whales such as the sperm whale, have a longer cycle that starts with a 12–17-month gestation and a 3-year, or longer, lactation period. Reproductive synchrony varies more among toothed whales, reflecting the range of habitats they occupy. Species inhabiting temperate waters, such as the harbor porpoise, typically have relatively discrete calving seasons that fit with peak ecosystem productivity. Species inhabiting tropical waters, such as spinner dolphins, have more variable seasons with calves born year round.

Longevity

Cetaceans are relatively long-lived, like many large mammal species. Among baleen whales, longevity ranges from 60 years for minke whales to up to 100 years for fin whales and more than 100 years for bowhead whales. Bowheads are the longest-living cetacean. A weapon fragment recovered from a whale in 2007 gave a minimum age estimate for that whale of 115–130 years. Recent research suggests that bowheads may live for more than 200 years. Among toothed whales, longevity ranges from around 20 years for harbor porpoises to 70 years for sperm whales.

Bowhead Whale

Baleen whale two-year cycle

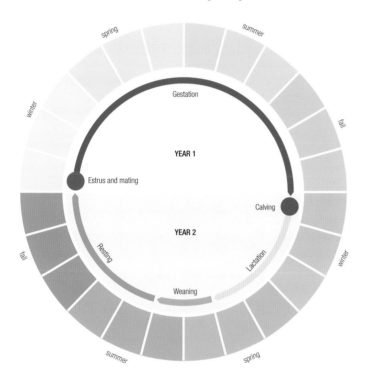

spring

summer

winter

fall

Gestation

YEAR 1

Estrus and mating

Calving

YEAR 2

Resting

fall

Lactation

winter

Weaning

summer

spring

Toothed whale three-year cycle

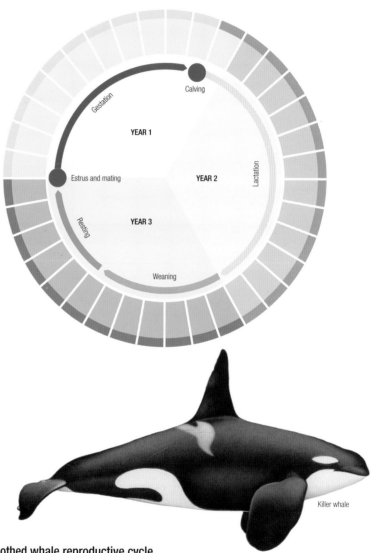

Gestation

YEAR 1

Calving

Estrus and mating

YEAR 2

Lactation

Resting

YEAR 3

Weaning

North Atlantic right whale

Killer whale

Baleen whale reproductive cycle

The reproductive cycle of baleen whales is a minimum of two years, and the reproductive events are correlated with their annual migratory cycle. Most mating and calving takes place on the wintering grounds, and calves are weaned when they arrive on the feeding grounds during their first summer. Females complete one migratory cycle while pregnant and the first half of the next cycle with a nursing calf.

Toothed whale reproductive cycle

The reproductive cycle of toothed whales typically lasts a minimum of three years. In many species, the cycle includes a one-year gestation and an approximately two-year lactation period. Mating, calving, and weaning—when calves begin foraging but are still dependent—are distinct events in the cycle, with each occurring over several months.

Life History

Growth rate in calves

All newborn cetacean calves grow rapidly, but baleen whale calves grow much faster than toothed whale calves. The faster growth of baleen whale calves is in part driven by the differences in the reproductive cycles and migration patterns of the two groups, and the ability to do so is fueled by high-energy content milk. Baleen whales have milk with three to five times higher energy content than that of toothed whales. The rapid growth of calves together with their large size and ability to swim immediately after birth increases their probability of survival. This is vital for baleen whale calves to maximize their chances of making a successful trip to feeding grounds within months of birth. Toothed whales typically have lactation periods that last a year or more beyond weaning. This additional period of investment likely further increases a calf's chance of survival (see below).

Calf survival

To enhance their chances of survival, calves of toothed whales, such as this short-beaked common dolphin calf, spend two or more years with their mother learning to forage and socialize.

Population dynamics

Knowledge of a species' life-history strategy provides the foundation for understanding their population dynamics—changes in abundance due to population-level patterns of reproduction and survival (see below right). Comparing life-history characteristics among populations of a species contributes to this knowledge by revealing how traits vary. For example, pantropical spotted dolphins have different breeding seasons on either side of the Equator and different characteristics of individual growth on either side of the Pacific Ocean: adult dolphins in the western Pacific are 1½–2¾ in (40—70 mm) longer than those in the eastern Pacific. These two examples reveal traits that largely reflect adaptations to the habitats occupied by these discrete populations. There are also examples that illustrate how cetacean populations respond to increases in prey availability. For example, the onset of sexual maturity at younger ages in populations of fin, sei, and minke whales has been interpreted as a response to increased prey availability fueling faster individual growth rates after commercial whaling reduced the whale's population numbers.

The reproductive output of populations varies from year to year, reflecting annual variability in environmental conditions. Long-term studies are particularly valuable for understanding a population's natural variability in reproductive output and which environmental parameters influence those changes. One example is the correlation of annual calf production in eastern North Pacific gray whales with seasonal access to the feeding grounds for newly pregnant females (see right).

Population monitoring

Monitoring select parameters, particularly those associated with reproduction, help us understand the natural variability in population dynamics and provide indicators of how they change. The low reproductive rates—late maturation and long inter-birth intervals—of cetaceans make them vulnerable to habitat degradation and exploitation. Identifying and monitoring parameters that reflect critical aspects of a population's dynamics are essential components to include in conservation and management plans for assessing population health and protecting populations at risk.

Vital rates

The number of individuals in a population fluctuates over time as a result of natural fluctuations in the combined ability of individuals to survive and reproduce. A population's survival rates are revealed by knowing its age structure. A population's reproductive rates are revealed by knowing the age at which females first give birth, the interval between births, and how many females are breeding. Rates of survival and reproduction vary by age and sex in cetaceans. In combination, a population's survival and reproductive rates define its dynamics.

Gray whale calf production

Gray whale calf production is tightly correlated with the extent of spring ice covering their Arctic feeding grounds. In years when seasonal ice breaks up early in the Arctic spring, pregnant females are able to feed and put on enough fat to support pregnancy and lactation for new calves. During these years, calf production is high. Conversely, when seasonal ice extends far to the south, newly pregnant females temporarily lack access to foraging areas and fewer calves are born. Counts of calves made during their first migration to the Arctic feeding grounds are made from a site in central California and used to estimate annual calf production. As the Arctic environment continues to change, this data set will provide a valuable time series for interpreting the influence of climate change on future gray whale calf production.

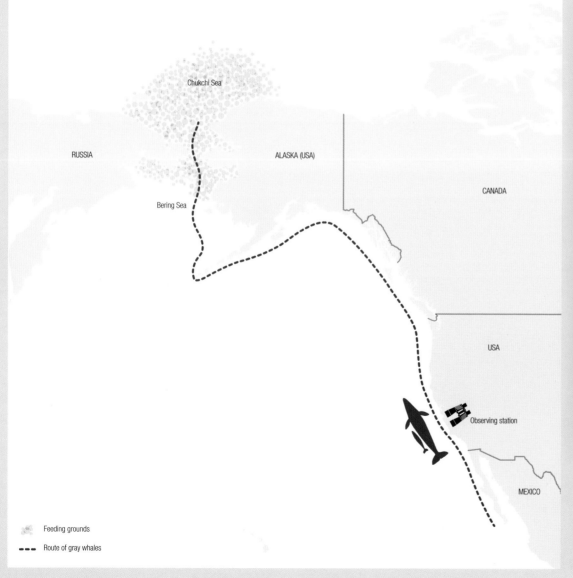

Chukchi Sea

RUSSIA

ALASKA (USA)

CANADA

Bering Sea

USA

Observing station

MEXICO

Feeding grounds

Route of gray whales

Gray whale adult female with calf
Most calves are just a few months old when they embark on their first journey to the feeding grounds. Adult females nurse their calves until they arrive and repeat the journey with a new calf every few years throughout their life.

Range

The range of a given species is defined as the spatial and temporal extent of the distribution of all of the resources required for survival. Cetaceans are found throughout nearly all of our planet's oceanic and coastal habitats—from warm equatorial waters to freshwater river systems and the ice-covered polar regions. As mobile animals, cetaceans range over broad areas and have the longest migratory routes of any mammals. Alternatively, some cetacean species have small home ranges and spend their entire lives in refined locations where they are able to feed, breed, and avoid predation.

The range of an individual or species can be measured in a number of ways. One of the simplest ways is through the use of photographic identification. For nearly 40 years, researchers have used photographs of the distinct markings or scarring patterns unique to individual animals to capture information on where and when animals are seen. Over time, by matching photographs of individual animals, an animal's range can be defined. This method is particularly valuable for animals that inhabit coastal areas or are easily photographed. Some of the most fundamental information on the range of migratory species such as humpback whales has been made based on the matching of photographs by researchers in both breeding and feeding grounds.

Alternatively, for animals that range over large areas that are not easily accessible to researchers, the use of satellite telemetry has been critical to our understanding of the range and movement of cetaceans. Tracking devices that provide information on the geographic position (latitude and longitude) of an animal can provide critical information on the location, movement patterns, and ultimately the range of individual animals. By replicating this over a number of individuals, ultimately the range of a species can be defined by the locations in which animals are found. The core area in which an individual or species is found is commonly referred to as a home range, and contains the necessary resources for its survival (for example, habitat).

Fluke scarring
Natural pigmentation patterns and scars acquired throughout their lives create unique "fingerprints" that researchers use to identify individual whales.

Variation in home ranges: rolling stones vs. home-bodies

For some cetaceans, finding suitable habitat for critical life-history events means traveling great distances over long periods of time. Both humpback and gray whales exhibit extremely long seasonal migrations (the longest migration for any mammalian species). Both species spend summer months in high latitudes, where prey is abundant, feeding almost constantly to replenish energy stores. In the fall, most whales migrate to more tropical calving and breeding grounds where they fast. These two spatially discrete portions of their range can be separated by more than 3,000 miles (5,000 km). After spending winter months giving birth and rearing young in warm, shallow, and secluded waters, the whales once again travel back toward feeding grounds in the spring. The migratory corridors that these whales use tend to be close to shore, in part to protect small calves from predation. The total home range for these whales can span more than 6,000 miles (10,000 km) from feeding to breeding grounds and back again.

This pattern is in contrast to several species of dolphins that maintain small home ranges for their entire lives. Hector's dolphins, endemic to New Zealand, are known to have home ranges that are less than 20 square miles (50 square km). Similarly, some river dolphin species, including the tucuxi, range over areas less than 8 square miles (20 square km). Unlike large baleen whales, these dolphins are able to find the necessary resources to satisfy their energetic and reproductive demands in a small area.

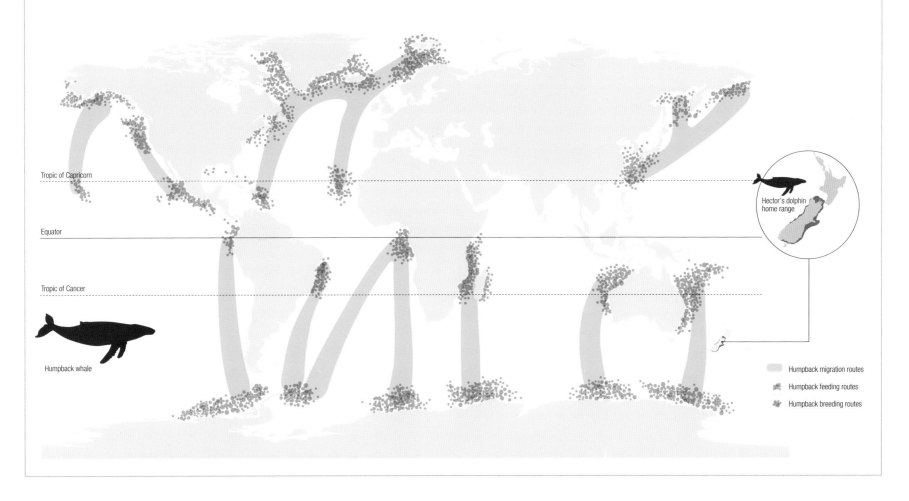

Tropic of Capricorn

Equator

Tropic of Cancer

Hector's dolphin home range

Humpback whale

Humpback migration routes
Humpback feeding routes
Humpback breeding routes

Habitat

Cetaceans must balance the many demands of living in a dynamic marine environment in order to survive. As large, warm-blooded predators, cetaceans must be able to find and acquire enough food to satisfy enormous energetic needs. Conversely, there are thermal challenges to living in a marine environment. A large, well-insulated body means that cetaceans must live in environments that allow them to dissipate heat to maintain suitable body temperatures. With increasing body size comes increased energetic demands. Therefore the largest cetaceans, the baleen whales, must find regions with sufficient, predictable prey resources in order to satisfy their enormous appetites. Conversely, young baleen whales face substantial predation risk from killer whales, and the migratory destinations of humpback and gray whales may reflect areas that are safe for small animals to grow.

While baleen whales feed on small prey that aggregate into dense and often dispersed patches, odontocetes catch larger prey items one at a time. Baleen whales typically feed near the ocean's surface—around the upper 980 ft (300 m)—beaked whales are extreme divers that routinely dive and feed at more than 5,000 ft (1,500 m) deep, sometimes reaching depths of more than 8,000 ft (2,500 m) to feed on squid and deep-water fish.

In order to characterize the habitat of cetaceans, scientists typically measure many features of the ocean environment to determine which parameters best predict where a certain species is likely to be found. Most often, cetacean habitat is defined as having greater than average densities of prey. However, the oceanographic features and mechanisms that promote enhanced productivity and prey can vary regionally. For cetaceans to thrive, they must avoid competition with each other, and certain species that live in close proximity to each other often prefer slightly different combinations of environmental conditions—niches—to avoid competition.

The habitats in which cetaceans are found are as diverse as the animals themselves. While narwhal and bowhead whale habitat is characterized by frigid water and narrow leads between ice floes, several species of river dolphin prefer the murky freshwater deltas of the world's great rivers. Habitat can also vary depending on the behavior of the cetacean. During the feeding season, humpback whales are known to change their distribution and habitat preferences as the distribution of their prey changes. On a seasonal scale, many baleen whales change their habitat from areas with rich prey resources for feeding to areas in warm water and close to shore in order to give birth and raise their young.

Spinner dolphins
Spinner dolphins are known to feed in open pelagic waters at night and then move into sheltered bays to rest during the day. Spinner dolphins are an example of a species that has very different foraging and resting habitats.

Cetacean habitats

Sea ice

For a number of cetaceans, the seasonal sea ice found in both the Arctic and Antarctic provides suitable habitat for both foraging and avoiding predation. In the Arctic, bowhead, beluga, and narwhal all rely on sea ice, while in the Antarctic, both minke and killer whales use this habitat.

Lagoons

Gray whales utilize the warm and shallow lagoons of Baja California as calving habitat. The whales use these waters to avoid predation from killer whales but also to provide calves with calm, warm water in which they can nurse and learn to swim.

Rivers

Many of the largest rivers in the world provide habitat for a number of dolphin species. These murky and fast-moving waters are home to a variety of fish species that support the dolphins.

Offshore waters

The offshore waters of the world's oceans constitute the largest single habitat on the planet. Many pelagic dolphin and beaked whale species utilize this environment. For dolphins living in large social groups, sometimes in the thousands, they must find schools of small bait fish that associate with oceanographic features such as gyres and eddies. For more solitary beaked whales that feed on deep-water squid, habitats associated with offshore seamounts and submarine canyons are preferred.

Warm tropical waters

Warm tropical waters often around islands are a haven for many dolphin species. While these clear waters lack the primary productivity found in colder, higher latitude waters, they are rich in reefs and features that aggregate fish. In the case of spinner dolphins around the Hawaiian islands, the sheltered bays also provide critical resting habitat where the dolphins can avoid predation from sharks and rest between foraging bouts.

Sea ice

Lagoons

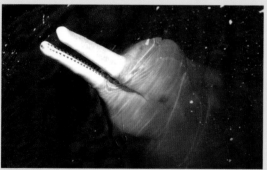

Rivers

Offshore waters

Warm tropical waters

Conservation & Management

Conservation, as defined by the International Union for the Conservation of Nature (IUCN), is the "protection, care, management and maintenance of ecosystems, habitats, wildlife species and populations, within or outside of their natural environments, in order to safeguard the natural conditions for their long-term permanence." Cetaceans, as large, wide-ranging species, are important "umbrella" species whereby efforts to conserve them may also protect other species in the ecosystem. Furthermore, as top predators, cetaceans can be used as indicator species of ecosystem health.

Conservation status

The Earth is in the midst of the sixth mass extinction and cetaceans are not immune to this. Several species, such as the vaquita and the North Pacific right whale hover on the brink of extinction with small and dwindling populations. The conservation status of all cetaceans is regularly evaluated with the IUCN Red List of Threatened Species (see table below).

Vaquita

Endangered and extinct species
The Chinese river dolphin or baiji (above) and vaquita (above right) are critically endangered. The baiji is most likely extinct as no individuals were sighted during the last comprehensive survey of its habitat in the Yangtze River. Habitat loss and bycatch are the causes of this species' decline. Bycatch has also been identified as a reason for the decline of the vaquita in the Sea of Cortez, where accidental drowning in gill-net fisheries has caused the population to fall to fewer than 100 individuals at the time of writing.

IUCN Red List classification of cetaceans

Red List classifications are made by a team of experts on a given species using the best available data. Classifications reflect the risk of extinction based on the major threats facing a species. These classifications are regularly updated as new data become available. The table below reflects the classifications at the time of writing (Fall 2014).

Category	Definition	No. cetacean species
Critically Endangered	Species faces an extremely high risk of extinction in the wild	2
Endangered	Species faces a very high risk of extinction in the wild	7
Vulnerable	Species faces a high risk of extinction in the wild	6
Near Threatened	Species does not qualify for status as Critically Endangered, Endangered, or Vulnerable but is close to or likely to qualify in the near future	5
Least Concern	Species is widespread and abundant and does not qualify for Critically Endangered, Endangered, Vulnerable, or Near Vulnerable status	22
Data Deficient	Inadequate information available to make an assessment of a species' extinction risk	45

Threats

Threats to cetaceans may be direct—causing a direct loss of life—or indirect, causing a threat to life through events occurring in the environment. Direct effects include hunting for subsistence use, whaling, poaching, and drive fisheries; entanglement in fishing gear and marine debris; bycatch; ship strikes; habitat loss due to coastal development; and mortality from underwater sonar testing. Indirect effects include altered migration routes and prey availability due to climate change, decreased reproductive success due to prey depletion, increased mortality risk due to bioaccumulation of contaminants, increased risk of disease from coastal pollution, and cumulative effects of stress from noise pollution and boat traffic.

Conservation actions

Cetaceans have a global distribution and individuals of highly migratory species may occur in international waters in addition to the waters of several nations. Thus, cetaceans are protected by a number of international and national policies and regulations. The International Whaling Commission, established under the International Convention for the Regulation of Whaling in 1946, marked the first international effort to manage cetaceans through the establishment of whaling catch limits. Further acts passed in 1973 and 1979 offered additional international protections to cetaceans. However, international conservation acts are difficult to enforce and passage of domestic laws such as the US Marine Mammal Protection Act in 1972 and the Endangered Species Act in 1973 provide enforceable and comprehensive measures for protecting cetaceans.

Cetaceans are also protected through the actions of non-governmental organizations (NGOs) and individuals. Many NGOs lobby for conservation action, support research, and educate the public about conservation issues. Individual actions such as beach clean-ups and taking measures to reduce one's carbon footprint—for example through recycling and using public transportation—also help to conserve cetaceans.

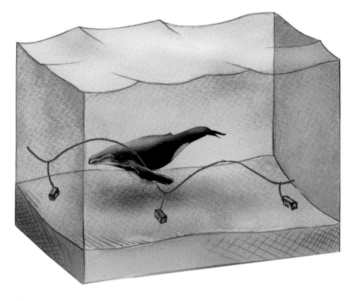

Floating line
Traditionally, lobster pots are strung together on the seafloor using a floating ground line. This line may imperil whales as they swim along the seafloor.

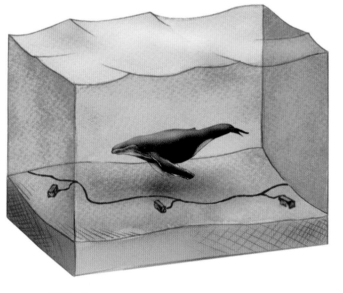

Sinking line
A new type of lobster-pot ground line is weighted so that it sinks along the seafloor. This development minimizes the risk of entanglement.

"Whale-safe" lobster lines
A major threat to humpback and North Atlantic right whales in the Northwest Atlantic is entanglement in fishing gear. Efforts to alleviate entanglement may have a reactive or proactive approach. A reactive approach involves skilled disentanglement teams cutting the lines off a whale. A proactive approach involves the modification of fishing gear, for example, through sinking ground lines, as shown here.

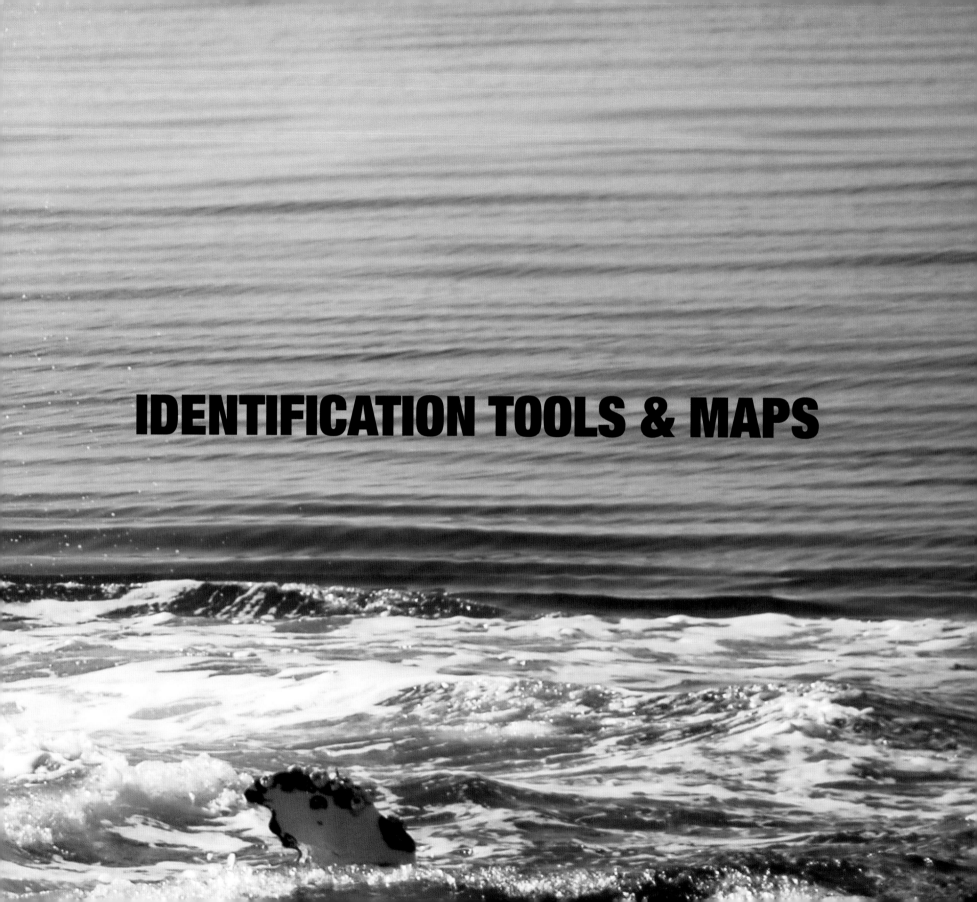

IDENTIFICATION TOOLS & MAPS

Identification Keys

The following keys can be used to identify species of whales, dolphins, and porpoises in the field using sets of physical characteristics. Only portions of the bodies of cetaceans can be seen as the animals surface to breathe; however, there are several observable features that are diagnostic. Those features that are most diagnostic include the shape and size of the blow (exhaled water vapor) in the big whales, as well as external features, including shapes of the head, dorsal fin, and tail flukes, and overall body size.

Blow: 29+ ft (9+ m) high

Blue whale *Balaenoptera musculus*

Blows (spouts)

Most small whales and dolphins produce small blows that are difficult to see. The blows of large whales are identified by shape.

Single (columnar) blow

Double ("V"-shaped) blow

Sperm whale
Physeter macrocephalus
6½ ft (2 m) high

Sei whale
Balaenoptera borealis
10 ft (3 m) high

Humpback whale
Megaptera novaeangliae
8–10 ft (2.4–3 m) high

Gray whale
Eschrichtius robustus
10–15 ft (3–4.6 m) high

Bryde's whale
Balaenoptera edeni
10–13 ft (3–4 m) high

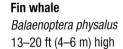

Fin whale
Balaenoptera physalus
13–20 ft (4–6 m) high

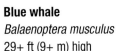

Blue whale
Balaenoptera musculus
29+ ft (9+ m) high

Northern/Southern right whales
Eubalaena spp.
16½ ft (5 m) high

Bowhead whale
Balaena mysticetus
23 ft (7 m) high

External features

Whales and dolphins can be distinguished by head shape (being either beaked, beakless, or flat), coloration (uniform, counter-shaded, or with complex markings), the shape and position of the dorsal fin, and the shape of the tail fluke.

Head shape

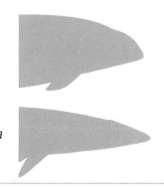

Short-beaked common dolphin
Delphinus delphis
prominent beak

Harbor porpoise
Phocoena phocoena
no beak, rounded head

Common minke whale
Balaenoptera acutorostrata
flattened head

Dorsal fin shape

Pantropical spotted dolphin
Stenella attenuata
curved

Killer whale
Orcinus orca
straight and tall

Hector's dolphin
Cephalorhynchus hectori
rounded

Amazon River dolphin
Inia geoffrensis
small hump

Body coloration

Beluga
Delphinapterus leucas
uniform coloration

Common bottlenose dolphin
Tursiops truncatus
counter-shaded (bottom lighter than top)

Atlantic white-sided dolphin
Lagenorhynchus acutus
complex markings

Dorsal fin position

Andrews' beaked whale
Mesoplodon bowdoini
posterior

Striped dolphin
Stenella coeruleoalba
middle

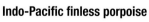

Indo-Pacific finless porpoise
Neophocaena phocoenoides
absent

Tail flukes

Bowhead whale
Balaena mysticetus
pointed tips, slight middle notch, slightly concave trailing edges

Gray whale
Eschrichtius robustus
deep middle notch, convex and ragged trailing edges

Humpback whale
Megaptera novaeangliae
deep middle notch, sigmoidal trailing edges with bumps

Sperm whale
Physeter macrocephalus
broad and triangular, straight trailing edges, deep middle notch

Narwhal
Monodon monoceros
distinctly convex trailing edges, deep middle notch

Short-beaked common dolphin
Delphinus delphis
pointed tips, slight middle notch, distinctly concave trailing edges

Harbor porpoise
Phocoena phocoena
pointed tips, distinct middle notch, slightly concave trailing edges

Dall's porpoise
Phocoenoides dalli
rounded tips, slight middle notch, distinctly convex trailing edges

Body size

The whales and dolphins in this book are separated into general batches based on body size and further divided by the presence or absence of a beak and dorsal fin.

Length up to 10 ft (3 m), beak and dorsal fin present

Franciscana
Pontoporia blainvillei;
s. hemisphere;
4¼–5¾ ft (1.3–1.7 m), page 254

Tucuxi
Sotalia fluviatilis;
n. and s. hemisphere;
4¼–6 ft (1.3–1.8 m), page 156

Clymene dolphin
Stenella clymene;
n. hemisphere;
5¾–6½ ft (1.7–2 m), page 170

Spinner dolphin
Stenella longirostris;
n. and s. hemisphere;
4¼–7 ft (1.3–2.1 m), page 176

Guiana dolphin
Sotalia guianensis;
s. hemisphere;
7 ft (2.1 m), page 158

Atlantic spotted dolphin
Stenella frontalis;
n. and s. hemisphere;
5¾–7½ ft (1.7–2.3 m), page 174

Pantropical spotted dolphin
Stenella attenuata;
n. and s. hemisphere;
5¾–8 ft (1.7–2.4 m), page 168

Long-beaked common dolphin
Delphinus capensis;
n. and s. hemisphere;
6¼–8 ft (1.9–2.4 m), page 116

Short-beaked common dolphin
Delphinus delphis;
n. and s. hemisphere;
5¾–8 ft (1.7–2.4 m), page 118

Striped dolphin
Stenella coeruleoalba;
n. and s. hemisphere;
6–8¼ ft (1.8–2.5 m), page 172

Atlantic humpback dolphin
Sousa teuszii;
n. and s. hemisphere;
6½–8¼ ft (2–2.5 m), page 166

**Indian Ocean bottlenose
dolphin** *Tursiops aduncus*;
n. and s. hemisphere;
8½ ft (2.6 m), page 180

Rough-toothed dolphin
Steno bredanensis;
n. and s. hemisphere;
7–8½ ft (2.1–2.6 m), page 178

Indo-Pacific humpback dolphin
Sousa chinensis;
n. and s. hemisphere;
6½–9¼ ft (2–2.8 m), page 160

Indian humpback dolphin
Sousa plumbea;
n. and s. hemisphere;
8–9¼ ft (2.4–2.8 m), page 162

Australian humpback dolphin
Sousa sahulensis;
s. hemisphere;
6½–9 ft (2–2.7 m), page 164

Common bottlenose dolphin
Tursiops truncatus;
n. and s. hemisphere;
6¼–12¾ ft (1.9–3.9 m),
page 182

Length up to 10 ft (3 m), beak present, dorsal fin absent or reduced

Baiji
Lipotes vexillifer;
n. hemisphere;
4¾–8¼ ft (1.4–2.5 m), page 252

Ganges River dolphin
Platanista gangetica;
n. hemisphere;
5–8¼ ft (1.5–2.5 m), page 258

Amazon River dolphin
Inia geoffrensis;
n. and s. hemisphere;
6–8¼ ft (1.8–2.5 m), page 256

Southern right whale dolphin
Lissodelphis peronii;
s. hemisphere;
6–9½ ft (1.8–2.9 m), page 144

Northern right whale dolphin
Lissodelphis borealis;
n. hemisphere;
6½–10 ft (2–3 m), page 142

50 ft (15.25 m)

Length up to 10 ft (3 m), beak absent, dorsal fin present or absent

Hector's dolphin
Cephalorhynchus hectori;
s. hemisphere;
4–5 ft (1.2–1.5 m), page 114

Vaquita
Phocoena sinus;
n. hemisphere;
4–5 ft (1.2–1.5 m), page 270

Commerson's dolphin
Cephalorhynchus commersonii;
s. hemisphere;
4¼–5¾ ft (1.3–1.7 m), page 108

Heaviside's dolphin
Cephalorhynchus heavisidii;
s. hemisphere;
5¼–5¾ ft (1.6–1.7 m), page 112

Chilean dolphin
Cephalorhynchus eutropia;
s. hemisphere;
4–5¾ ft (1.2–1.7 m), page 110

Hourglass dolphin
Lagenorhynchus cruciger;
s. hemisphere;
5¼–6 ft (1.6–1.8 m), page 136

Indo-Pacific finless porpoise
Neophocoena phocaenoides;
n. and s. hemisphere;
4–6¼ ft (1.2–1.9 m), page 264

Narrow-ridged finless porpoise
Neophocaena asiaeorientalis;
n. hemisphere;
6¼ ft (1.9 m), page 262

Harbor porpoise
Phocoena phocoena;
n. hemisphere;
4¾–6¼ ft (1.4–1.9 m), page 268

Burmeister's porpoise
Phocoena spinipinnis;
s. hemisphere;
4¾–6½ ft (1.4–2 m), page 272

Dusky dolphin
Lagenorhynchus obscurus;
s. hemisphere;
5¼–7 ft (1.6–2.1 m), page 140

Spectacled porpoise
Phocoena dioptrica;
s. hemisphere;
4¼–7¼ ft (1.3–2.2 m), page 266

Dall's porpoise
Phocoenoides dalli;
n. hemisphere;
5¾–7¼ ft (1.7–2.2 m), page 274

Peale's dolphin
Lagenorhynchus australis;
s. hemisphere;
6½–7¼ ft (2–2.2 m), page 134

Pacific white-sided dolphin
Lagenorhyhchus obliquidens;
n. hemisphere;
5¾–8 ft (1.7–2.4 m), page 138

Atlantic white-sided dolphin
Lagenorhynchus acutus;
n. hemisphere;
6¼–8¼ ft (1.9–2.5 m), page 130

Fraser's dolphin
Lagenodelphis hosei;
n. and s. hemisphere;
6½–8½ ft (2–2.6 m), page 128

Irrawaddy dolphin
Oracella brevirostris;
n. and s. hemisphere;
7–8½ ft (2.1–2.6 m), page 146

Pygmy killer whale
Feresa attenuata;
n. and s. hemisphere;
7–8½ ft (2.1–2.6 m), page 120

Dwarf sperm whale
Kogia simus;
n. and s. hemisphere;
7–9 ft (2.1–2.7 m), page 192

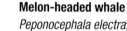

Melon-headed whale
Peponocephala electra;
n. and s. hemisphere;
7–9 ft (2.1–2.7 m), page 152

Australian snubfin dolphin
Oracella heinsohni;
s. hemisphere;
6–9¼ ft (1.8–2.8 m), page 148

White-beaked dolphin
Lagenorhynchus albirostris;
n. hemisphere;
8¼–9¼ ft (2.5–2.8 m), page 132

Pygmy sperm whale
Kogia breviceps;
n. and s. hemisphere;
9–11¼ ft (2.7–3.4 m), page 190

Risso's dolphin
Grampus griseus;
n. and s. hemisphere;
8½–12½ ft (2.6–3.8 m), page 126

50 ft (15.25 m)

Length 10–33 ft (3–10 m), beak and dorsal fin present

Pygmy beaked whale
Mesoplodon peruvianus; n. and s. hemisphere;
11¼–12¼ ft (3.4–3.7 m), page 240

Hector's beaked whale
Mesoplodon hectori; n. and s. hemisphere;
13–15 ft (4–4.6 m), page 230

Perrin's beaked whale
Mesoplodon perrini; n. hemisphere;
13–14¾ ft (4–4.5 m), page 238

Andrews' beaked whale
Mesoplodon bowdoini; s. hemisphere;
13–15½ ft (4–4.7 m), page 218

Deraniyagala's beaked whale
Mesoplodon hotaula; n. and s. hemisphere;
12¾–15¾ ft (3.9–4.8 m), page 232

Sowerby's beaked whale
Mesoplodon bidens; n. and s. hemisphere;
13–16½ ft (4–5 m), page 216

Gervais' beaked whale
Mesoplodon europaeus; n. and s. hemisphere;
14¾–17 ft (4.5–5.2 m), page 224

Ginkgo-toothed beaked whale
Mesoplodon ginkgodens; n. and s. hemisphere;
15½–17 ft (4.7–5.2 m), page 226

True's beaked whale
Mesoplodon mirus; n. and s. hemisphere;
16–17½ ft (4.9–5.3 m), page 236

Stejneger's beaked whale
Mesoplodon stejnegeri; n. hemisphere;
16½–17½ ft (5–5.3 m), page 242

Hubbs' beaked whale
Mesoplodon carlhubbsi; n. hemisphere;
16½–17½ ft (5–5.3 m), page 220

Spade-toothed beaked whale
Mesoplodon traversii; s. hemisphere;
16–18 ft (4.9–5.5 m), page 244

Gray's beaked whale
Mesoplodon grayi; s. hemisphere;
14¾–18½ ft (4.5–5.6 m), page 228

Blainville's beaked whale
Mesoplodon densirostris; n. and s.
hemisphere; 14¾–20 ft (4.5–6 m), page 222

Strap-toothed whale
Mesoplodon layardii; s. hemisphere;
16½–20¼ ft (5–6.2 m), page 234

Cuvier's beaked whale
Ziphius cavirostris; n. and s. hemisphere;
18–23 ft (5.5–7 m), page 248

Shepherd's beaked whale
Tasmacetus shepherdi; s. hemisphere;
20–23 ft (6–7 m), page 246

Southern bottlenose whale
Hyperoodon planifrons; s. hemisphere;
20–24½ ft (6–7.5 m), page 212

Longman's beaked whale
Indopacetus pacificus; n. and s. hemisphere;
22–26 ft (6.7–8 m), page 214

Northern bottlenose whale
Hyperoodon ampullatus; n. hemisphere;
23–29½ ft (7–9 m), page 210

Arnoux's beaked whale
Berardius arnuxii; s. hemisphere;
25½–31¾ ft (7.8–9.7 m), page 206

50 ft (15.25 m)

Length 10-33 ft (3–10 m), beak and dorsal fin absent

Beluga
Delphinapterus leucas;
n. hemisphere; 9¼–16½ ft
(2.8–5 m), page 200

Narwhal
Monodon monoceras;
n. hemisphere; 12½–16½ ft
(3.8–5 m), page 196

Length 10–33 ft (3–10 m), beak absent, dorsal fin present

Long-finned pilot whale
Globicephala melas;
n. and s. hemisphere;
12½–20 ft (3.8–6 m),
page 124

Killer whale
Orcinus orca;
n. and s. hemisphere;
18–32¼ ft (5.5–9.8 m),
page 150

False killer whale
Pseudorca crassidens;
n. and s. hemisphere;
14–20 ft (4.3–6 m), page 154

Common minke whale
Balaenoptera acutorostrata;
n. and s. hemisphere;
23–33 ft (7–10 m), page 86

Pygmy right whale
Caperea marginata;
s. hemisphere; 18–21½ ft
(5.5–6.5 m), page 78

Antarctic minke whale
Balaenoptera bonaerensis;
s. hemisphere;
24–35 ft (7.3–10.7 m),
page 88

Omura's whale
Balaenoptera omurai;
n. and s. hemisphere;
31½–37¾ ft (9.7–11.5 m),
page 98

Short-finned pilot whale
Globicephala macrorhynchus;
n. and s. hemisphere;
12–21½ ft (3.6–6.5 m),
page 122

50 ft (15.25 m)

Length over 33 ft (10 m), beak and dorsal fin present

Baird's beaked whale *Berardius bairdii*;
n. hemisphere; 35–42 ft (10.7–12.8 m), page 208

Length over 33 ft (10 m), beak absent, dorsal fin absent or reduced

Gray whale *Eschrichtius robustus*;
n. hemisphere; 39½–56 ft (12–17 m), page 82

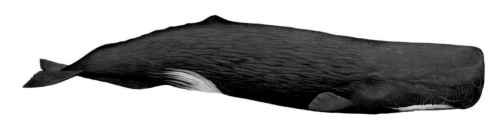

Sperm whale *Physeter macrocephalus*;
n. and s. hemisphere; 36–59 ft (11–18 m), page 186

Southern right whale *Eubalaena australis*;
s. hemisphere; 49 ft (15 m), page 68

Bowhead whale *Balaena mysticetus*;
n. hemisphere; 46–59 ft (14–18 m), page 76

North Atlantic right whale *Eubalaena glacialis*;
n. hemisphere; 43–52 ft (13–16 m), page 70

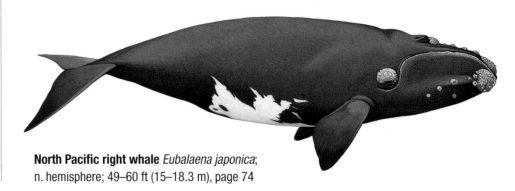

North Pacific right whale *Eubalaena japonica*;
n. hemisphere; 49–60 ft (15–18.3 m), page 74

100 ft (30.5 m)

Length over 33 ft (10 m), beak absent, dorsal fin present

Bryde's whale *Balaenoptera edeni*;
n. and s. hemisphere; 37¾–47½ ft (11.5–14.5 m), page 92

Sei whale *Balaenoptera borealis*;
n. and s. hemisphere; 39½–52 ft (12–16 m), page 90

Humpback whale *Megaptera novaeangliae*;
n. and s. hemisphere; 37¾–49 ft (11.5–15 m), page 102

Fin whale *Balaenoptera physalus*;
n. and s. hemisphere; 59–72¼ ft (18–22 m), page 100

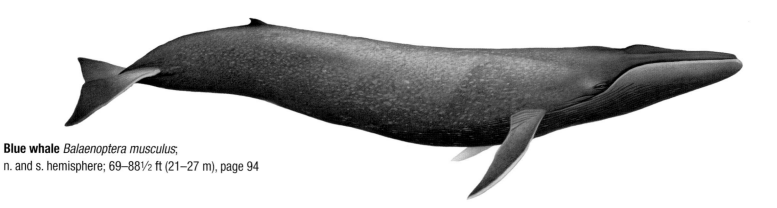

Blue whale *Balaenoptera musculus*;
n. and s. hemisphere; 69–88½ ft (21–27 m), page 94

100 ft (30.5 m)

Surface Behaviors

Cetacean behaviors at the surface can range from simply breaking the surface of the water to breathe, to highly conspicuous leaps. When a cetacean exhales, carbon dioxide is released from the lungs via the blowhole(s). When this warm exhaled air meets the cooler outside air, it condenses and forms a spout or blow. In large whales, the size and shape of the blows can be used to identify species from a distance. The rate of exhalation can also be used to infer an individual's activity state and energy expenditure.

Fluking is a common surface behavior in large cetaceans. This behavior occurs when the flukes break the surface of the water at the beginning of a dive. The angle at which the flukes break the surface may be used as an indication of dive depth. When a whale flukes high and straight up in the air, this indicates the whale is using the full force of its body to propel itself to depth. Shallower dives are indicated by lower, more angled flukes.

Other types of surface behaviors can be categorized into three groups: high-speed traveling, aerial behavior, and surface-feeding behavior. In general, dolphins tend to exhibit the widest variety of surface behaviors. Despite their massive size, baleen whales may also display an astonishing variety of surface behaviors including breaches and lunges. Porpoises have a tendency to be more cryptic at the surface (with the exception of Dall's porpoise, see right below).

Whale blows

The blows of large whales are highly visible from a distance. The blue whale is the largest cetacean and, accordingly, has the tallest blow. The right whale blow has a "V"-shape, which reflects its angled blowholes. The humpback whale blow is slightly shorter and has a "bushy" appearance. The sperm whale blow points forward and to the left, which reflects the high degree of asymmetry in the skull and nasal passages.

Large whale blows

Blue whale

Right whale

Humpback whale

Sperm whale

Traveling

Porpoising occurs during high-speed traveling. Porpoising is characterized by rapid swimming just below the surface of the water, punctuated by quick, lateral leaps out of the water. Porpoising is often a group behavior, with all individuals porpoising in a coordinated fashion. Despite its name, porpoising occurs in dolphins as well as porpoises (and even in sea otters, fur seals, and sea lions). The function of porpoising is that it enables an animal to travel more efficiently. Water is more than 800 times denser than air, and so for the brief moment during which a cetacean is out of the water, the drag force is markedly reduced. This enables dolphins and porpoises to travel more quickly and with less energy expenditure. High-speed travel may also form "rooster-tails." This occurs when dolphins and porpoises skim or "run" along the surface of the water, creating a "V"-shaped wave.

Bow-riding is another form of surface traveling behavior. During bow-riding, dolphins, porpoises, and other small odontocetes ride the pressure wave formed at the bow of a vessel. Individuals are lifted up and pushed forward as water is circulated in front of an advancing vessel. A bow-riding individual frequently leaps in order to catch a breath quickly. The function of bow-riding is unclear since bow-riding individuals are often observed approaching traveling vessels, bow-riding for some time, and then returning to their original location. Thus, enhancement of energetic efficiency during traveling is not likely to be the main function of bow-riding. Instead, it may be a form of play behavior.

Porpoising and rooster-tailing
Surface traveling behaviors such as porpoising and rooster-tailing enable small cetaceans to swim more efficiently. On the left, a group of southern right whale dolphins porpoises off Kaikoura, New Zealand. On the right, a Dall's porpoise in Southeast Alaska forms the characteristic "rooster-tail" of this species, which occurs during high-speed travel.

Surface Behaviors

Aerial behavior

One of the most conspicuous aerial behaviors is leaping—known as breaching in large cetaceans. Leaps may serve several functions including: removal of ectoparasites; enhancement of foraging, social, and mating strategies; communication; and play. For example, the spinning leaps of spinner dolphins may create enough force to dislodge a remora (suckerfish). Leaps performed during cooperative foraging enable dolphins to take a breath quickly while gaining momentum to return to depth swiftly. Leaps performed during socialization may occur during mating chases, male displays, and male-male competition (see page 22). Leaps may also serve as visual forms of in-air communication or percussive forms of underwater communication. At other times, cetaceans appear to leap for the fun of it and so leaps may serve as a form of play behavior. Leaps may also be a by-product of an intense level of activity or excitement, sometimes serving as an "exclamation point" to accentuate the message that is being communicated.

Other aerial behaviors include spyhops, lobtailing (also called fluke- or tail-slapping), and flipper-slaps. During spyhopping, a cetacean rises vertically so that the eyes are above the surface of the water, allowing an individual to scan the surface of the water. Spyhopping is often performed in a social or play context. Both lobtailing and flipper-slapping are percussive behaviors that create sounds that travel in air and underwater. During lobtailing and flipper-slapping, the flukes or flippers, respectively, are slapped against the surface of the water, often repeatedly. These behaviors may serve to communicate an internal state of aggression or annoyance. Lobtailing may also serve to herd and stun fish during foraging.

Acrobatic cetaceans
Spinner dolphins and dusky dolphins are two of the most acrobatic cetaceans. Spinner dolphins (left) get their name from their high, spinning leaps. As many as 14 complete rotations may be performed during a single leap. Dusky dolphins (above left) perform four different types of leaps, including the coordinated leap shown here. Coordinated leaps performed in synchrony by two individuals may function to form and reinforce social bonds.

Spyhopping and lobtailing

Spyhopping
Spyhopping involves projection of the rostrum out of the water. This enables the individual to see above the surface of the water, perhaps to spot potential prey (for killer whales), predators, or conspecifics.

Lobtailing
During lobtailing, the animal—for example, a humpback whale—throws its fluke into the air. The flukes are slapped back against the surface creating a splash.

Surface feeding behavior

Cetacean foraging strategies may involve surface behaviors. In addition to the aerial behaviors described left, cetacean foraging strategies may involve lunge-feeding, skim-feeding, and beach-hunting.

Lunge-feeding is performed by rorquals (such as blue, Bryde's, fin, and humpback whales) and occurs when a rapidly swimming individual breaks the surface of the water with the mouth open. During surfacing, the ventral pleats expand as the animal engulfs massive amounts of water— up to 70 percent of the animal's body weight—and prey. The mouth then slowly closes as the water filters out of the mouth, leaving the prey items trapped inside by the baleen plates. In some cases, an individual or group herds prey toward the surface, using the air–water interface as an impenetrable boundary against which the prey are trapped (see page 29).

During skim-feeding, cetaceans slowly travel along the surface of the water with their mouths open, trapping small prey items occurring just below the surface of the water. This behavior is typically performed by right and pygmy right whales. The highly arched rostrum of these species holds long, fine baleen plates that are capable of trapping small prey items, such as copepods and euphasiids (krill).

Beach-hunting occurs in certain populations of killer whales and bottlenose dolphins. This specialized foraging strategy involves intentional stranding. In Patagonia at the southern end of South America and the Crozet archipelago in the southern Indian Ocean, killer whales beach themselves to capture seal and sea lion pups. In northwest Australia and the southeastern United States, bottlenose dolphins beach themselves to trap fish against the shore. Beach-hunting requires a learning period and the behavior may be passed down through generations.

Beach-hunting (above left)

A killer whale intentionally strands itself to capture a South American sea lion pup. This specialized foraging strategy requires years of practice. Studies indicate that older pod members teach this behavior to calves.

Skim-feeding (above right)

A southern right whale skim-feeds at the surface. The baleen plates are visible; they act like a sieve by trapping small prey items and expelling water. Also noticeable are the highly arched upper and lower jaws, and the hardened patches of skin called callosities.

How & Where to Watch

Watching whales, dolphins, and porpoises in the wild can be one of the most enjoyable experiences of a lifetime. Knowing where to watch cetaceans, and how to watch them, are the keys to success. The first step is to select a location. There are many magnificent locations around the world to watch cetaceans, from land, sea, or air. The next step is to prepare for the experience. Bringing proper equipment and gear will increase your viewing capabilities and ensure your personal comfort. It is also important to do some background research ahead of time regarding the most appropriate viewing platform, the natural history of the species, and whale-watching guidelines.

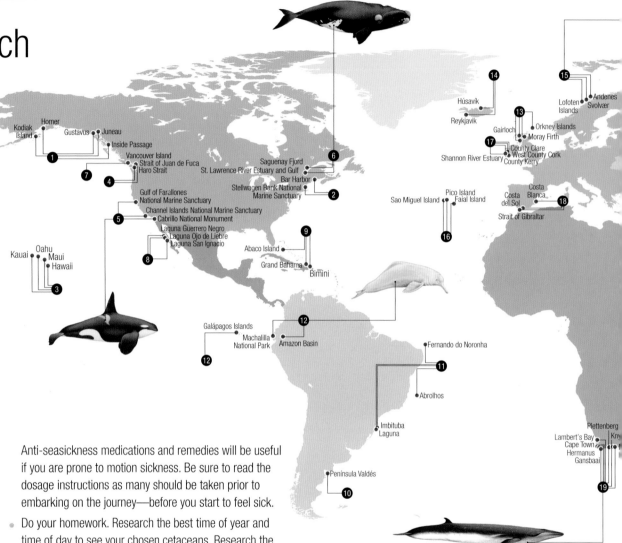

To have a successful and enjoyable cetacean-viewing experience, consider the following factors:

- Select your viewing platform. There are a variety of ways to view cetaceans. The most common platforms are boats, which may range from kayaks or small inflatable boats to large vessels carrying 100 or more passengers. Cetaceans may also be viewed opportunistically from cruise ships and ferries. In some places, where permissible by law, it is possible to swim with cetaceans. Aerial platforms, such as small planes and helicopters, provide a unique perspective for viewing large whales and large groups of dolphins. Cetaceans may also be viewed from shore in many places.

- Bring proper equipment. Marine mammal laws and protections (see below) regulate how closely cetaceans may be viewed, so binoculars and a camera with a telephoto lens are important for obtaining close-up views of cetaceans. If viewing cetaceans from shore, a spotting scope is useful. It is also important to be prepared for the weather. Bring plenty of protection from the cold, wind, rain, and sun. It is always a few degrees cooler on the water, and weather can change quickly, so layers are essential. It is also a good idea to bring water and snacks.

Anti-seasickness medications and remedies will be useful if you are prone to motion sickness. Be sure to read the dosage instructions as many should be taken prior to embarking on the journey—before you start to feel sick.

- Do your homework. Research the best time of year and time of day to see your chosen cetaceans. Research the various tour operators to see if you would prefer a smaller, private charter tailored to your specific interests or a larger, more general tour. Inquire about the educational and conservation-oriented aspects of the tour. Is there a naturalist on board or are other educational opportunities available? Does the operator partner with a research or non-profit organization?

- Learn to identify species. This book provides useful information for identifying cetacean species according to overall body size, shape, and color. It is also important to be able to recognize distinctive features that may be viewed from afar such as the shape of the blow, dorsal fin, and fluke.

- Abide by local, regional, and federal marine mammal laws and protections. These help to avoid disturbing the natural behavior of the animals and help to maintain a sustainable cetacean-viewing industry. These laws and protections may restrict the distance at which cetaceans may be viewed, the length of time that they may be viewed, and/or the number of vessels that may be near an individual cetacean or group at one time.

Where to watch

The map above is not an all-inclusive account of places and species but shows some of the best locations for viewing cetaceans, based on factors such as biodiversity, abundance, predictability, and accessibility.

Approach zones

Whale watching guidelines help to minimize disruption of whales' natural behaviors by restricting how vessels approach whales. Speed limits help to ensure that vessels slow down as they approach the whales. No head-on approach zones prevent vessels cutting off a whale's direction of movement.

North America

1. USA, Alaska: *humpback whale, killer whale, gray whale, harbor porpoise, Dall's porpoise, Pacific white-sided dolphin*

2. USA, New England: *North Atlantic right whale, humpback whale, fin whale, minke whale, Atlantic white-sided dolphin, harbor porpoise*

3. USA, Hawaii: *humpback whales, spinner dolphin, common bottlenose dolphin, short-finned pilot whale, false killer whale*

4. USA, Washington: *killer whale, gray whale, minke whale, humpback whale, Dall's porpoise, harbor porpoise*

5. USA, California: *gray whale, humpback whale, blue whale, long-beaked common dolphin, short-beaked common dolphin, Risso's dolphin, Dall's porpoise, Pacific white-sided dolphin, common bottlenose dolphin, killer whale, northern right whale dolphin*

6. Canada, Québec: *beluga, blue whale, fin whale, minke whale, humpback whale, North Atlantic right whale, Atlantic white-sided dolphin, harbor porpoise, white-beaked dolphin*

7. Canada, British Columbia: *killer whale, Dall's porpoise, gray whale, minke whale, humpback whale, harbor porpoise, Pacific white-sided dolphin*

8. Mexico, Baja California: *gray whale, common bottlenose dolphin*

Caribbean

9. Bahamas: *Atlantic spotted dolphin, common bottlenose dolphin, sperm whale, humpback whale, Blainville's beaked whale, Risso's dolphin, short-finned pilot whale, false killer whale, killer whale, dwarf sperm whale, pygmy sperm whale*

South America

10. Argentina: *southern right whale, killer whale, dusky dolphin*

11. Brazil: *spinner dolphin, southern right whale, common bottlenose dolphin, humpback whale, minke whale, rough-toothed dolphin*

12. Ecuador: *humpback whale, Bryde's whale, sperm whale, Amazon River dolphin (boto), common bottlenose dolphin, killer whale, pantropical spotted dolphin, short-finned pilot whale, tucuxi*

Europe

13. United Kingdom, Scotland: *common bottlenose dolphin, harbor porpoise, minke whale*

14. Iceland: *minke whale, humpback whale, blue whale, white-beaked dolphin, harbor porpoise, killer whale, northern bottlenose whale, sperm whale*

15. Norway: *killer whale, sperm whale, minke whale, white-beaked dolphin, harbor porpoise, humpback whale, long-finned pilot whale*

16. Portugal, Azores: *short-beaked common dolphin, sperm whale, short-finned pilot whale, common bottlenose dolphin, Atlantic spotted dolphin, Risso's dolphin, blue whale, fin whale, sei whale, false killer whale, killer whale, striped dolphin, northern bottlenose whale, Cuvier's beaked whale*

17. Ireland: *fin whale, minke whale, common bottlenose dolphin, short-beaked common dolphin, harbor porpoise*

18. Spain: *Risso's dolphin, common bottlenose dolphin, long-finned pilot whale, fin whale, minke whale, Cuvier's beaked whale, short-beaked common dolphin, harbor porpoise, sperm whale, striped dolphin, killer whale*

Africa

19. South Africa, Western Cape Province: *southern right whale, humpback whale, Bryde's whale, Heaviside's dolphin, bottlenose dolphin (Note: it is unclear if this species is the common or Indo-Pacific bottlenose dolphin), long-beaked common dolphin*

Asia

20. Japan, Rausu, Hokkaido: *minke whale, sperm whale, Baird's beaked whale, killer whale, short-finned pilot whale, Pacific white-sided dolphin, Dall's porpoise, harbor porpoise*

21. Japan, Okinawa main island and Zamami, Okinawa: *humpback whale*

22. Japan, Ogasawara Islands: *bottlenose dolphin, spinner dolphin, sperm whale, humpback whale*

Australia, New Zealand

23. New Zealand, South Island: *dusky dolphin, killer whale, sperm whale, Hector's dolphin, humpback whale*

24. New Zealand, North Island: *short-beaked common dolphin, common bottlenose dolphin, Bryde's whale, killer whale*

25. Australia, Western Australia: *bottlenose dolphin (Note: it is unclear if this species is the common or Indo-Pacific bottlenose dolphin)*

26. Australia, Queensland: *humpback whale, Indo-Pacific bottlenose dolphin, Indo-Pacific humpback dolphin*

27. Australia, New South Wales: *humpback whale, Indo-Pacific bottlenose dolphin*

THE SPECIES DIRECTORY

How to Use the Species Directory

Following the introductory material that highlights the biology and natural history of cetaceans, the main portion of the book is The Species Directory, organized according to the major groups (suborder and family-level classification) as well as particular species.

Group introduction

Each family or family group is described including common name, number of species, interesting biology, and illustrations of a representative member of the family that shows key characteristics and typical dive or feeding sequences.

Araguaian river dolphin
Inia araguaiaensis

Species accounts

Every entry gives detailed information about the biology of a particular species. This guide follows the scientific names of species and subspecies and the taxonomic arrangement of the Marine Mammal Society's Committee on Taxonomy (2014). The authoritative list is updated annually and can be consulted at www.marinemammalscience.org. Some described species have been omitted because of concerns about their distinctiveness. For example, two new species of the Amazon river dolphin have been proposed, *Inia boliviensis* and *Inia araguaiaensis* but they are not listed here since there are questions about their species-level distinctiveness given the limited sampling of individuals from different geographic populations. Future work is needed to clarify their taxonomy.

Each species account begins with the currently accepted common name of the species. Some species names have alternate common names and those are also given. Common names can vary but each species has only one scientific name, which is listed below the common name. Next, under taxonomy, subspecies or different populations and closest relatives are listed. Any similar species are also listed, covering other whales that share some characteristics of appearance of the featured species. Features that can be used to help distinguish it are given. This is followed by measurements of birth and adult weight, and a brief mention of diet and group size. The habitat preference is listed and a range map showing the geographical distribution of the species provided. A list of major historic and current threats to the species is followed by the species' current status in the wild according to the largest global conservation organization, the International Union for the Conservation of Nature (IUCN).

An identification checklist of key features is provided for quick reference. This is followed by a paragraph on anatomy, which presents the animal's physical characteristics that are most useful for identification. The coloration pattern is described, as well as differences between males and females or between different populations. The behavior paragraph describes the social organization of the species, including group size, and whether it is solitary or social. Also included are typical behaviors to look out for, such as swimming, diving, and vocalizations. Next, the food and foraging paragraph describes the typical foraging behaviors and prey location in relation to diving behavior.

Life history is also covered. This paragraph describes the species mating system and behavior, breeding season, lactation and weaning, and life span. Next, under the

Proposed new species

Inia araguaiaensis was described as a new river dolphin species from the Araguaia River basin of Brazil. However, the study examined samples from two extremes of the range of *Inia* so it is unclear if the DNA differences represent species level differences or were due to sampling from two widely separated locations. Also, the diagnostic anatomical differences reported in the skulls were based on examination of very few specimens.

conservation and management heading, information on population estimates, threats, and efforts to protect the species is provided.

Finally, each account includes a main image of the species in profile. Some accounts also include illustrations showing animals at different life stages, or special features or variations such as distinctive flukes or flipper shapes. Illustrations of characteristic features of the skull and teeth or baleen for some species are included. In most cases a typical dive sequence or typical surface behaviors are also illustrated.

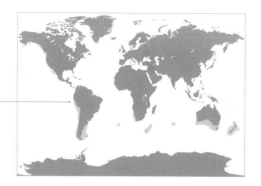

Information chart

Chart of at-a-glance information about the species including scientific name, common names, taxonomy, weight, diet, major threats, and IUCN status.

Distribution map

Shows the known distribution range of the species, including information on depth and habitat.

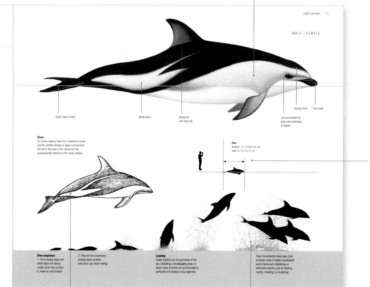

Main profile illustration

A typical example of the species with annotation highlighting key features and characteristics. The main image typically applies to both males and females except as noted.

Measurements

A silhouette of the species is shown to scale, along with information about adult and newborn length.

Identification checklist

A list of key features of the species.

Species description

Main narrative text describing anatomy, behavior, food and foraging, life history, and conservation and management.

Distinctive features

Smaller illustrations here focus on distinctive features of the species, as well as other interesting characteristics or variations.

Dive sequence and surface behavior

A description of a typical dive sequence and surface behaviors where known.

BALEEN WHALES
Right Whales

Balaenidae (right whales) includes one species of bowhead and three species of right whales. All four balaenid species are large-bodied cetaceans, ranging in adult body lengths of around 43–59 ft (13–18 m). All species possess a strongly arched rostrum with the longest baleen of all baleen whales. Right whales can be distinguished from bowheads by distinctive growths known as callosities on their snouts and above their eyes. Balaenids are found in both hemispheres and occupy temperate and polar waters.

Right whale characteristics

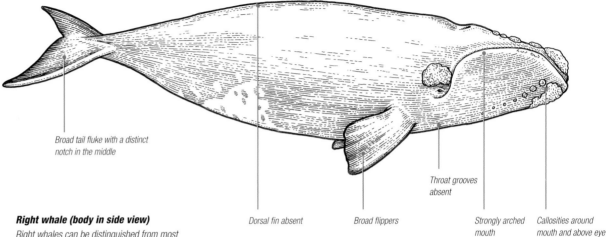

Broad tail fluke with a distinct notch in the middle

Throat grooves absent

Dorsal fin absent

Broad flippers

Strongly arched mouth

Callosities around mouth and above eye (absent in bowheads)

Right whale (body in side view)
Right whales can be distinguished from most other large cetaceans by the absence of the dorsal fin and their extremely thick bodies in profile. Callosities around the mouth are white on account of whale lice.

- The southern right whale (*Eubalaena australis*) is the only balaenid species to occupy waters in the southern hemisphere. The North Pacific (*Balaena japonica*) and Atlantic right whales (*Eubalaena glacialis*) occupy their respective ocean basins, and the northern extents of their ranges overlap with that of the bowhead whale (*Balaena mysticetus*).

- Balaenids are among the largest cetacean species. The sperm whale is the only toothed whale that approaches the size of the balaenids.

- The diet of balaenids consists of small- to medium-sized crustaceans, including krill and copepods. Prey items are trapped within the baleen, which may achieve lengths of 10 ft (3 m) in bowhead whales.

- Balaenids do not have the throat grooves and pouches that characterize rorqual whales, which engage in bulk engulfment feeding. Rather, balaenids skim-feed at the water's surface by swimming slowly with their mouths open and do not lunge forward toward their prey.

- All balaenid species lack dorsal fins. In this regard, balaenids are distinct from all other baleen whale species, except for the gray whale. Balaenids can be distinguished from gray whales by size, general color, and the shape of the head.

- The callosities (patches of rough, calcified skin) of right whales can be used to distinguish those species from the bowhead whale, which lacks the distinctive growths. In addition, the flippers of bowheads tend to be narrower relative to body size than the right whale species.

- The bowhead and southern right whales have been assigned a conservation status of Least Concern but both northern species of right whale are Endangered. All species of balaenid have been hunted in the past, beginning in the 16th century with the decimation of bowhead and North Pacific right whale populations.

North Atlantic right whale skull

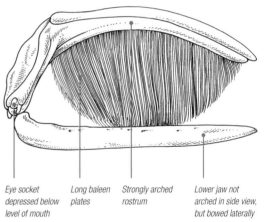

Eye socket depressed below level of mouth

Long baleen plates

Strongly arched rostrum

Lower jaw not arched in side view, but bowed laterally

Skull
The rostrum of the right and bowhead whales is greatly arched to accommodate the long plates of baleen hanging from the edges of the upper jaw. The eye socket is depressed below the level of the mouth, further accentuating the curve of the rostrum.

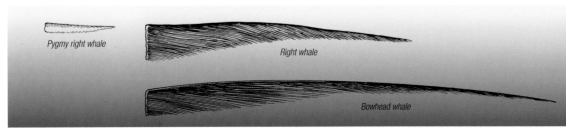

Pygmy right whale

Right whale

Bowhead whale

Baleen plates
Baleen is developed as long plates that hang from the edges of the roof of the mouth like vertical blinds. The "fringe" of the baleen is located on the inner edge in contact with the tongue. The baleen of balaenids is narrow, long, and dark in color. The baleen of the pygmy right whale is cream in color and narrow, but nowhere near the length of the plates in balaenids. The baleen of rorquals ranges in color from light to dark.

SOUTHERN RIGHT WHALE

Family Balaenidae

Species *Eubalaena australis*

Other common names Black right whale

Taxonomy No subspecies recognized

Similar species North Atlantic and North Pacific right whales

Birth weight 1,760–2,200 lb (800–1,000 kg)

Adult weight 44,000–220,500+ lb (20,000–100,000+ kg)

Diet Zooplankton—such as calanoid copepods—also krill, and lobster-krill

Group size 1–2, up to 30 or more in temporary feeding or social aggregations

Major threats Collisions with ships, entanglement in fishing gear

IUCN status Least Concern

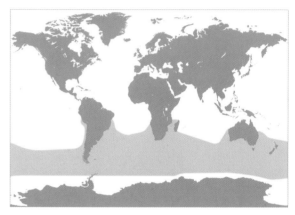

Range and habitat This species is found all around the southern hemisphere—in nearshore waters at lower latitudes during the winter calving and breeding season, and far offshore mainly between 40°S and 50°S during the summer feeding season.

Identification checklist

- Smooth, black body, some animals have white belly patches
- Rotund body
- No dorsal fin
- Large head with strongly arched mouthline
- Square-shaped flippers
- Rough, whitish patches on head

Anatomy

Southern right whales are morphologically similar to North Atlantic or North Pacific right whales. Slight differences between North Atlantic and Southern right whales in the cranial bones near the eye socket are likely due to age rather than species differences.

Behavior

Southern right whale behavior is likely to be the same as North Atlantic or North Pacific right whales. During their summer feeding season, they occur far offshore—research has focused on the near-shore calving grounds. Large breeding populations are known from South Africa, Argentina–Brazil, and Australia, with others known from New Zealand, Chile–Peru, Tristan da Cunha, and Madagascar. The separations are maintained through habitat fidelity.

Food and foraging

These whales migrate annually between low-latitude, nearshore winter calving grounds and high-latitude summer feeding grounds far offshore around the Antarctic. Stomach contents from 20th-century Soviet whaling were mainly copepods north of 40°S and mainly krill south of 50°S, with a mix in between. Other prey include mysid shrimp and pelagic crab juveniles. Prey selection involves decision-making about relative costs and benefits. Larger prey contains more energy, but the whale spends more energy during feeding—because faster swimming results in more drag.

Life history

Females mature at 9–10 years old, and give birth to single calves in winter every third year. All life-history parameters should be expected to be similar across all three right whale species.

Conservation and management

The Chile–Peru population is listed as Critically Endangered, with fewer than 50 breeding adults. The three larger populations are growing at 6–7 percent annually, suggesting that the population could have grown from 7,500 in 1997 to 20,000–25,000 today. This species was depleted by more than 150,000 animals, which were killed by whalers between the 18th century and the 1930s, and another 3,000 by illegal Soviet whaling in the 1950–60s. Sources of anthropogenic mortality are likely similar to the two northern hemisphere species, but lower rates of human activity and relatively remote feeding grounds keep the mortality rates lower.

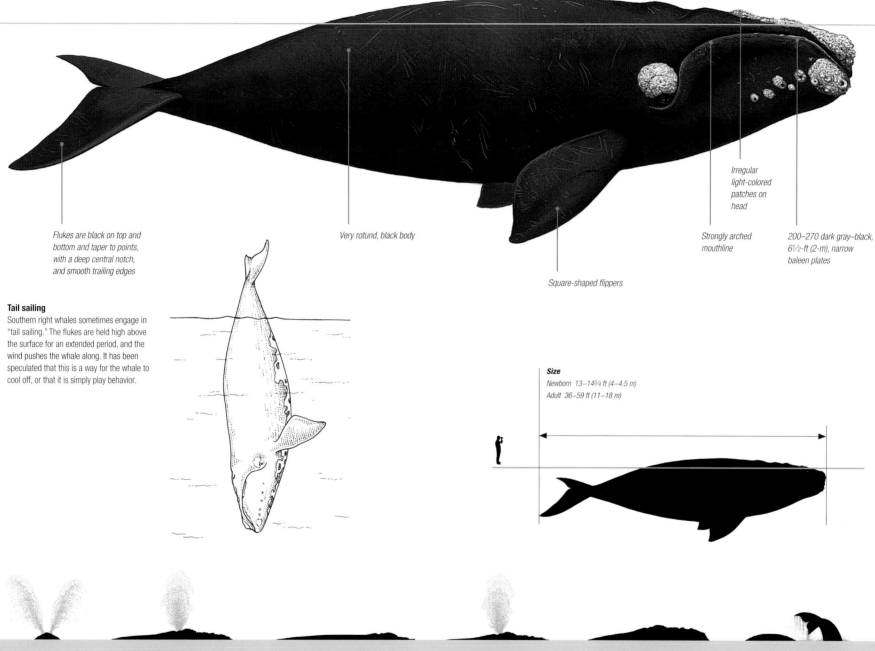

MALE / FEMALE

Flukes are black on top and
bottom and taper to points,
with a deep central notch,
and smooth trailing edges

Very rotund, black body

Irregular
light-colored
patches on
head

Strongly arched
mouthline

200–270 dark gray–black,
6½-ft (2-m), narrow
baleen plates

Square-shaped flippers

Tail sailing
Southern right whales sometimes engage in
"tail sailing." The flukes are held high above
the surface for an extended period, and the
wind pushes the whale along. It has been
speculated that this is a way for the whale to
cool off, or that it is simply play behavior.

Size
Newborn 13–14¾ ft (4–4.5 m)
Adult 36–59 ft (11–18 m)

Blow
The blow is about 6½–10 ft
(2–3 m) high, bushy, and
clearly "V"-shaped when
seen from ahead or behind.

Dive sequence
1. The blow appears oval
and bushy from the side.

2. The head disappears below
the surface as the whale moves
forward, showing only the
broad, smooth back.

3. A typical surfacing
includes about 4–6 blows in
succession, approximately
10–30 seconds apart.

4. After the last blow, the
head is lifted much higher
out of the water as the
whale inhales deeply.

5. The head is pushed down and the
whale dives at a very steep angle, the
body bends, and more of the back is
visible. The tail flukes emerge as the
whale wheels forward and down,
finishing more or less vertical above
the surface and then sliding down.

NORTH ATLANTIC RIGHT WHALE

Family Balaenidae

Species *Eubalaena glacialis*

Other common names Black right whale, northern right whale

Taxonomy No subspecies recognized

Similar species North Pacific and southern right whales

Birth weight 1,760–2,200 lb (800–1,000 kg)

Adult weight 44,000–220,500 lb (20,000–100,000 kg)

Diet Zooplankton—such as calanoid copepods—also krill, barnacle larvae, and pteropods

Group size 1–2, up to 30 or more in temporary feeding or social aggregations

Major threats Collisions with ships, entanglement in fishing gear

IUCN status Endangered

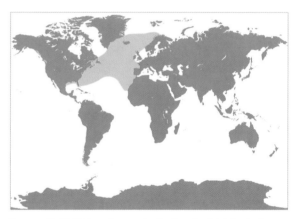

Range and habitat This species occurs in primarily continental shelf waters from Florida to eastern Canada. There are a few far offshore sightings, and occasional to rare occurrences in northern and eastern portions of its historical range.

Identification checklist

- Smooth, black body—some animals have white belly patches
- Rotund body
- No dorsal fin
- Large head with strongly arched mouthline
- Square-shaped flippers
- Rough, whitish patches on head

Anatomy

North Atlantic right whales are robust, with broad, smooth backs, and no dorsal fin. The color is usually black, and some animals have irregular white patches on the belly. The head comprises about a quarter or third of the body length. The rostrum is narrow and arched, and the curve of the mouth opening is strongly arched. There are irregular whitish patches called callosities on the rostrum, on the chin, along the lower jaw, and over the eye—usually behind the blowholes, and sometimes on the lower lips. The callosities are patches of thickened, keratinized skin inhabited by dense populations of light-colored whale lice. The flippers are up to 5¾ ft (1.7 m) long, and the flukes are up to 20 ft (6 m) across. Right whale baleen plates are up to 9 ft (2.7 m) long.

Behavior

North Atlantic right whales are relatively asocial and do not form stable pods or herds. They are capable of diving for 15–20 minutes or more, and can easily reach near-bottom depths in their continental shelf feeding grounds. A typical right whale foraging dive involves a rapid and steep descent powered by continuous fluke strokes to overcome buoyancy, a prolonged bottom excursion at a relatively constant depth, and a rapid and shallow buoyant glide back to the surface. Vigorous behaviors at the surface are common, including breaching, lobtailing, and flipper-slapping.

Food and foraging

North Atlantic right whales are strongly migratory, moving annually between high-latitude feeding grounds and low-latitude calving grounds. Known feeding grounds in the North Atlantic are in the Gulf of Maine and adjacent waters. The calving ground is in coastal waters off the southeastern United States, but occasional births have occurred in New England waters. The location of breeding is unknown, although recent observations suggest that some mating may occur in the central Gulf of Maine. Feeding is by "skimming." The whales simply swim slowly forward with the mouth open. Water flows in through the opening at the front, then passes through the baleen, straining prey organisms from the water and collecting them on the inside. Feeding can occur at or just below the surface where it can be observed easily, or more often at depth and out of sight. North Atlantic right whales are obligate planktivores, with the principal prey being large, late-stage juveniles and adults of the copepod *Calanus finmarchicus* (crustaceans approximately the

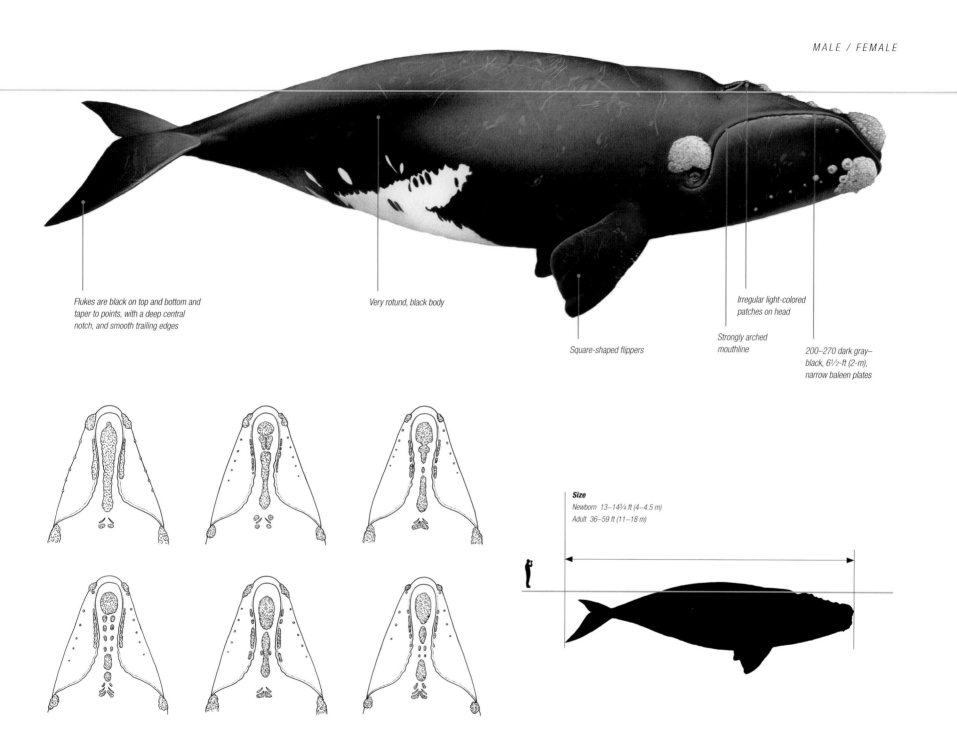

MALE / FEMALE

Flukes are black on top and bottom and taper to points, with a deep central notch, and smooth trailing edges

Very rotund, black body

Irregular light-colored patches on head

Square-shaped flippers

Strongly arched mouthline

200–270 dark gray–black, 6½-ft (2-m), narrow baleen plates

Size
Newborn 13–14¾ ft (4–4.5 m)
Adult 36–59 ft (11–18 m)

Callosity patterns
Callosity patterns—here viewed looking down on the top of the head—develop in the first year and serve as "fingerprints" that enable repeated identification of individual animals from photographs.

NORTH ATLANTIC RIGHT WHALE

size of a grain of rice). At times they also feed on other zooplankton, including smaller copepods, euphausiids ("krill"), barnacle larvae, and pteropods (planktonic snails). The sensory mechanisms involved in prey detection and foraging probably include at least sight and touch, if not also sound and possibly taste.

Life history

Females give birth to single calves in winter after a 12–13 month gestation—most births occur between December and February. Most calves are weaned around the age of one. Following weaning, the female typically takes a year to "rest"—feeding and rebuilding blubber stores before mating the following winter. The result is a three-year inter-birth interval under good conditions with adequate prey resources. By the time of weaning about a year after birth, North Atlantic right whales have about doubled in length and reached nearly 11,000 lb (5,000 kg) in weight. North Atlantic right whales can live to be at least 65–70 years old.

Conservation and management

North Atlantic right whales are considered to be one of the most endangered mammals in the world, with about 500 living as of 2012. Separate eastern (European) and western (North American) populations likely existed historically. If the two populations were listed separately, the western population would be Endangered and the eastern population would be Critically Endangered, probably Extinct. Substantial anthropogenic mortality is continuing, and is suspected to be retarding recovery of the population. The two most significant sources of mortality are collisions with ships and entanglement in commercial fishing gear. Other hypothesized anthropogenic impacts on right whales include toxic contaminants, habitat loss, noise, and global climate change.

Surface active group

This image shows a surface-active group (SAG) of right whales, among which courtship and breeding take place. The female is upside-down in the center, with her flippers pointing upward, surrounded by multiple males. The number of whales in a SAG ranges from 2 to more than 30, averaging 5. They form in response to a vocalization by the female, who then mostly stays belly-up in the center of the group. Males cluster around, and those that are adjacent to the female attempt to copulate when she rolls over to breathe. SAGs are observed in all North Atlantic right whale feeding grounds—these occur at the wrong time of year for conception, and may represent adult females assessing potential future mates.

Entanglement scarring

More than three-quarters of all North Atlantic right whales carry scars from prior entanglements in fishing gear, particularly at the junction of the flukes with the tail stock. Some individuals have survived multiple entanglements.

Blow
The blow is about 6½–10 ft (2–3 m) high, bushy, and clearly "V"-shaped when seen from ahead or behind.

Dive sequence
1. The blow appears oval and bushy from the side.

2. Very little of the whale is seen above the surface after most blows in a surfacing sequence.

3. The head disappears below the surface as the whale moves forward, showing only the broad, smooth back.

Breaching and lobtailing

Breaching and lobtailing by right whales are seen commonly. Both behaviors produce loud underwater sounds and may have a signaling function.

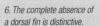

4. A typical surfacing includes about 4–6 blows in succession, approximately 10–30 seconds apart.

5. Their very low profile while surfacing makes right whales difficult to see from the bridge of a ship.

6. The complete absence of a dorsal fin is distinctive.

7. With the final blow, the whale is preparing for a longer dive.

8. After the last blow, the head is lifted much higher out of the water as the whale inhales deeply.

9. The head is pushed down and the whale dives at a very steep angle, the body bends, and more of the back is visible.

10. The tail flukes emerge as the whale wheels forward and down, finishing more or less vertical above the surface and then sliding down.

NORTH PACIFIC RIGHT WHALE

Family Balaenidae

Species *Eubalaena japonica*

Other common names Black right whale, northern right whale

Taxonomy No subspecies recognized

Similar species North Atlantic and southern right whales

Birth weight 1,760–2,200 lb (800–1,000 kg)

Adult weight 44,000–220,500 lb (20,000–100,000 kg)

Diet Zooplankton—such as calanoid copepods—and krill

Group size 1–2, up to 30 or more in temporary feeding or social aggregations

Major threats Collisions with ships, entanglement in fishing gear

IUCN status Endangered

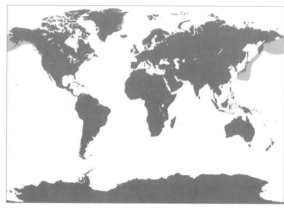

Range and habitat This species was formerly widespread between 40°N and 60°N latitudes on both sides of the Pacific Basin but is presently observed only in and near the Sea of Okhotsk, the southeastern Bering Sea, and the Gulf of Alaska.

Identification checklist

- Smooth, black body—some animals have white belly patches
- Rotund body
- No dorsal fin
- Large head with strongly arched mouthline
- Square-shaped flippers
- Rough, whitish patches on head

Anatomy

North Pacific right whales are morphologically very similar to North Atlantic and southern right whales. It is possible that North Pacific right whales grow somewhat larger than their North Atlantic relatives, however the maximum lengths reported in the literature for the two species differ by very little—60 ft (18.3 m) in the North Pacific and 59 ft (18 m) in the North Atlantic.

Behavior

The behavior of North Pacific right whales is very likely to be essentially the same as in the North Atlantic species. They more often occur in deeper water farther offshore, and may be deeper divers, although there are no quantitative data.

Food and foraging

Based on the patterns of whaling catches, these whales migrate between higher latitudes in summer and lower latitudes in winter. No calving or breeding ground is known—calving and mating may occur far offshore. These whales are skim-feeders on zooplankton, with a broader menu of large copepods to choose from than in the North Atlantic. Some of these copepods are substantially larger than the *Calanus finmarchicus* eaten in the North Atlantic, which may confer energetic advantages. The whales may benefit from the differing peaks in abundance of different prey species.

Life history

Although quantitative data are not available, one would expect North Pacific right whale life history to be the same as in the North Atlantic right whale. A very few sightings of mother-calf pairs have occurred in the southeastern Bering Sea in recent years.

Conservation and management

The northeast Pacific population is listed as Critically Endangered. Present numbers are estimated at several hundred in the northwest and substantially fewer than 100 in the northeast. North American whalers reached the North Pacific in the 1830s, and in less than 20 years killed at least 11,000 right whales. Whaling continued until protective treaties in the 1930–40s. Soviet whalers illegally killed 372 whales from the northeast population in 1963–66, and probably came close to extirpating the stock. There is no recent information on anthropogenic mortality, although they are likely susceptible to the same impacts as North Atlantic right whales.

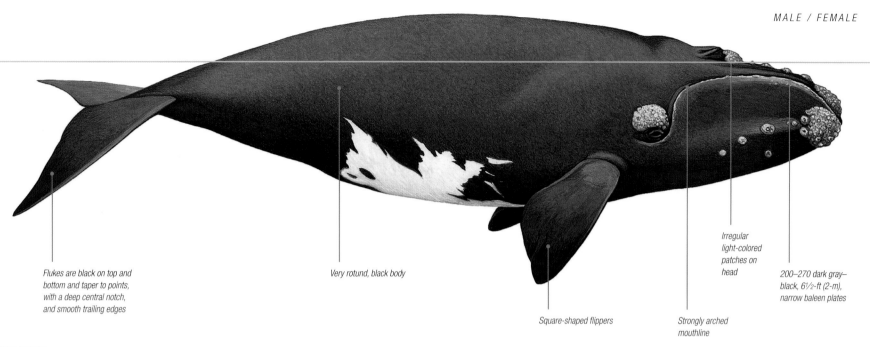

MALE / FEMALE

Flukes are black on top and
bottom and taper to points,
with a deep central notch,
and smooth trailing edges

Very rotund, black body

Square-shaped flippers

Strongly arched
mouthline

Irregular
light-colored
patches on
head

200–270 dark gray–
black, 6½-ft (2-m),
narrow baleen plates

Skim-feeding
A feeding right whale swims forward with the mouth
open, skimming zooplankton from the water as it
flows into the opening between the two rows of
baleen and then laterally through the baleen.

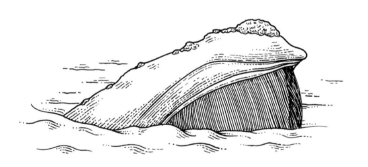

Size
Newborn 13–14¾ ft (4–4.5 m)
Adult 36–60 ft (11–18.3 m)

Blow
The blow is about 6½–10 ft
(2–3 m) high, bushy and
clearly "V"-shaped when
seen from ahead or behind.

Dive sequence
*1. The blow appears oval
and bushy from the side.*

*2. The head disappears below
the surface as the whale moves
forward, showing only the broad,
smooth back.*

*3. A typical surfacing
includes about 4–6 blows in
succession, approximately
10–30 seconds apart.*

*4. After the last blow, the head is
lifted much higher out of the water
as the whale inhales deeply.*

*5. The head is pushed down and the
whale dives at a very steep angle, the
body bends, and more of the back is
visible. The tail flukes emerge as the
whale wheels forward and down,
finishing more or less vertical above
the surface and then sliding down.*

BOWHEAD WHALE

Family Balaenidae

Species *Balaena mysticetus*

Other common names Greenland right whale, Arctic right whale, Great polar whale

Taxonomy No subspecies recognized—closest relatives are right whales

Similar species Superficially resemble right whales

Birth weight 2,000–2,200 lb (900–1,000 kg)

Adult weight 66,000–220,500+ lb (30,000–100,000+ kg)

Diet Zooplankton—including copepods—also krill, mysids, and amphipods

Group size 1 to a few, up to 30 or more in temporary feeding or social aggregations

Major threats Global warming, collisions with ships, and entanglement in fishing gear

IUCN status Least Concern: Okhotsk population is Endangered; Svalbard/Barents is Critically Endangered

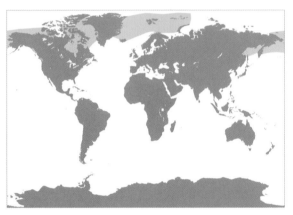

Range and habitat The only large whale living in Arctic waters around the globe, with five distinct populations—Okhotsk Sea, Bering/Chukchi/Beaufort Seas, Hudson Bay/Foxe Basin, Davis Strait/Baffin Bay, and Svalbard/Barents Sea.

Identification checklist

- Smooth, black body with extensive white on the chin—some animals have pale areas on the tail stock
- Rotund body
- No dorsal fin
- Very large head with a distinct peak and a strongly arched mouthline
- Square-shaped flippers

Anatomy

Bowhead whales resemble right whales, but are even more robust, heavier, and lack the callosities on the head. Bowheads possess the thickest blubber layer of any whale, up to 1½ ft (50 cm) thick. The head is proportionately larger, up to 40 percent of body length, with much longer baleen plates.

Behavior

Bowhead whales are slow swimmers generally feeding at the surface but they are capable of long dives, possibly more than an hour—an adaptation for navigating under ice. A bowhead is capable of breaking through ice as thick as 2 ft (60 cm). As in right whales, surface behaviors such as breaching, lobtailing, and flipper-slapping are common.

Food and foraging

Bowhead whales are migratory, with summering and wintering grounds in cold, high-latitude waters—following the advance and retreat of the sea ice. For example, the Bering/Chukchi/Beaufort stock occurs in feeding grounds in the Beaufort Sea in May– September, migrates through the Chukchi Sea in October– November, winters in the northern Bering Sea in November–March, and returns through the Chukchi in March–June. Like right whales, bowheads skim-feed on very small zooplankton prey.

Life history

Bowhead whale life history is likely related to living in a lower-productivity habitat. They grow more slowly and mature later than right whales—at 12–18 years old and possibly as old as 25. Their gestation period is slightly longer at 13–14 months. Calving intervals are similar at three to four years. Bowheads may be the longest-lived mammals. Several taken by Inuit hunters were more than 100 years old, and one was estimated to be 211.

Conservation and management

Whaling depleted bowhead numbers, especially during the 18–19th centuries. Two populations are listed as threatened. Inuit hunts are permitted in Alaska, Russia, and Canada, but have not affected growth. Their remote habitats have insulated bowheads from most anthropogenic impacts. Climate change and the disappearance of sea ice is a concern, as this could alter the ecosystem and allow increases in shipping, fishing, and industrial activity.

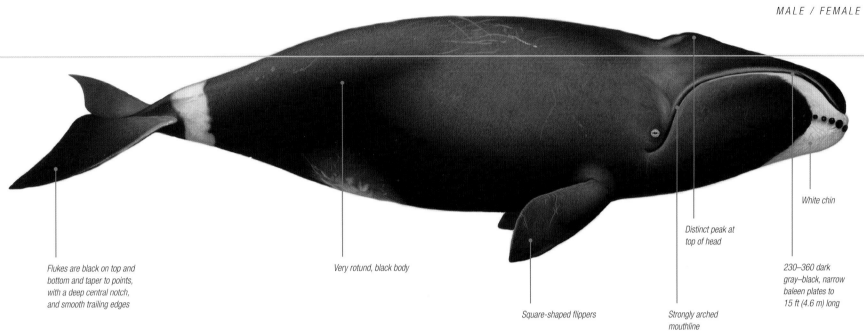

MALE / FEMALE

Flukes are black on top and
bottom and taper to points,
with a deep central notch,
and smooth trailing edges

Very rotund, black body

Square-shaped flippers

Distinct peak at
top of head

Strongly arched
mouthline

White chin

230–360 dark
gray–black, narrow
baleen plates to
15 ft (4.6 m) long

Bowhead anatomy
The bowhead's anatomy is adapted for survival in a frigid,
low-productivity habitat—thick blubber provides insulation
and allows long periods without feeding, an immense mouth
enables feeding on relatively low concentrations of prey, and
a very large tail creates the force to overcome drag from
skim-feeding with the mouth wide open.

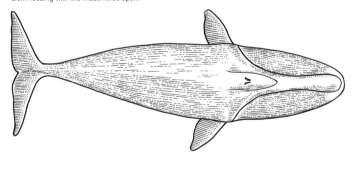

Size
Newborn 13–14¾ ft (4–4.5 m)
Adult 39–59 ft (12–18 m)

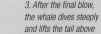

Dive sequence
*1. The blow appears oval
and bushy from the side,
and "V"-shaped from
ahead or behind.*

*2. The broad back with no
dorsal fin is distinctive.*

*3. After the final blow,
the whale dives steeply
and lifts the tail above
the surface.*

*4. The tail flukes lifted high above
the surface at the end of a dive
or before smashing them back
down again (lobtailing).*

*5. A breaching whale
typically twists and
lands back on its side.*

PYGMY RIGHT WHALE

Family Neobalaenidae

Species *Caperea marginata*

Other common names None

Taxonomy Closest relatives are either right whales (based on anatomical evidence) or rorquals (based on DNA evidence)

Similar species Owing to its small size, the pygmy right whale is sometimes confused with the minke whale

Birth weight Unknown

Adult weight Around 7,000 lb (3,200 kg) female; 6,400 lb (2,900 kg) male

Diet Copepod crustaceans and occasional krill

Group size Typically 1–2 individuals, but pods of up to 100 have been observed

Major threats Little information exists for this species and there may be no direct anthropogenic threats

IUCN status Data Deficient

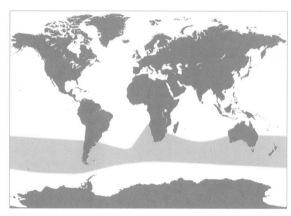

Range and habitat This species is circumpolar in temperate waters of the southern hemisphere. Sightings of live individuals in the ocean are rare.

Identification checklist

- Dark gray dorsally and light gray ventrally
- Smallest baleen whale
- Arched rostrum
- Small dorsal fin positioned posteriorly
- Two throat grooves

Anatomy

Pygmy right whales share features in common with right whales. Most notable is an arched rostrum, although it is not as high as in right and bowhead whales. However, pygmy right whales possess a small, posteriorly positioned dorsal fin and two throat grooves similar to gray whales. Recent fossil evidence suggests that pygmy right whales are the last living relative of a group of baleen whales once thought to be extinct. This whale possesses between 210–230 yellowish-white baleen plates per upper jaw, up to 27 in (69 cm) in length. Pygmy right whales have 17 flattened ribs, which is more than any other baleen whale.

Behavior

Sightings of live pygmy right whales are rare in the wild and most information is gathered from stranded animals. Thus, little is known of their behavior. They do not appear to engage in activities such as breaching or spyhopping and they rarely display their tails.

Food and foraging

Pygmy right whales are strong and fast swimmers, and they employ a skim-type feeding similar to that of right and bowhead whales. Pygmy right whales feed mostly on copepod crustaceans although some stranded whales contained krill in their stomachs.

Life history

Owing to the elusive nature of the species, very little is known about the reproductive behavior and life histories of the pygmy right whale. It is likely that only one calf is born at a time.

Conservation and management

Pygmy right whales are not actively hunted, although a small number of individuals have been taken by accident. Pygmy right whales are covered by international protection measures for all whales, although there are no specific conservation measures for the species at this time.

MALE / FEMALE

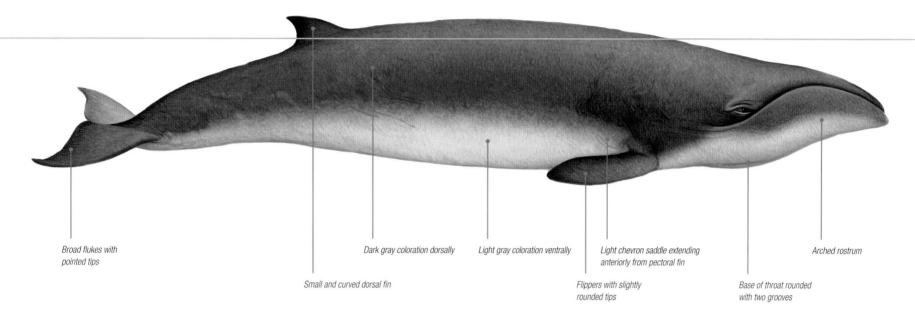

Broad flukes with
pointed tips

Small and curved dorsal fin

Dark gray coloration dorsally

Light gray coloration ventrally

Light chevron saddle extending
anteriorly from pectoral fin

Flippers with slightly
rounded tips

Base of throat rounded
with two grooves

Arched rostrum

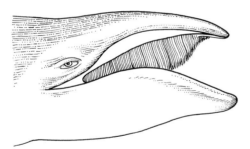

Baleen plates
The rostrum of the pygmy right whale is
arched, but not as strongly as in bowhead
and right whales. There are 210–30 pale
yellow to white baleen plates per side that
are up to 27 in (69 cm) long.

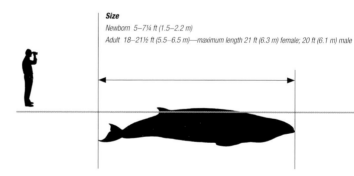

Size
Newborn 5–7¼ ft (1.5–2.2 m)
Adult 18–21½ ft (5.5–6.5 m)—maximum length 21 ft (6.3 m) female; 20 ft (6.1 m) male

Dive sequence
Dives are short, lasting
a few minutes at a time
with only brief periods
at the surface.

*1. The head and blowhole
emerge before the dorsal
fin, although sometimes
the head and dorsal fin are
exposed simultaneously.*

*2. The blow is small
and inconspicuous.*

*3. Next the head
disappears and the back
becomes visible.*

*4. Then the sickle-shaped
dorsal fin emerges.*

*5. The tail stock is
rounded out of the water
as the whale dives.*

*6. The tail fluke typically
does not surface during
the dive. Dives are likely
to be shallow owing to
the short duration.*

BALEEN WHALES

Rorqual Whales & Gray Whale

Rorquals are the most diverse of the baleen whales. They range in size from the small 33 ft (10 m) common minke whale to the giant blue whale more than 100 ft (33 m) in length, the largest mammal to have ever lived. Gray whales, though not rorquals, are closely related and share many similar characteristics. Most are large and slender-bodied with females being a little larger than males in all species. Distribution is worldwide and generally oceanic. Most rorquals make long north to south migrations between warm-water breeding grounds and cold-water summer feeding grounds.

Bryde's whale (right)
This species has its throat pleats expanded after feeding in sardine-laden waters off Baja California, Mexico. Lunge-feeding in rorquals is said to be the largest biomechanical event on Earth.

Rorqual characteristics

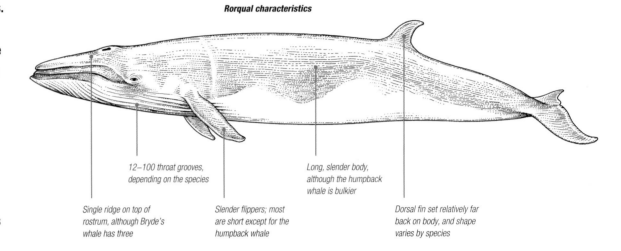

12–100 throat grooves, depending on the species

Long, slender body, although the humpback whale is bulkier

Single ridge on top of rostrum, although Bryde's whale has three

Slender flippers; most are short except for the humpback whale

Dorsal fin set relatively far back on body, and shape varies by species

- Rorquals are in the family Balaenopteridae, one of four families of baleen whales. The group includes two genera: *Balaenoptera* with eight species and *Megaptera* with a single species, the humpback whale. More species may be distinguished with genetic analyses.

- The name rorqual is from the Norwegian word *røyrkval*, meaning "furrow whale," with reference to their numerous throat grooves.

- This group includes Bryde's whale, which uniquely spends the entire year in tropical–subtropical waters; the newly discovered Omura's whale; and the humpback whale, which has the longest and most complex songs known in the animal kingdom.

- No other cetaceans have so many or such well-developed throat grooves. They extend from below the lower jaw to behind the flippers and even to the navel. The throat grooves, also called pleats, allow enormous expansion of the mouth cavity during feeding.

- Rorquals are engulfment-feeders, exerting huge amounts of energy and lunging open-mouthed through prey-laden water. Food is sieved from these enormous gulps of seawater using the baleen plates that entangle prey. The baleen plates are short to moderate in length at ⅓–3½ ft (10–100 cm), and the density and bristle diameter vary

among species, inferring their preferred prey. Rorquals have a unique sensory organ embedded within the connection between the lower jaws, which consists of a bundle of mechanoreceptors that helps the brain to coordinate engulfment-feeding.

- The larger species of rorquals have massive heads that make up nearly a quarter of the whale's body length. They exhibit the most diverse range of body sizes within any cetacean group.

- Rorquals produce the loudest sustained known sounds of any animal on Earth, with whales moaning sounds in the 10 Hz to 40 kHz range (compared to human sound frequencies that range from 18 Hz to 15kHz), and which may travel hundreds of miles.

- Rorqual whales are oceanic and found in all the world's oceans, though populations may range over more limited areas. Most are highly migratory, breeding in tropical waters from November to March, and then migrating to polar regions.

- Gray whales are the only species of a separate family of baleen whales, Eschrichtiidae. Their baleen is the shortest and thickest of all baleen whales, well adapted to their unique feeding technique of sucking up invertebrates from bottom sediments.

Skull (side view)

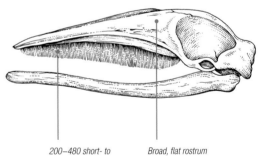

200–480 short- to moderate-length baleen plates per side depending on the species

Broad, flat rostrum

Rorqual feeding
After diving to a depth of up to 330 ft (100 m), the whale drives through the water at high speed so as to engulf a large volume of prey-laden water. The drag that is generated by the gulped water forces the whale's mouth to open to full capacity, four times greater than at rest, as a consequence of the expanded throat grooves. With the mouth closed, the large tongue forces engulfed water through the sieve-like baleen plates, entangling the prey in the bristles. Over several hours of feeding, a whale can ingest more than a ton of fish or invertebrates.

GRAY WHALE

Family Eschrichtiidae

Species *Eschrichtius robustus*

Other common names Grayback whale, scrag, devil fish

Taxonomy Gray whales are the only species in the family Eschrichtiidae—related to the rorqual whales (Balaenopteridae)

Similar species The distinctive mottled gray color, and the "knuckles" on the tail stock should make identification easy at close examination

Birth weight 1,500–2,030 lb (680–920 kg)

Adult weight 35,300–99,200 lb (16,000–45,000 kg)

Diet Crustaceans such as amphipods, copepods, mysid shrimp, krill, and larval and red crabs, and small fish

Group size 1–3, up to 12 when breeding

Major threats Ocean debris, ocean pollution, noise pollution, entanglement, global warming, and ship strikes

IUCN status The western population: Critically Endangered; eastern population: Least Concern

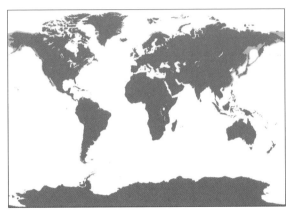

Range and habitat This species is found primarily in shallow coastal waters of the North Pacific Ocean. Two populations are recognized—the eastern stock along North America's coast and western stock along the eastern coast of Asia.

Identification checklist

- Mottled gray body
- Dorsal hump
- Bumps along top of the peduncle
- Barnacles and whale lice on the body
- Low heart-shape blow

Anatomy

Gray whales have a large and robust body. The head is narrow and triangular when seen from above. The skull is relatively small at approximately 20 percent of their total length. The mouth appears slightly arched and contains 130–80 coarse, cream–pale-yellow baleen plates per side. Gray whales possess between two and seven short, deep throat grooves. Vibrissae emerge from follicles on the rostrum and chin areas. Instead of a dorsal fin, gray whales have a hump of variable size and shape, which is followed by a series of fleshy knobs, or "knuckles," along the dorsal ridge of the tail stock. Their flippers are relatively short and paddle shape. The flukes of adults are broad, 10–12 ft (3–3.6 m) wide.

Behavior

Gray whales do not form lasting associations. They frequently travel alone or in small unstable groups. Migrating gray whales move steadily in one direction, breathing and diving in predictable patterns. Breaching is relatively common in this species, and also it also regularly spyhops (raises its head out of the water). Gray whales are known for approaching boats and their "friendly" nature.

Food and foraging

Their principal prey consists of a wide variety of seabed-dwelling amphipods, which they filter from sediment in shallow shelf or coastal waters. Foraging whales often leave long trails of mud in their wake on the ocean bottom. Gray whales also feed on midwater prey, probably more than has usually been assumed. These whales travel around 12,500 miles (20,000 km) every year between their winter calving lagoons in the warm waters surrounding Mexico and their summer feeding grounds in the cold Arctic seas. Gray whales are long-range explorers capable of extensive migrations in search of suitable habitats and resources. This was demonstrated in 2010 when a gray whale migrated from the North Pacific across the Arctic to the Mediterranean Sea, and in 2013 another wandering gray whale found its way to the coast of Namibia, Africa.

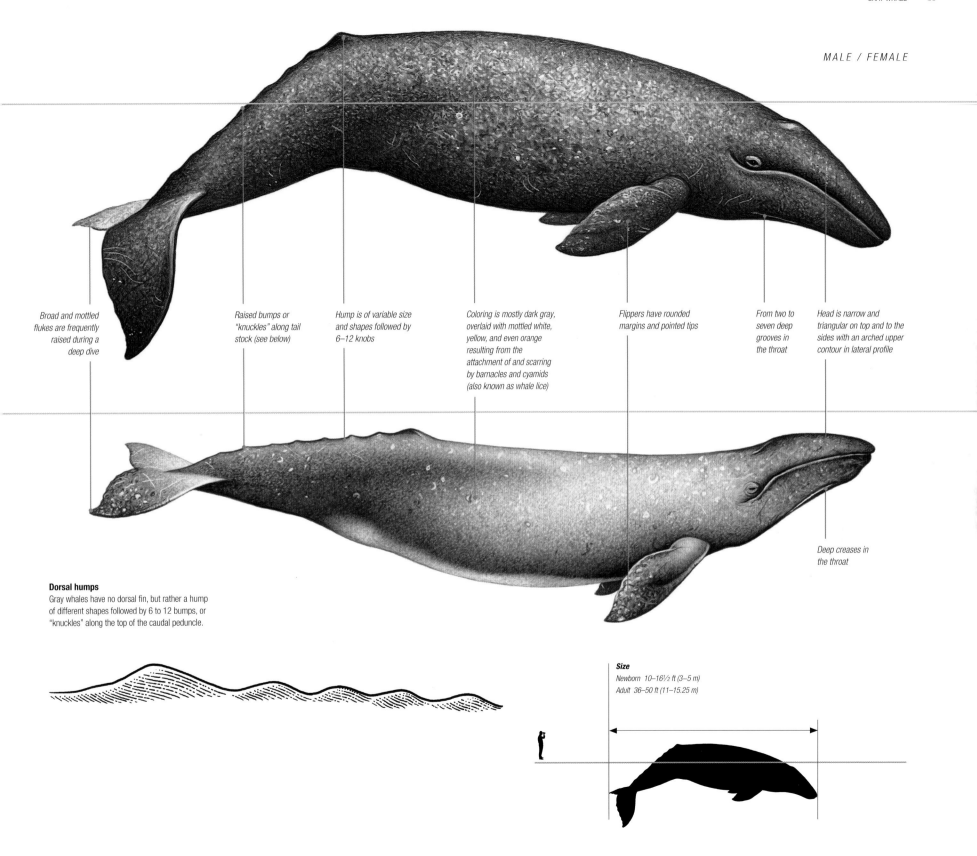

MALE / FEMALE

Broad and mottled
flukes are frequently
raised during a
deep dive

Raised bumps or
"knuckles" along tail
stock (see below)

Hump is of variable size
and shapes followed by
6–12 knobs

Coloring is mostly dark gray,
overlaid with mottled white,
yellow, and even orange
resulting from the
attachment of and scarring
by barnacles and cyamids
(also known as whale lice)

Flippers have rounded
margins and pointed tips

From two to
seven deep
grooves in
the throat

Head is narrow and
triangular on top and to the
sides with an arched upper
contour in lateral profile

Deep creases in
the throat

Dorsal humps
Gray whales have no dorsal fin, but rather a hump
of different shapes followed by 6 to 12 bumps, or
"knuckles" along the top of the caudal peduncle.

Size
Newborn 10–16½ ft (3–5 m)
Adult 36–50 ft (11–15.25 m)

GRAY WHALE

Life history

Gray whales are thought to have a promiscuous mating system. Because multiple inseminations can occur, it is proposed that sperm competition may be taking place—meaning that sperm from two or more males compete to fertilize the female's single ovum. Reproduction in gray whales is strongly seasonal. The female reproductive cycle lasts two years and most females bear young in alternate years. Conception occurs primarily in late November and December while the whales are migrating south from the feeding grounds, but some do not conceive until they are in the breeding lagoons. Length of gestation is disputed, but is generally thought to last 11–13 months. The birth season lasts from about late December to early March. Cows bear a single calf, unattended, and provide sole parental care. Calves drink about 400 pints (189 liters) of rich milk—containing about 53 percent fat and 6 percent protein—each day They grow rapidly, reaching 28½ ft (8.7 m) when weaned, eight months later, in the summer feeding areas. Lifespan is estimated to be between 60 and 70 years.

Conservation and management

The status of the two extant populations differs greatly. The eastern population has been a model of stock recovery since its protection in 1937. Its abundance is estimated at 22,000 animals. This population is the target of a small aboriginal hunt of the Chukotka Peninsula in Russia. The protected western population is one of the most critically endangered whale stocks in the world, with close to 130 animals remaining. However, photographs of western gray whales obtained within the Baja California breeding lagoons demonstrate that both populations mix to some degree and they may interbreed.

Mother and calf (above)
Mothers are very protective of their young and establish strong bonds that are needed to keep them together on the migration northward. Calves nurse for about eight months.

Fluking (opposite)
The flukes can be more than 10 ft (3 m) across and they have pointed tips with a marked notch in the center. When preparing for long, deep dives the whales often display their flukes clearly out of the water.

Spyhopping (left)
Spyhopping is a behavior in which the whales are in a vertical position with their heads partially or completely out of the water and can last few minutes.

Blow
The blow is heart-shaped and 10–13 ft (3–4 m) in height.

Dive sequence
1. The whale arches and shows its dorsal hump and bumps along the top of the peduncle.

2. In deep waters the flukes may appear above the surface of the water.

Fluking
The flukes are scalloped with a big notch in the middle and are often scarred with barnacles or killer whale tooth marks.

Spyhopping
Spyhopping is a characteristic behavior of the gray whale.

Breaching
Gray whales commonly breach, especially during migration, and in or near the breeding lagoons in Baja California.

COMMON MINKE WHALE

Family Balaenopteridae

Species *Balaenoptera acutorostrata*

Other common names Lesser rorqual, sharp-snouted fin whale, little piked whale, and pygmy whale

Taxonomy Three subspecies: North Atlantic (*B.a. acutorostrata*), North Pacific (*B.a. scammoni*), and dwarf in southern hemisphere (unnamed)—most closely related to the Antarctic minke whale

Similar species Bryde's whale but smaller size, distinct white band across flipper

Birth weight 330–660 lb (150–300 kg)

Adult weight 11,000–22,000 lb (5,000–10,000 kg)

Diet Small fish, squid, and krill

Group size 1–3, up to a few hundred in highly abundant feeding areas

Habitat Coastal and continental shelf, around islands, and offshore banks where upwelling occurs

Major threats Ocean pollution, noise pollution, fisheries interactions, climate change

IUCN status Least Concern, though some populations are not well known

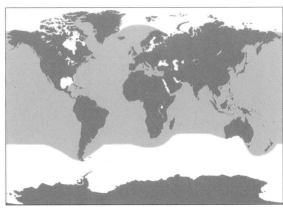

Range and habitat This species is found in all ocean basins and enclosed bays worldwide. Primarily coastal waters, over continental shelves, around islands, and seamounts where there is consistent upwelling.

Identification checklist

- Dark gray body, bright white belly
- Small, streamlined body
- Falcate dorsal fin
- Short flippers with white band
- Narrow, pointed head
- Blazes of lighter color across the back
- Fin position: center, two-thirds of the way along the back

Anatomy

Common minke whales are the smallest of the rorqual whales, one-third the size of fin whales. They are easily identified by their small, sleek size, white countershading from chin to belly, sharp-pointed head, and white band across each flipper.

Behavior

With a global distribution, researchers have noted a wide range of behaviors in common minke whales. Their diminutive size allows for leaping fully out of the water. Common activities include breaching and bursts of speed when chasing prey. Swim speeds approach 22 mph (35 kmph), outpacing a persistent predator, orcas. Some regional individuals are curious and follow alongside boats.

Food and foraging

Worldwide, minke whales feed on a range of schooling fish (cod, herring, juvenile salmon, sardine, and anchovy) and invertebrates (krill, copepods, squid, and pteropods). Depending on the prey, they use a range of techniques such as lunging, gulping, and skimming surface waters while rolling sideways. Specialists called "bird ball feeders" take advantage of seabird feeding flocks that amass prey.

Life history

Minke whales lead a solitary life, or they occur in groups of two to three with mother-calf and attending male. Male mating calls are distinctive, including "boings" and "duck quacks." Gestation is around 11 months, but calving intervals fluctuate by region. Migration patterns vary globally and are poorly understood. In temperate latitudes of the Northeast Pacific and the Canary Islands, some individuals are resident year round. Other populations follow a typical migratory cycle of foraging in spring–summer in subpolar regions and breeding in winter in the tropics. Female minke whales are sexually mature as young as 6 years, and may live for 30–50 years.

Conservation and management

The smallest of rorqual whales were hunted as early as the 11th century, and extensively hunted until commercial whaling was suspended by the IWC. Minke whales are one of the few whales now collected for "scientific studies." Since minke whales inhabit coastal waters, they are often exposed to pollution and fisheries entanglement. Nevertheless, they are the most abundant of all baleen whales.

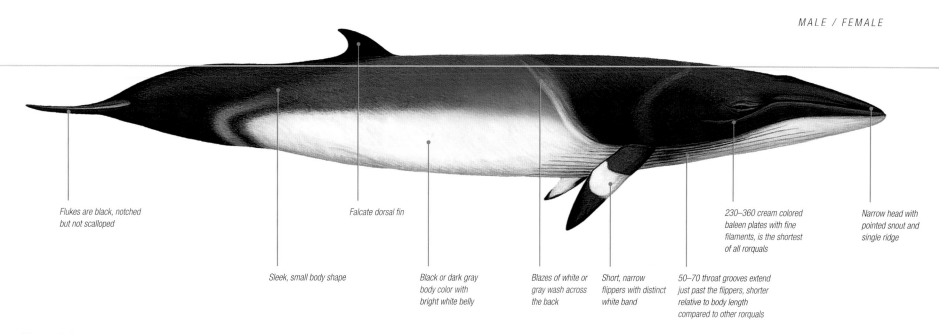

MALE / FEMALE

Flukes are black, notched but not scalloped

Falcate dorsal fin

230–360 cream colored baleen plates with fine filaments, is the shortest of all rorquals

Narrow head with pointed snout and single ridge

Sleek, small body shape

Black or dark gray body color with bright white belly

Blazes of white or gray wash across the back

Short, narrow flippers with distinct white band

50–70 throat grooves extend just past the flippers, shorter relative to body length compared to other rorquals

Flipper variations

The common minke whale was only recently distinguished as genetically separate from the larger Antarctic minke whale. The even smaller dwarf minke whale of the southern hemisphere, though, remains for now recognized as a subspecies of the common minke whale. The dwarf minke whale has a flipper mark that extends onto the shoulder and gray/brown baleen. The other two subspecies have a single white flashing band across the each flipper.

Antarctic minke flipper
Mostly gray

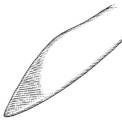

Dwarf minke flipper
White patch on flipper/shoulder area

Common minke flipper
Distinct white band

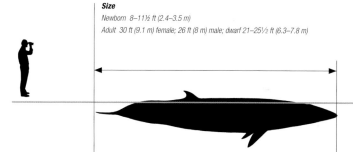

Size
Newborn 8–11½ ft (2.4–3.5 m)
Adult 30 ft (9.1 m) female; 26 ft (8 m) male; dwarf 21–25½ ft (6.3–7.8 m)

Blow
The blow is light and usually not seen; it is 6½ ft (2 m) high and dissipates quickly

Dive sequence
1. The head with a small splash guard surfaces at a low slant before back emerges.

2. The dorsal fin is exposed as the splash guard submerges.

3. The body then rolls and arches, accentuating the distinctive dorsal fin.

4. The dorsal fin is still visible as the whale begins to dive with arched tail stock.

5. The tail stock is exposed at the end of a dive. Common minke whales almost never raise their flukes above the surface when diving.

Breaching
Common minke whales can lift their entire bodies out of the water, and some feeders breach with a mouth full of fish.

ANTARCTIC MINKE WHALE

Family Balaenopteridae

Species *Balaenoptera bonaerensis*

Other common names Southern minke whale

Taxonomy Closely related to the common minke whale

Similar species Unnamed subspecies of the common minke whale also known as dwarf minke whale; it can be distinguished from the latter based on differences in flipper-shoulder patch

Birth weight 441–550 lb (200–250 kg)

Adult weight 18,700–24,250 lb (8,500–11,000 kg)

Diet Krill, occasionally copepods, and small fish

Group size 2–5 individuals, hundreds in large feeding aggregations

Major threats Ocean pollution, climate change, and whaling

IUCN status Data Deficient

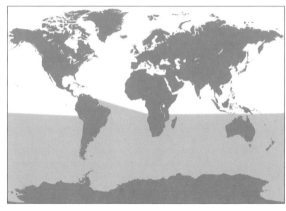

Range and habitat This species inhabits oceanic habitats from the Equator to the Antarctic, with some whales found in the North Atlantic Ocean. The summer range is between 55°S and the ice edge and the winter range is between 5° and 35°S.

Identification checklist

- Black or dark gray back, white belly
- Triangular-shaped head, with a single head keel
- Tall and falcate dorsal fin positioned at the aft (rear) third of the back
- Small, gray flippers
- Slender and streamlined body

Anatomy

The Antarctic minke whale is one of the smaller rorqual species, rarely reaching more than 33 ft (10 m) in length. Females are slightly larger than males. The dark gray or black back is separated from the almost all-white ventral area by a light gray flank that undulates along the length of the body. Often, a light gray chevron extends from the flipper toward the back and the head.

Behavior

These whales are fast swimmers, reaching around 20 mph (32 kmph) and occasionally porpoising or breaching. Their narrow, vertical, 6–10 ft (1.8–3 m) tall blows are more visible in colder environments. Their vocal behavior is poorly known, but a mysterious quacking sound (known as the "bio-duck") recorded by sonars in the Southern Ocean has been recently attributed to this species.

Food and foraging

Long migrations occur between summer feeding grounds and tropical breeding areas. Their primary prey is the Antarctic krill, but occasionally their diet includes other zooplankton and small fish. They are lunge-feeders, trapping large quantities of water in their mouth and filtering their food with their baleen plates.

Life history

Mating occurs in the winter. Calves are born after a 10-month gestation and are weaned 4–5 months after birth. Antarctic minke whales become sexually mature at seven to eight years and are thought to reproduce nearly every year after that. They live for 50–60 years. Killer whales are their main predator.

Conservation and management

The Antarctic minke whale became a primary target of whaling after the collapse of the larger rorqual stocks in the 1970s. They were protected when the whaling moratorium was implemented in the mid 1980s, but thousands have been killed by the Japanese scientific whaling program in the Southern Ocean. Abundance estimates recently reviewed by the IWC show a decline of yet unknown reasons from the late 1980s (estimate of 720,000 individuals) to the mid 1990s (estimate of 515,000 whales).

MALE / FEMALE

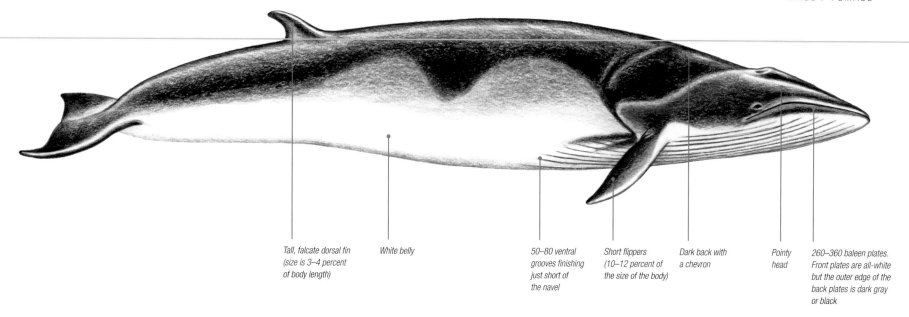

Tall, falcate dorsal fin
(size is 3–4 percent
of body length)

White belly

50–80 ventral
grooves finishing
just short of
the navel

Short flippers
(10–12 percent of
the size of the body)

Dark back with
a chevron

Pointy
head

260–360 baleen plates.
Front plates are all-white
but the outer edge of the
back plates is dark gray
or black

Dorsal chevron
The Antarctic minke may have a
distinctive light gray chevron marking
on its dorsal surface.

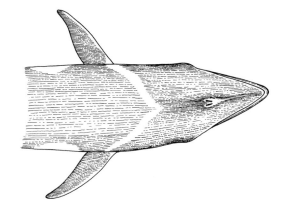

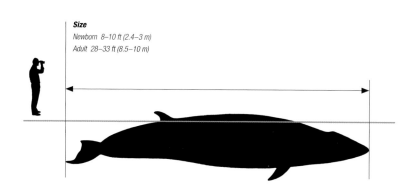

Size
Newborn 8–10 ft (2.4–3 m)
Adult 28–33 ft (8.5–10 m)

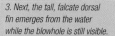

Dive sequence
1. When surfacing, the
minke whale head comes
out of the water first and is
followed by a slender and
vertical blow 6½–10 ft
(2–3 m) in height.

2. As the head rolls
beneath the surface, the
dark back is exposed.

3. Next, the tall, falcate dorsal
fin emerges from the water
while the blowhole is still visible.

4. Toward the end of the
surfacing, the back
arches pronouncedly in
preparation for the dive.

5. Lastly, the dorsal fin goes
into the water and the
peduncle is occasionally
exposed but, unlike in other
baleen whale species, the
fluke almost never comes out
of the water.

Spyhopping
Antarctic minke whales
can place their bodies
vertically in the water
and expose their heads
above the surface.

Breaching
Antarctic minke whales
breach regularly,
sometimes exposing
almost the whole body.

SEI WHALE

Family Balaenopteridae

Species *Balaenoptera borealis*

Other common names Sei whale is a variation on the Norwegian name for pollock whale; other names are coalfish whale, sardine whale, northern rorqual, lesser fin whale

Taxonomy Two subspecies recognized: *B.b. borealis* (northern sei whale) and *B.b. schlegellii* (southern sei whale)—most closely related to Bryde's whale and fin whale

Similar species Can be confused with fin whale and Bryde's whale but can be distinguished by differences in head shape (see below)

Birth weight 1,322 lb (600 kg)

Adult weight 30,900–60,000 lb (14,000–27,000 kg)

Diet Copepods, other invertebrates, small fish

Group size 1–5, rare up to 50 in rich feeding areas

Habitat open ocean, coastal waters

Major threats Ocean pollution, climate change

IUCN status Endangered

Range and habitat Found in all major oceans, except the north Indian Ocean, ranging from subpolar to temperate in summer to subtropical in winter. This species is associated with deep water in open ocean, submarine canyons, and continental slopes but avoids enclosed ocean basins.

Identification checklist

- Dark gray to black body, white sides and belly
- Large but slim, sleek body
- Tall upright, falcate dorsal fin
- Short pointed flippers
- Rostrum is pointed, downward arched profile
- Single narrow ridge on head
- Sides pock-marked
- Fin position: center of the lower back

Anatomy

The "most graceful of all the whales," sei whales are sleek for their large size. The shape and position of the dorsal fin are the best features for identifying them. The fin shape is tall, upright, and falcate, though the curve can vary. The tall dorsal fin may be confused with that of Bryde's whales, but sei whales have a single prominent ridge along the narrow, "V"-shaped head compared to the triple-ridged, broader-shaped Bryde's whale. Baleen bristles are extremely fine and can be used to identify the species. Females are larger than males, and southern hemisphere whales are larger than northern ones.

Behavior

Their sleek shape enables sei whales to swim in bursts of 16–35 mph (25–56 kmph), among the fastest of all whales. They are not acrobatic, and seldom raise their flukes or breach. Sei whales avoid boats, which may explain why they are rarely observed. The term "sei whale years" refers to their irregular appearance on feeding grounds.

Food and foraging

The fine bristles of sei whales allow them to feed on the smallest invertebrates, copepods, the "fleas of the sea." They also feed on krill and schooling fish such as herring, anchovies, hake, and saury, but copepods are the main prey. They lunge-feed on fish but swim on their sides to skim the surface for weak-swimming copepods.

Life history

Sei whales usually are alone, in pairs, or groups of three—a cow and calf with a male. Larger groups of up to 50 are may occur where food is highly abundant, but this is rare. Males pair up with females with calves during the winter mating season. Calving occurs a year later in subtropical waters. As with most other rorqual whales, sei whales skip calving years. They are sexually mature at 5–15 years and may live to 65 years or longer.

Conservation and management

Sei whales were intensively hunted by whalers, declining worldwide by 80 percent, and populations have not recovered. Death by ship strike has been documented, but human interactions are low because they occur mostly offshore.

MALE / FEMALE

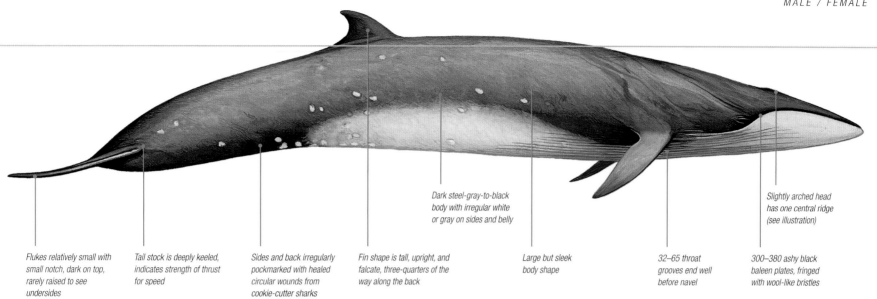

Flukes relatively small with small notch, dark on top, rarely raised to see undersides

Tail stock is deeply keeled, indicates strength of thrust for speed

Sides and back irregularly pockmarked with healed circular wounds from cookie-cutter sharks

Fin shape is tall, upright, and falcate, three-quarters of the way along the back

Dark steel-gray-to-black body with irregular white or gray on sides and belly

Large but sleek body shape

32–65 throat grooves end well before navel

Slightly arched head has one central ridge (see illustration)

300–380 ashy black baleen plates, fringed with wool-like bristles

Dorsal fin and head
The dorsal fin is tall and slightly falcate compared to all other large whales, except for Bryde's whales. Sei whales have a single central ridge along the rostrum, compared to the three ridges of Bryde's whale.

Dorsal fin

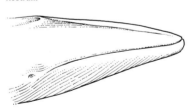

Rostrum

Size
Newborn 13–16 ft (4–4.9 m)
Adult 47½–64 ft (14.5–19.5 m) female; 45–61 ft (13.7–18.6m) male

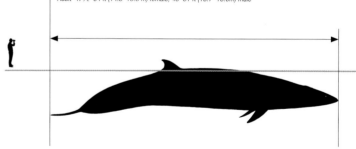

Blow
Rising 6½–10 ft (2–3 m) the blow is high, broad, and diffuse. It is similar to that of the fin whale but is shorter.

Dive sequence
1. Much of the back is exposed when the sei whale is surfacing and diving.

2. The head and blow occur at surface at the same time as the tall, falcate dorsal fin.

3. The animal slips or sinks below surface rather than arching before a dive, with the dorsal fin seen last.

4. Unlike other rorquals, this species does not hyperventilate before dives since dives are relatively shallow.

5. Tail flukes are rarely raised above the surface because dives are usually shallow.

Breaching
Sei whales rarely breach, and usually belly flop when they do.

BRYDE'S WHALE

Family Balaenopteridae

Species *Balaenoptera edeni*

Other common names None

Taxonomy The nomenclature and taxonomy of Bryde's whales remains unresolved with a lack of clarity about how many Bryde's-like whale species exist—*B. edeni* and *B. brydei* are commonly collectively known as Bryde's whales and *B. omurai* as Omura's whale. Two subspecies are recognized: *B. e. brydei*, which occupies offshore waters, and *B. e. edeni* (Eden's whale).

Similar species Omura's whale (*Balaenoptera omurai*)

Birth weight Not known

Adult weight 26,450–44,100 lb (12,000–20,000 kg)

Diet Schooling fish, krill, and plankton

Group size 1–2 whales

Major threats Habitat degradation, vessel strike

IUCN status Data Deficient

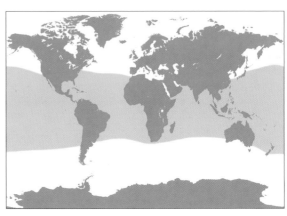

Range and habitat This species is distributed throughout temperate and tropical inshore and offshore waters, typically between the latitudes 40°N and 40°S.

Identification checklist

- Dark gray back
- Three rostral ridges
- Long ventral throat grooves
- Small, falcate dorsal fin, about three-quarters of the way along the back

Anatomy

Bryde's whales have three distinctive rostral ridges running from the rostrum to the blowhole, and long throat grooves, extending beyond the navel. They are slender, dark gray on the upper side and along the pectoral fins, and lighter underneath. Sometimes they are pink on their underside, as a result of the warm water and increased activity. They have a thin blubber layer compared to other balaenopterids. Their dorsal fin is typically falcate although there is variation in fin-shape between individuals.

Behavior

Bryde's whales are typically found alone or in pairs but may be in loose aggregations when foraging. They do not appear to cooperatively forage, instead individually catching their prey. Because Bryde's whales breed year round, they do not form large breeding aggregations. They are most active when foraging—either lunging through prey or traveling to another prey patch. Bryde's whales have similar low-frequency tonal and swept calls to other balaenopterids, with variation between areas.

Food and foraging

They feed on schooling fish and plankton. Resident Bryde's whales will change their diet with seasonal shifts in prey availability. They lunge-feed to capture prey and will rapidly turn onto their side when feeding. Sometimes solitary whales will bubble net.

Life history

Bryde's whales probably have a similar lifespan to that of other balaenopterids. They calve year round as their warm-water habitat provides favorable temperatures for newborns. They have an 11-month gestation period, lactation lasts around 6 months, and calves are weaned when about 23 ft (7 m) in length—although this varies depending on whether they are inshore or offshore whales. Females typically give birth to a single calf every two to three years.

Conservation and management

Japan has killed more than 1,000 Bryde's whales for their scientific whaling program. Vessel-strike mortality is reported throughout the world and is a threat to some populations. Entanglement in fishing gear, noise pollution, and overfishing of prey are likely threats. For coastal populations accumulation of toxins may pose a threat.

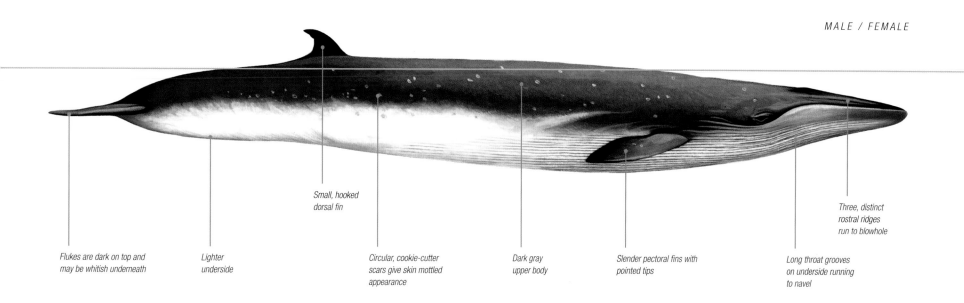

MALE / FEMALE

Small, hooked
dorsal fin

Three, distinct
rostral ridges
run to blowhole

Flukes are dark on top and
may be whitish underneath

Lighter
underside

Circular, cookie-cutter
scars give skin mottled
appearance

Dark gray
upper body

Slender pectoral fins with
pointed tips

Long throat grooves
on underside running
to navel

Rostral ridges
Bryde's whales are unique among the baleen
whales as they have three distinctive rostral
ridges on the top of their head. They sometimes
lift their head, and on occasion most of their
body, out of the water and slap it on the surface.

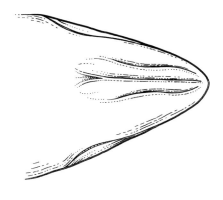

Size
Newborn 10–16½ ft (3–5 m)
Adult 36–50 ft (11–15.5 m)

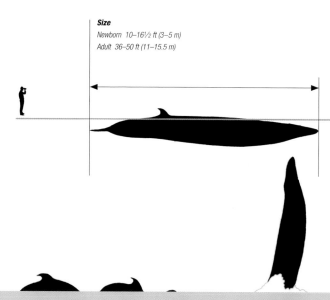

Blow
Bryde's whales have a
small, bushy blow when
they surface to breathe.

Dive sequence:
1. After the whale has
surfaced the blowhole
becomes visible.

2. Next, the back
emerges and comes
into view.

3. This is followed by
the dorsal fin, which is
further along the back.

4. Finally, the whale
arches its back slightly
as it dives beneath the
surface of the water.

Fluking
These whales do not
typically show their
flukes when they dive.

Breaching
Sometimes Bryde's will raise
most of their body almost
straight out of the water and
splash back into the water.

BLUE WHALE

Family Balaenopteridae

Species *Balaenoptera musculus*

Other common names Blue rorqual, Sibbald's rorqual, sulfur-bottomed whale

Taxonomy Four proposed subspecies based on regional differences: *B. m. musculus* (northern hemisphere); *B. m. intermedia* (Antarctic); *B. m. brevicauda* (sub-Antarctic Indian Ocean); and *B. m. indica* (northern Indian Ocean)

Similar species Their large size with their tiny dorsal fin make blue whales distinctive but they may be confused with fin or sei whales

Birth weight 8,800–11,000 lb (4,000–5,000 kg)

Adult weight Average 15,500–300,000 lb (7,000–136,000 kg); maximum 390,218 lb (177,000 kg) female

Diet Krill, rarely fish

Group size 1–3, rarely 50–80 in rich feeding areas

Major threats Ship strikes, fisheries interactions, ocean pollution, noise pollution, and climate change

IUCN status Endangered

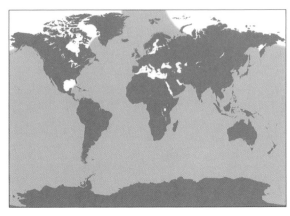

Range and habitat The blue whale's distribution is cosmopolitan, occurring in all world oceans with depths 260–12,140 ft (80–3,700 m); it congregates along shelf and shelf break, seamounts, and islands.

Identification checklist

- Blue-gray with light or dark spots
- Large, sleek body shape
- Miniature dorsal fin with a variable shape
- Flippers short and pointed
- Flukes notched with straight trailing edge
- Head broad with prominent middle ridge
- Fin position: center and far back

Anatomy

Blue whales are the largest animal that has ever lived, reaching lengths in adult females of up to 95 ft (29 m) in the larger Antarctic subspecies. The skin has a bluish tint, which when seen underwater may appear turquoise. They are quickly identified from all other whales by their extremely large size, contrasted with a tiny dorsal fin, located far back on the body.

The ventral throat pleats allow the mouth of blue whales to expand to capture 20,000 lb (9,000 kg) of krill and seawater. The lower jaws are long and heavy with special muscular attachments to the skull to accommodate this "largest biomechanical action in the animal kingdom." Observed from above, the rostrum is broad and flat, and the tail stock is deep but narrow, enabling powerful strokes for lunging and fast swimming.

Behavior

Though not as graceful as humpbacks, blue whales can pirouette 360 degrees when lunging through dense krill patches. They fluke when diving, but rarely breach. When cruising, they travel at 12 mph (20 kmph) with bursts to 31 mph (50 kmph). Blues have distinct regional vocalizations mostly below 20 Hz, and also the loudest sustained sounds from any known living animal. The sounds are the loudest of any animal at 188 decibels, which are blasted hundreds of miles away, and likely vital for communication between these rare whales.

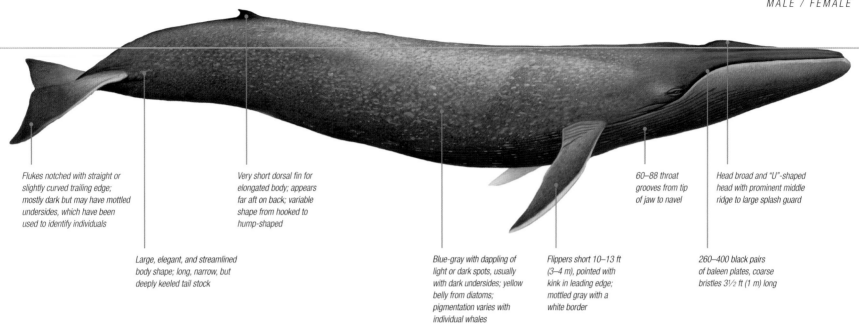

Flukes notched with straight or slightly curved trailing edge; mostly dark but may have mottled undersides, which have been used to identify individuals

Very short dorsal fin for elongated body; appears far aft on back; variable shape from hooked to hump-shaped

60–88 throat grooves from tip of jaw to navel

Head broad and "U"-shaped head with prominent middle ridge to large splash guard

Large, elegant, and streamlined body shape; long, narrow, but deeply keeled tail stock

Blue-gray with dappling of light or dark spots, usually with dark undersides; yellow belly from diatoms; pigmentation varies with individual whales

Flippers short 10–13 ft (3–4 m), pointed with kink in leading edge; mottled gray with a white border

260–400 black pairs of baleen plates, coarse bristles 3½ ft (1 m) long

Aerial view

The long, streamlined body contrasts with the enormous body mass. The broad, flat head accounts for one-quarter of the body length, with a long back tapering to the tail. Note the tiny dorsal fin far aft and the turquoise color of the body underwater.

Size

Newborn 20–26 ft (6–8 m)

Adult 75½–100 ft (23–33 m); Antarctic subspecies larger, maximum 110 ft (33.6 m)

Dorsal fin variations

The blue whale's small dorsal fin varies in shape and size from hooked to more rounded, to a triangular hump.

Falcate   **Rounded**   **Triangular**

BLUE WHALE

Food and foraging

This largest of mammals feeds almost exclusively on one of the smallest invertebrates, krill. Pairs of blue whales have been noted lunging side by side, but usually they lunge-feed alone. They eat other invertebrates such as copepods, but rarely small schooling fish. Blue whales seasonally migrate to higher latitudes to forage in upwelling areas near continental shelves, at the edge of polar ice and in open oceans. They forage alone or in small groups of 2–3, and loose groups of 50–80 distributed widely where prey is super concentrated. During a single dive, they can lunge up to six times on swarms of krill, and hold their breath for 30 minutes. Each day, 8,000 lb (3,600 kg) of krill are swallowed to sustain this behemoth.

Life history

Little is known of mating behavior, which occurs in fall and winter. One year later a calf is born in the subtropics, weighing the equivalent of an adult hippopotamus. Cows nurse calves for several months, skipping one to two years between births. They become sexual mature at 7–12 years, and have a life span of 80 years or more. Because of their enormous size and swift speed, predation by orcas is rare.

Conservation and management

Blue whales almost became extinct because of intensive whaling. Most populations have not recovered, and they remain endangered. Some recovery has occurred in the northeast Pacific Ocean population, but globally there are less than 10,000 individuals.

Fluking

Fluking exposes the triangular-shaped, notched tail. The tail stock from the front appears narrow, but from the side is broad, which explains the powerful thrust of the tail (see dive sequence opposite).

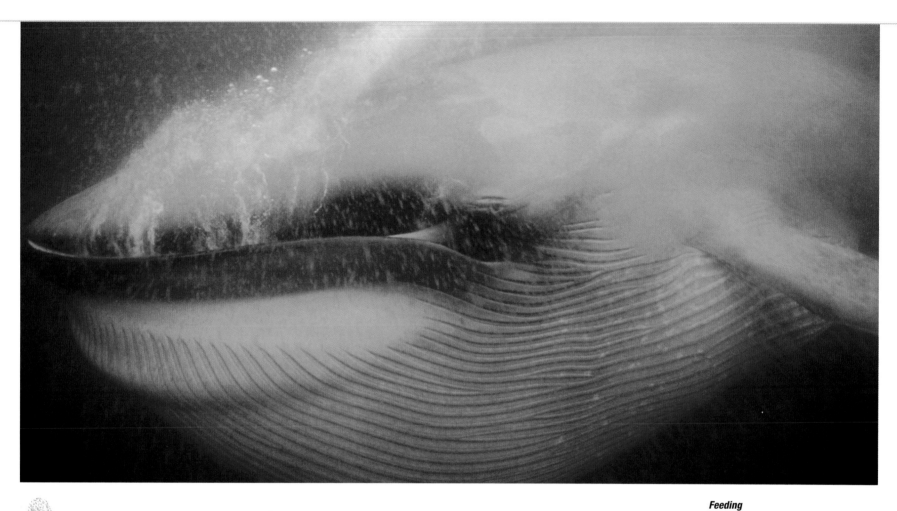

Feeding

Lunge-feeding is the largest biomechanical action in the animal kingdom. The sleek body is transformed into a "bloated tadpole" by the enormous expansion of the throat pleats, capturing a mouth full of krill and seawater.

Blow

Tall, straight, and dense cloud, the blow rises 29½–39½ ft (9–12 m.) On calm days, the blow looks like a column.

Dive sequence

1. The animal surfaces with its broad-shaped head and distinctive splash guard that appears as a hump.

2. The long, broad back appears before exposing the tiny dorsal fin.

3. It often settles below the surface after exposing the back and dorsal fin rather than arching before diving.

4. If a whale arches, the massive, deeply keeled tail stock is exposed as the dorsal fin slips below the surface.

5. The tail flukes may then appear above the surface, and can be used for identifying individuals.

Breaching

Blue whales do breach, though not often, exposing much of their body out of the water and creating an enormous splash.

OMURA'S WHALE

Family Balaenopteridae

Species *Balaenoptera omurai*

Other common names Pygmy Bryde's whale, dwarf fin whale

Taxonomy Although it was originally considered a pygmy form of Bryde's whale it is now recognized as outside the clade formed by the blue, Bryde's and sei whales

Similar species Can be confused with the fin whale, Bryde's whale, or the minke whale (see below)

Birth weight Unknown

Adult weight Estimated less than 44,000 lb (20,000 kg)

Diet Krill, very limited information

Group size 1–4

Habitat Open ocean and shallow coastal waters

Major threats Fisheries interactions, climate change

IUCN status Data Deficient

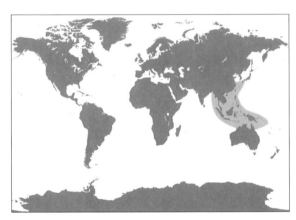

Range and habitat The range is likely broader than the current estimate, which is based on a few confirmed sightings and specimens collected in the Indo-Pacific Ocean (Sea of Japan, eastern Indian Ocean, Philippines, and Solomon Sea). Habitats are tropical–subtropical, both nearshore and open ocean over continental shelves.

Identification checklist

- Streamlined body shape
- Dark back with light belly
- Asymmetrical coloration of jaw and throat
- Dorsal fin hooked
- Broad head with one ridge
- Flippers slender; dark on top with white border
- Flukes are straight-edged
- Fin position: center, lower back

Anatomy

Omura's whale, so named to honor a Japanese cetologist, was only recently recognized as a distinct species. They are most often confused with Bryde's whales, but Omura's whales have a single ridge on the broad head compared to three ridges on Bryde's whales. The Omura's whale's distinctive asymmetrical coloration of the head (similar to fin whales) is the best identifying feature, and is why they have been called dwarf fin whale.

Behavior

There are so few confirmed sightings of Omura's whale that little is known of their behavior, and because of misidentification, descriptions may be misleading. Males are not known to call during mating, though surface rolling during mating has been reported. Fluking has not been described, though breaching has.

Food and foraging

Omura's whales have been observed lunge-feeding in coastal waters and krill have been identified in the stomachs of dead individuals. Fish is likely a larger part of their diet, though, because of the small number and size of baleen plates.

Life history

Groups of one to four Omura's whales, including females with calves, have been reported, but information on mating behavior or even migration is unknown. Limited knowledge suggests that they occur year round in the Indian and eastern Pacific Oceans and may not have a seasonal breeding cycle. Gestation is likely one year, and as with all baleen whales, females are larger than males. From the few animals aged, the oldest female was estimated to be 29 years and the oldest male 38.

Conservation and management

Current conservation concerns are constrained by a paucity of information, though artisanal whale fisheries and bycatch have been documented. Population status, trend, and distribution are unknown.

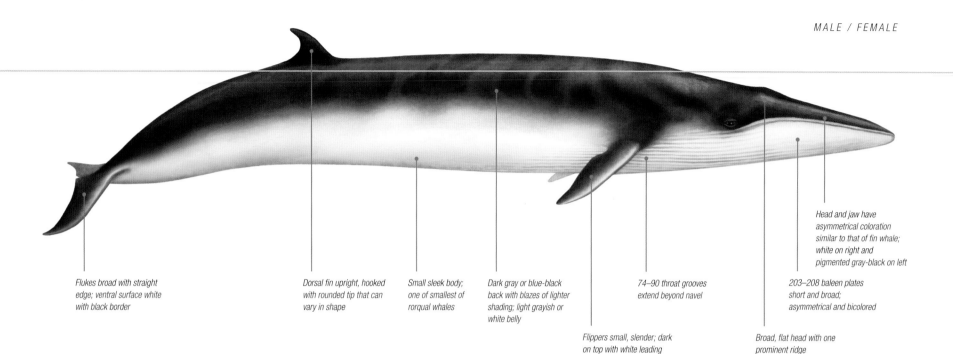

MALE / FEMALE

Flukes broad with straight edge; ventral surface white with black border

Dorsal fin upright, hooked with rounded tip that can vary in shape

Small sleek body; one of smallest of rorqual whales

Dark gray or blue-black back with blazes of lighter shading; light grayish or white belly

74–90 throat grooves extend beyond navel

Flippers small, slender; dark on top with white leading edge; white underneath

203–208 baleen plates short and broad; asymmetrical and bicolored

Head and jaw have asymmetrical coloration similar to that of fin whale; white on right and pigmented gray-black on left

Broad, flat head with one prominent ridge

Baleen plates bicoloration
Baleen plates are both asymmetrical and bicolored with white the dominant color on the right side and black on the left. On the right side (shown), the anterior plates are white, the middle bicolored, and the posterior black; on the left side (not shown), the anterior and middle plates are bicolored and the posterior are black.

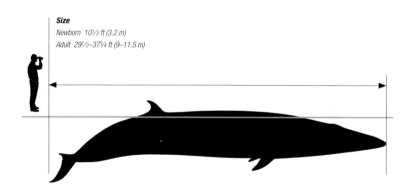

Size
Newborn 10½ ft (3.2 m)
Adult 29½–37¾ ft (9–11.5 m)

Dive sequence
1. The head surfaces at a low angle. There are few descriptions but the blow is noted as conspicuous.

2. The body then rolls, and as the splash guard submerges, the tall dorsal fin appears.

3. As the dorsal fin drops below the surface, the body rolls sharply and the whale begins to dive.

4. The tail stock arches strongly but the flukes do not appear above the surface. Omura's whales never fluke when diving, although this is based on few confirmed observations.

Breaching
Omura's whales are not reported to lift their entire bodies out of the water; however, they are hypothesized to lunge vertically out of the water when feeding.

FIN WHALE

Family Balaenopteridae

Species *Balaenoptera physalus*

Other common names Finback whale, fin-backed whale, finner, common rorqual, herring whale, razorback

Taxonomy Northern (*B. p. physalus*) and southern (*B. p. quoyi*) hemisphere subspecies are recognized. A third subspecies is also recognized, the pygmy fin whale (*B. p. patachonica*). Closely related to other species of rorquals, especially the humpback (*M. novaeangliae*).

Similar species Most likely to be confused with sei or blue whale but at close range the asymmetrical pigmentation pattern is a useful identification feature.

Birth weight 2,200–3,300 lb (1,000–1,500 kg)

Adult weight 99,200 lb (45,000 kg) males and 110,250 lb (50,000 kg) females, in northern hemisphere; 132,250 lb (60,000 kg) males and 154,350 lb (70,000 kg) females, in southern hemisphere

Diet Krill and other planktonic crustaceans, schooling fish

Group size Single individuals and small groups of 2–7 whales

Major threats Commercial whaling, collisions with vessels, bycatch, and pollution

IUCN status Endangered (although most populations are recovering)

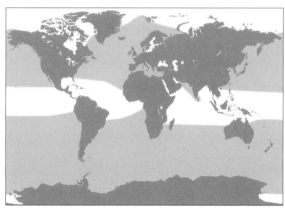

Range and habitat This species has a worldwide distribution, from the Equator to the polar regions. It is a pelagic species, found usually outside the continental shelf.

Identification checklist

- Large size
- Asymmetrical right side of head (especially lower jaw) and body coloration (paler on the right side)
- Yellowish pigmentation of front right baleen plates
- Falcate dorsal fin
- Fin position is three-quarters along the length of the body

Anatomy

Despite being the second largest whales, fin whales have a slender figure. They have a narrow rostrum with a single, well-developed, longitudinal ridge. Their typically asymmetrical coloration— with a light gray pattern on the right side of the head and front of the body—is a feature used together with the dorsal fin shape for identifying individuals. Fin whales are sexually dimorphic, with males usually slightly smaller than females.

Behavior

Fin whales are not a gregarious species and the only known social bond is the mother-calf pair. They form larger feeding aggregations only sporadically, though they often associate with blue whales, and sometimes dolphins or pilot whales. Normal cruising speed is 5–8 knots, with short dives lasting 3-10 minutes at 330–660 ft (100–200 m) deep. Breaching, lobtailing, and other aerial behaviors are rare. Vocalizations consist of low-frequency moans and grunts.

Food and Foraging

Fin whales vary their diet according to prey availability, season, and locality. They mainly feed on krill, but other species of planktonic crustaceans, schooling fishes, and small squids are also eaten. They migrate between high-latitude feeding grounds in summer and low-latitude breeding grounds in winter.

Life history

Fin whales reach sexual maturity at 6—10 years. Mating occurs in winter, with birth taking place 11 months later. Lactation lasts about six months—after weaning females remain reproductively inactive for about six months and mate again thereafter. Lifespan is unknown but individuals of up to 84 years old have been reported.

Conservation and management

Fin whale populations have been severely depleted by commercial whaling since the late 19th century. Current catches are strictly regulated, and the main threats for fin whales are now vessel collisions, pollution, and incidental catch in fishing gear.

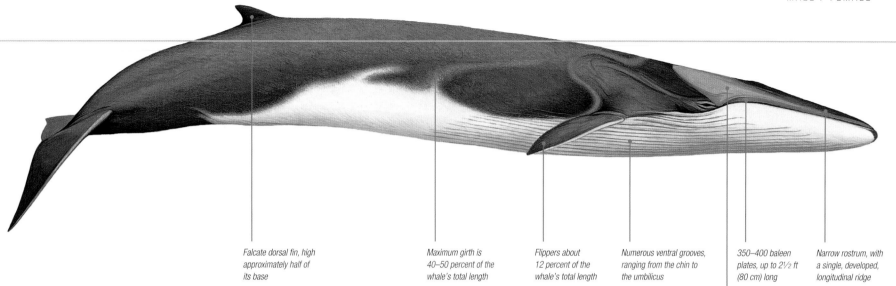

Falcate dorsal fin, high approximately half of its base

Maximum girth is 40–50 percent of the whale's total length

Flippers about 12 percent of the whale's total length

Numerous ventral grooves, ranging from the chin to the umbilicus

350–400 baleen plates, up to 2½ ft (80 cm) long

Narrow rostrum, with a single, developed, longitudinal ridge

Asymmetrical pigmentation of the head and the front right side of the body

Blaze coloration

Balaenoptera physalus has a striking asymmetrical coloration on the head and front right side of the body. Whereas the left side is dark slate, the right side shows a lighter coloration with a paler "blaze," the pattern of which is unique for each individual.

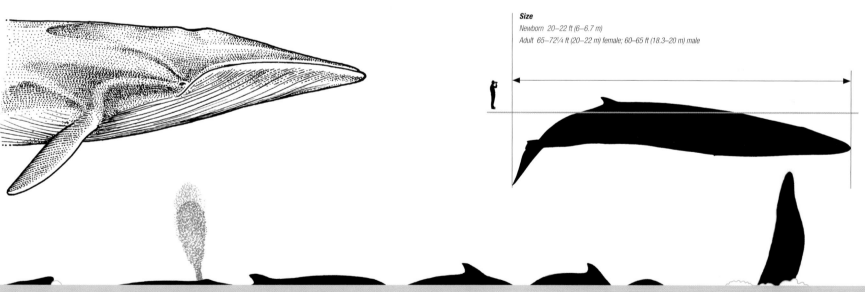

Size
Newborn 20–22 ft (6–6.7 m)
Adult 65–72¼ ft (20–22 m) female; 60–65 ft (18.3–20 m) male

Dive sequence
1. When the fin whale surfaces, it shows the rostrum first.

2. The frontal part of the back then surfaces, the dorsal fin still not being visible. Rising up to 20 ft (6 m) high, the blow is straight vertical and "V"-shaped.

3. The entire back of the whale then comes to surface.

4. When the whale starts diving, it arches and shows its caudal peduncle, and it then dives, almost never showing its caudal fin.

Breaching
Fin whales seldom breach, abruptly emerging from the water with their entire bodies and then landing on their back.

HUMPBACK WHALE

Family Balaenopteridae

Species *Megaptera novaeangliae*

Other common Names Hunchbacked whale, hump whale

Taxonomy This sole member of the genus *Megaptera* is most closely related to the fin whale (*Balaenoptera physalus*). Three subspecies are recognized: *M. n. australis* (southern humpback whale), *M. n. kuzira* (North Pacific humpback whale), and *M. n. novaeangliae* (North Atlantic humpback whale).

Similar Species Unmistakable with long pectoral flippers and dorsal hump

Birth weight 2,200–4,400 lb (1–2 tonnes)

Adult weight 55,000–66,0000 lb (25,000–30,000 kg)

Diet Diverse: krill, small schooling fishes

Group size 1–3, maximum 20 for breeding and feeding

Habitat Shallow coasts for breeding and feeding, open ocean during migration

Major threats Ship collisions, entanglements, noise pollution

IUCN status Least Concern

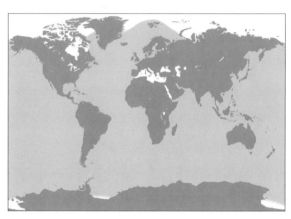

Range and habitat This species inhabits all major ocean basins. Primarily found in coastal waters and off continental shelves.

Identification checklist

- Black or dark-gray body
- Large, stocky body shape
- Low, stubby dorsal fin
- Very long flippers
- Knobby head and lower jaw
- Fin position: center rear

Anatomy

With their spindle-shaped body, elongate flippers, and large flukes, humpbacks are the most acrobatic of the great whales. The dorsal fin is stubby yet prominent, and a conspicuous hump forms when arching the back to dive. Two dozen or more ventral throat grooves or pleats are visible when rolling or feeding at the surface. Humpbacks' coloration is dark in the North Pacific, white-bellied in the southern oceans, and intermediate in the North Atlantic. The leading edge of the flippers are adorned with knobs that reduce drag, increase maneuverability, and assist with sensing of the environment. Individuals are identified by the shape and pigmentation on the underside of their flukes.

Behavior

Humpbacks exhibit spirited surface behaviors including lobtailing, flipper-slapping, spyhopping, head-lunging, and breaching. Complex social behaviors are underwritten by a brain three times larger than ours and wired with spindle neurons, which are active in processing emotion and social understanding. A strong bias toward right-handedness suggests considerable specialization of the brain hemispheres. This advanced cognition is shown in group hunting, bubble tool use, collective predator defense, and playful muggings of whale-watch vessels. Their astonishing acoustic repertoire includes songs, feeding calls, social sounds, wheezed blows, trumpet blasts, and surface-impacts sounds. With head down and eyes closed, the males broadcast beautiful songs that incorporate rhyme and syntax into ever-changing melodies. As the song cycle is repeated, other singers copy and improvise, allowing acoustic fads to ripple across our vast oceans.

Food and foraging

Humpback whales are generalist predators, using resourceful tactics to manipulate prey into exploitable configurations. The capture of elusive fish schools favors teamwork, enduring bonds, task specialization, and the deploying of communal bubble structures. Flipper movements and loud feeding calls may be employed to herd prey into bubble nets and trap it against the surface. With short, coarse, and unspecialized baleen, humpbacks exploit an astonishing diversity of prey. Adult krill is their preferred prey, though they target other midwater crustaceans including amphipods, pteropods, decapods, copepods, and mysid shrimp. Bait fish are also taken including herrings, sandlance, sardines,

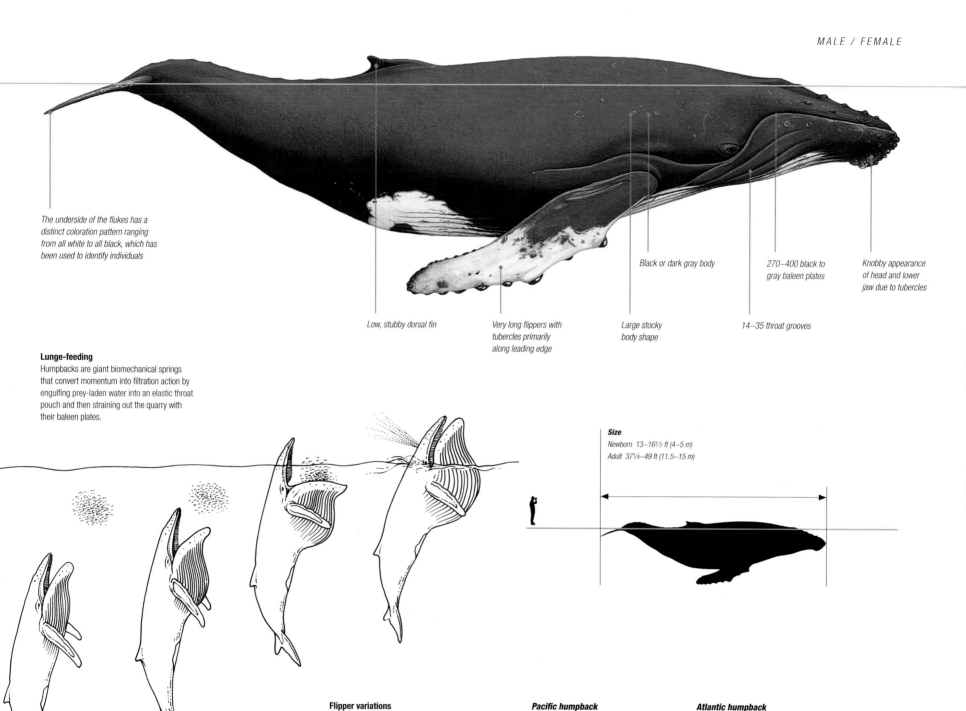

MALE / FEMALE

The underside of the flukes has a distinct coloration pattern ranging from all white to all black, which has been used to identify individuals

Low, stubby dorsal fin

Very long flippers with tubercles primarily along leading edge

Black or dark gray body

Large stocky body shape

14–35 throat grooves

270–400 black to gray baleen plates

Knobby appearance of head and lower jaw due to tubercles

Lunge-feeding

Humpbacks are giant biomechanical springs that convert momentum into filtration action by engulfing prey-laden water into an elastic throat pouch and then straining out the quarry with their baleen plates.

Size
Newborn 13–16½ ft (4–5 m)
Adult 37¼–49 ft (11.5–15 m)

Flipper variations

Humpbacks in the North Pacific generally have a black upper side of the flipper and a white underside. Humpbacks in the Atlantic and southern hemisphere usually have a flipper that is white on both sides, with some black markings.

Pacific humpback

Atlantic humpback

HUMPBACK WHALE

anchovies, capelin, eulachon, and mackerel. Even market squid are occasionally hunted. Commercially important fish are also taken including the young stages of pollock and cod. Some humpbacks have taken brazenly to raiding commercially reared salmon when they are released from net pens. When humpbacks are hunting krill, they become more asocial with fluid, transient groups consisting of one to three individuals.

Life history

Humpbacks undergo extensive seasonal migrations between summer foraging grounds in cold, productive, high latitudes and winter mating grounds in warm, low latitudes. These subtropical breeding grounds are situated near islands, coral reefs, and shallow coasts where the males engage in conspicuous communal displays known as leks. The males aggressively compete (and perhaps cooperate) for the primary escort position to mate with the asocial and elusive females. Females are impregnated in winter, and after a gestation period of nearly a year give birth to a single calf that is weaned after approximately eight months. Humpbacks are sexually mature at about eight years, and with a birth interval of two to three years. Life expectancy may approach that of human lifespans (60–70 years).

Conservation and management

Humpbacks were hunted intensively, but most worldwide populations are recovering robustly. However, the Japanese, Arabian Sea, New Zealand, and Fijian humpbacks remain Threatened. Worldwide conservation concerns include noise pollution, ship strikes, ocean acidification, fisheries depletion, and entanglement in recreation and commercial fishing gear. Many individuals exhibit scarring on the peduncle or the leading edge of the flukes from previous entanglements.

Humpback calf

A snowy-bellied calf from the Kingdom of Tonga reveals the white on the underside of the flippers, a possible adaptation for herding or "blazing" prey toward the mouth.

Breaching

The function of breaching is not known, but it is associated with play, aggression, hygienic maintenance, and communication. The nuisances of the impact sounds can reveal much about the leaper's temperament.

Blow
Rising 6½–10 ft (2–3 m) high, the blow is bushy and may appear "V"- or heart-shaped.

Dive sequence
1. Initially, the low dorsal fin is visible on the surface.

2. The body then arches, forming the characteristic humped back.

3. As the dorsal fin drops below the surface the tail is arched and the whale begins to dive.

4. The tail flukes appear above the surface, and they are lifted high on most dives.

Fluking
Humpbacks frequently raise the flukes when diving. Many have been photo-identified by their fluke markings.

Breaching
Humpbacks often lift their entire bodies out of the water, usually twisting on their side and landing on their back.

Pacific white-sided dolphin in motion (right)
During rapid swimming, delphinids leap completely
out of the water. Such porpoising behaviors, as
exhibited by the Pacific white-sided dolphin, are more
energetically efficient at high speeds than swimming.

TOOTHED WHALES
Oceanic Dolphins

Delphinidae, which includes oceanic dolphins and blackfish whales (such as *Orcinus orca*), is the most morphologically and taxonomically diverse group of cetaceans. Among the 38 species of delphinids is the smallest cetacean species, Hector's dolphin (*Cephalorhynchus hectori*). Some species of delphinid possess distinct beaks, such as the common bottlenose dolphin (*Tursiops truncatus*), while others are beakless, such as the long-finned pilot whale (*Globicephala melas*). Given that many delphinid species are abundant, social, and widespread geographically, dolphins are the most common cetaceans with which humans will come into contact.

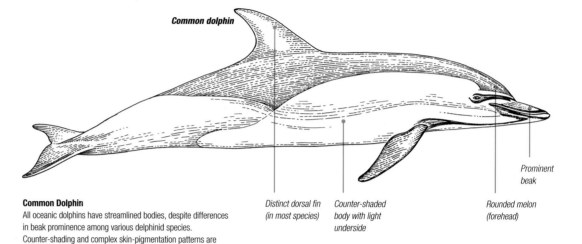

Common dolphin

Prominent beak

Distinct dorsal fin (in most species)   *Counter-shaded body with light underside*   *Rounded melon (forehead)*

Common Dolphin
All oceanic dolphins have streamlined bodies, despite differences in beak prominence among various delphinid species. Counter-shading and complex skin-pigmentation patterns are useful features to distinguish between different delphinid species.

- Dolphins occupy a broad range of body lengths from 4 ft (1.2 m)—Hector's dolphin (*Cephalorhynchus hectori*)—to 29½ ft (9 m)—(*Orcinus orca*).

- The skin coloration of delphinids varies a great degree from nearly uniform (*Sousa teuszii*), to counter-shaded (common bottlenose dolphin, *Tursiops truncatus*), to complex striping (hourglass dolphin, *Lagenorhynchus cruciger*), and spotting (Atlantic spotted dolphin, *Stenella frontalis*).

- The shape of the dorsal fin is variable among delphinids. The dorsal fin of the orca (killer whale, *Orcinus orca*) is tall and straight, the fin of others curve toward the tail (striped dolphin, *Stenella coeruleoalba*), while others are rounded (Hector's dolphin, *Cephalorhynchus hectori*).

- Most delphinid species are gregarious and live in pods consisting of hundreds to more than 1,000 individuals.

- As a family, Delphinidae is cosmopolitan and inhabits all of the world's oceans. Most species occupy shallow, coastal marine habitats although some dive to significant depths. Members of the genus *Sotalia* enter the Amazon River and associated tributaries.

- The diet of the various species of dolphins includes fish, squid, octopus, krill, and marine mammals, including other cetaceans.

- Delphinids tend to be very active at the surface and the smaller-bodied species often leap out of the water (porpoising) during periods of rapid swimming. Pilot whales have been observed spyhopping, lobtailing, and swimming on their sides.

- Although several species are assigned a conservation status of Least Concern, the Irrawaddy (*Orcaella brevirostris*), snubfin (*Orcaella heinsohni*), and some species of humpback dolphin (*Sousa* spp.) are either Near Threatened or Vulnerable. The Hector's dolphin is Endangered.

- Oceanic dolphins and porpoises (Phocoenidae) are easily confused, but the teeth of dolphins are conical as opposed to the spatulate teeth of porpoises.

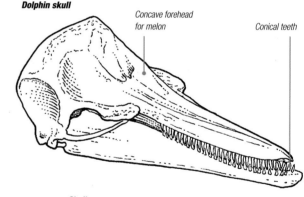

Dolphin skull

Concave forehead for melon   *Conical teeth*

Skull
Oceanic dolphins possess multiple conical teeth in both the upper and lower jaws. The skulls of oceanic dolphins can be distinguished from those of porpoises by the absence of distinct bumps near the blowhole.

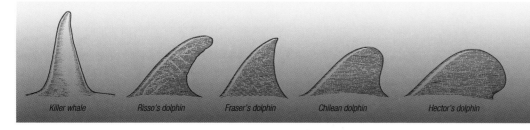

Killer whale   *Risso's dolphin*   *Fraser's dolphin*   *Chilean dolphin*   *Hector's dolphin*

All delphinids possess dorsal fins, except for the right whale dolphins. The dorsal fin of delphinids is variable in shape and can be useful in the identification of species in the field. The dorsal fin of the killer whale is triangular and slender, rising starkly from the back of the animal. The dorsal fins of other species tend to have a concave trailing edge, and the tips may be pointed (such as in Fraser's dolphin, Lagenodelphis hosei) or rounded (such as the Chilean dolphin, Cephalorhyncus eutropia). The dorsal fin of the Hector's dolphin is very rounded and lobe-like.*

COMMERSON'S DOLPHIN

Family Delphinidae

Species *Cephalorhynchus commersonii*

Other common names Black and white dolphin

Taxonomy Two subspecies are recognized *C.c. commersonii* and *C.c. kerguelenensis* (Kerguelen Islands Commerson's dolphin); the closest relative is the Chilean dolphin

Similar species The Chilean dolphin and Commerson's dolphin both occur near Cape Horn, and both have rounded dorsal fins—the sharply demarcated black-and-white color pattern of Commerson's dolphin make it distinctive

Birth weight 18–22 lb (8–10 kg)

Adult weight 88–110 lb (40–50 kg)

Diet Small fish, juvenile squid, shrimps

Group size 2–10, occasionally up to 35 when feeding

Major threats Entanglement in gill nets and trawls

IUCN status Data Deficient

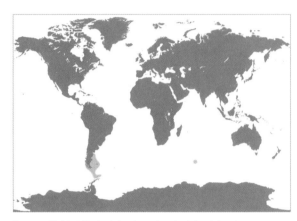

Range and habitat The South American subspecies is found along the Argentinean coast from the River Negro to Cape Horn, Straits of Magellan, and Falkland Islands. The Kerguelen subspecies inhabits coastal waters of the Kerguelen Islands. Both subspecies are usually seen at depths of less than 330 ft (100 m).

Identification checklist

- Small body—less than 5 ft (1.5m) long in South American subspecies
- Convex trailing edge of dorsal fin
- Striking coloration of black and white

Anatomy

Commerson's dolphins are small blunt-headed dolphins with striking black-and-white coloration, a convex trailing edge to the dorsal fin, and rounded flippers. Calves are born with a gray hue, which fades to the adults' light gray over about six months. In animals from Kerguelen, this darker hue is retained into adulthood. The Kerguelen animals are also larger, reaching more than 5¾ ft (1.7 m) and 190 lb (86 kg).

Behavior

Group size is usually small (2–10), though cooperative feeding bouts involving 30 or more dolphins have been observed. These dolphins are most often seen close inshore, sometimes feeding in the kelp or surfing on open coasts. Commerson's dolphins are curious and, in general, strongly attracted to boats, approaching to ride the bow or wake. Their most frequent leaps are low and horizontal, made while swimming at speed. Their sounds are high-frequency, narrow-band clicks. The dolphins in this genus do not whistle, as many other dolphins do.

Food and foraging

These dolphins show a diverse range of foraging behaviors, from single animals performing long dives to forage on the sea bottom, to large groups coordinating to herd fish at the surface. These are adaptable animals and will use any kind of barrier (moored boats, piers, rocky shores) to herd fish against. They eat a variety of fish, squid, and shrimp. Small schooling fish are sometimes pursued by groups working cooperatively. Individuals have small home ranges and localized populations rarely interbreed.

Life history

Commerson's dolphins reach maturity at five to eight years, and females produce one calf every two to four years. Gestation is about 12 months. The maximum recorded age is 18 years.

Conservation and management

Commerson's dolphins have a history of being hunted for crab bait. The most important threat is bycatch. In South America, they are caught in gill nets, and in midwater and bottom trawls for shrimp. Population size is unknown, but this is probably the most abundant member of the genus *Cephalorhynchus*.

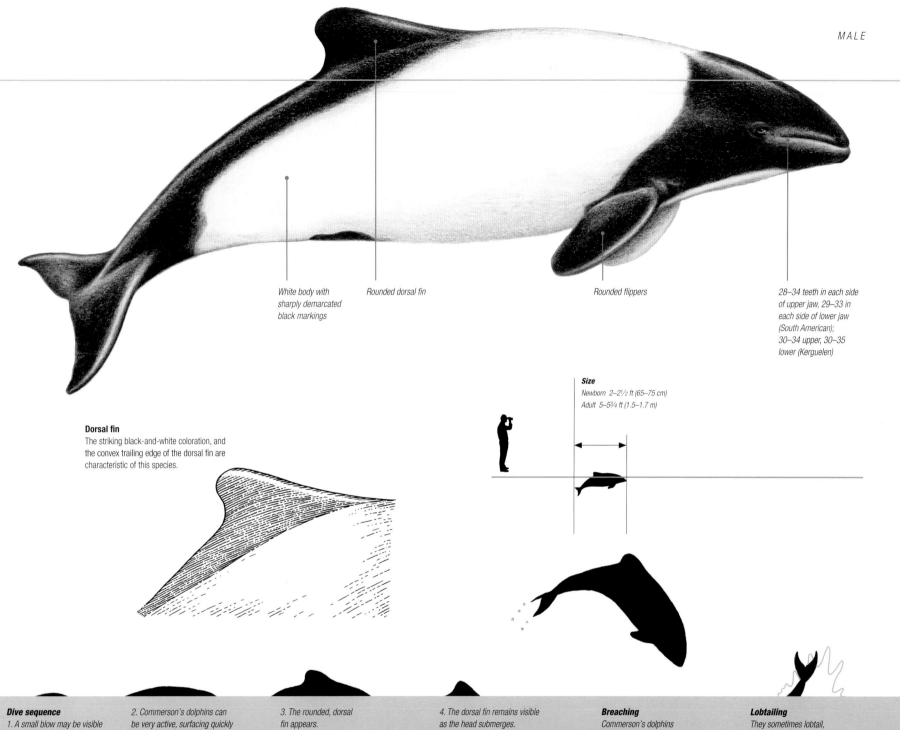

MALE

White body with
sharply demarcated
black markings

Rounded dorsal fin

Rounded flippers

28–34 teeth in each side
of upper jaw, 29–33 in
each side of lower jaw
(South American);
30–34 upper, 30–35
lower (Kerguelen)

Dorsal fin
The striking black-and-white coloration, and
the convex trailing edge of the dorsal fin are
characteristic of this species.

Size
Newborn 2–2½ ft (65–75 cm)
Adult 5–5¾ ft (1.5–1.7 m)

Dive sequence
*1. A small blow may be visible
in very cold conditions.*

*2. Commerson's dolphins can
be very active, surfacing quickly
with lots of splash, and rolling
slowly at the surface.*

*3. The rounded, dorsal
fin appears.*

*4. The dorsal fin remains visible
as the head submerges.*

Breaching
*Commerson's dolphins
sometimes show
vertical leaps.*

Lobtailing
*They sometimes lobtail,
which they do most
powerfully upside-down.*

CHILEAN DOLPHIN

Family Delphinidae

Species *Cephalorhynchus eutropia*

Other common names Black dolphin

Taxonomy Closest relative is Commerson's dolphin

Similar species Peale's dolphin and Burmeister's porpoise, but at close range the rounded dorsal fin of the Chilean dolphin is unmistakable

Birth weight 18–22 lb (8–10 kg)

Adult weight 133–154 lb (60–70 kg)

Diet Small fish, juvenile squid, shrimps

Group size 2–10, occasionally up to 25 when feeding

Major threats Entanglement in gill nets

IUCN Status Near Threatened

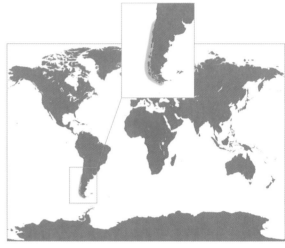

Range and habitat This species inhabits the fjords and open coasts of Chile from Valparaiso to Cape Horn.

Identification checklist

- Small body—less than 5¾ ft (1.7 m) long
- Convex trailing edge of dorsal fin
- Dark gray body, black flippers, flukes, and dorsal fin

Anatomy

Chilean dolphins show all the typical characteristics of their genus—they are small blunt-headed dolphins with rounded flippers, and as in the Hector's and Commerson's dolphin, the dorsal fin has a convex trailing edge. The color pattern broadly resembles a less well-defined and darker version of that of Hector's dolphins. The occasionally used name "black dolphin" is a misnomer—the main part of the body is a dark gray color, which darkens quickly to black after death.

Behavior

Group size is usually small (2–10), but larger associations can form when feeding. Chilean dolphins are somewhat shyer of boats than other *Cephalorhynchus* species, probably due to a history of being hunted. There are, however, isolated areas where they approach boats. Their sounds are high-frequency narrow-band clicks. The dolphins in this genus do not whistle, as many other dolphins do. At least around Isla Chiloé, individuals are highly philopatric.

Food and Foraging

Chilean dolphins are reported to prefer areas with rapid tidal flow, tide rips, and shallow waters over sills at the entrance to fjords. They eat a wide variety of fish species, squid, and shrimp. Small schooling fish such as sardines are sometimes pursued by a group of dolphins working cooperatively. There is no evidence of migration.

Life history

The limited information available suggests that like Hector's dolphins, Chilean dolphins reach maturity between seven to nine years, and thereafter females produce at least one calf every two to four years. Relatively few have been aged. Of these, the oldest was 19 years. They are likely to reach at least 25.

Conservation and management

Chilean dolphins have a history of being hunted for bait in crab fisheries, and are caught in gill nets throughout their range. Current abundance is unknown but likely to be low thousands at most. They are also displaced by intensive aquaculture, which is expanding dramatically in the northern fjords.

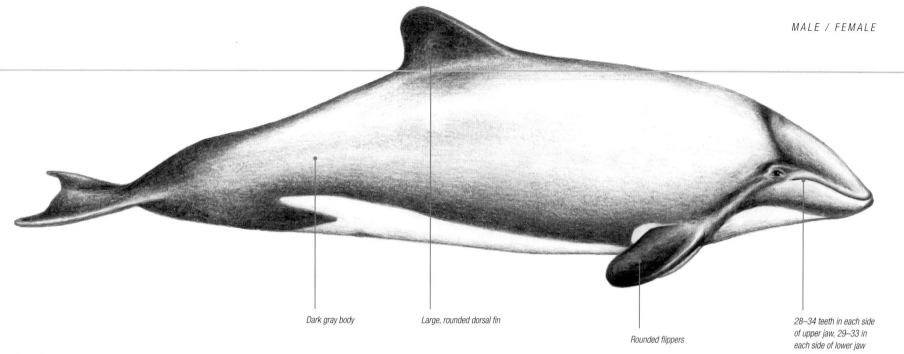

MALE / FEMALE

Dark gray body

Large, rounded dorsal fin

Rounded flippers

28–34 teeth in each side of upper jaw, 29–33 in each side of lower jaw

Dorsal fin
The shape of the dorsal fin is distinctive.
It is tall, with a convex trailing edge.

Size
Newborn 2¼–2½ ft (70–75 cm)
Adult 5–5¾ ft (1.5–1.7 m)

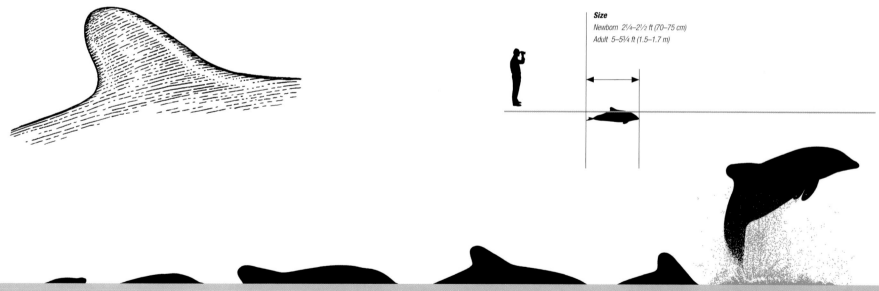

Dive sequence
1. The tip of the jaw grazes the surface as the dolphin breathes—a small blow is sometimes visible in very cold conditions.

2. Next, the dome of the melon becomes visible.

3. It is followed by the rounded, rather tall, dorsal fin.

4. The dorsal fin remains visible as the head submerges.

5. Finally, the animal disappears below the surface.

Breaching
Chilean dolphins jump infrequently. The most common type is a single vertical leap with head first re-entry.

HEAVISIDE'S DOLPHIN

Family Delphinidae

Species *Cephalorhynchus heavisidii*

Other common names Haviside's dolphin

Taxonomy Closest relative is Hector's dolphin

Similar species Superficially similar to the dusky dolphin but at close range the triangular dorsal fin of Heaviside's dolphin cannot be confused with the recurved fin of the dusky dolphin

Birth weight 18–22 lb (8–10 kg)

Adult weight 133–165 lb (60–75 kg)

Diet Bottom-dwelling fish, especially juvenile hake, octopus, and squid

Group size 2–10, occasionally up to 30

Major threats Entanglement in gill nets and trawls

IUCN status Data Deficient

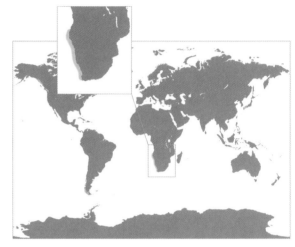

Range and habitat This species is found in waters around southwest Africa, to about 17°S and is usually seen at depths of less than 330 ft (100 m).

Identification checklist

- Small body—less than 6 ft (1.8 m) long
- Prominent, roughly triangular dorsal fin
- Light gray head and sides with a dark (almost black) cape extending from the blowhole along flanks

Anatomy

Heaviside's dolphin is the only member of this genus that does not have a rounded, convex trailing edge to its dorsal fin. Instead, the dorsal fin is roughly triangular, with a reasonably long base. Their coloration displays three shades—almost black dorsally, mid-gray head and thorax, and a complex pure white pattern ventrally.

Behavior

Group size is usually small (2–10), though larger groups have been seen. Heaviside's dolphins are curious and will approach boats—and like the other open-coast *Cephalorhynchus* dolphins, are fond of surfing. None of the members of this genus are as acrobatic as spinner or dusky dolphins, but in social contexts they show a variety of leaps and lobtails. Their sounds are high-frequency, narrow-band clicks.

Food and foraging

While they take a variety of prey, Heaviside's dolphins are specialists among other species of the genus, focusing particularly on juvenile hake. This fish migrates vertically to be close to the surface at night, and this is thought to drive diurnal movements of Heaviside's dolphins, which at least in the southern part of their range are seen close inshore during the day, moving offshore in the late afternoon in order to forage. This diurnal inshore-offshore movement is not seen in all populations, however. Individuals have relatively small home ranges of less than 50 miles (80 km) alongshore.

Life history

Specific information is scant, but it is likely that the life history of Heaviside's dolphins is broadly similar to that of the other members of its genus. Hence, maturity would be expected at between five to eight years, and calving interval every two to four years.

Conservation and management

Historically, Heaviside's dolphins have been hunted by fishermen. Bycatch occurs in gill nets, trawls, and purse seines, but at unknown levels. Total population size is also unknown. The exposed coast of southwest Africa offers little shelter to inshore fishing vessels, so it is possible that bycatch in fishing nets is less of a problem for this species than for the other *Cephalorhynchus* species.

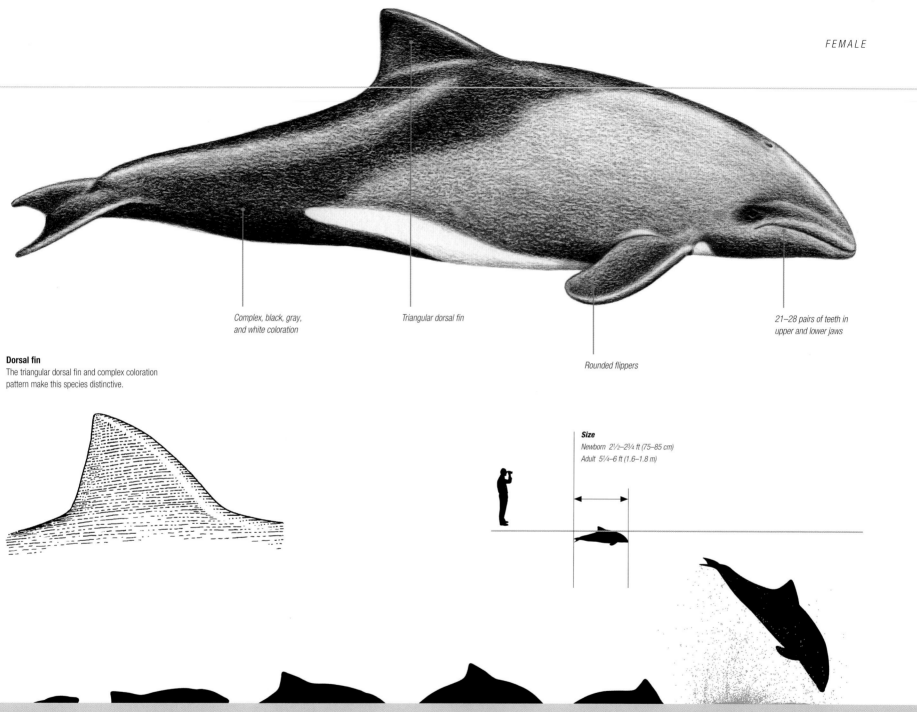

FEMALE

Complex, black, gray,
and white coloration

Triangular dorsal fin

21–28 pairs of teeth in
upper and lower jaws

Rounded flippers

Dorsal fin
The triangular dorsal fin and complex coloration
pattern make this species distinctive.

Size
Newborn 2½–2¾ ft (75–85 cm)
Adult 5¼–6 ft (1.6–1.8 m)

Dive sequence
1. No blow is visible as
this dolphin begins a slow
roll while surfacing.

2. The tip of the triangular
dorsal fin emerges very soon
after the dolphin breathes.

3. The body rolls relatively
high in the water.

4. Next, the head
submerges.

5. Finally, the dorsal fin
disappears below the surface.

Breaching
Vertical jumps with head first
re-entry are most common
aerial display.

HECTOR'S DOLPHIN

Family Delphinidae

Species *Cephalorhynchus hectori*

Other common names New Zealand dolphin, Maui's dolphin (North Island subspecies)

Taxonomy Two subspecies: South Island (*C. h. hectori*), North Island west coast (*C. h. maui*). South Island Hector's dolphins form three genetically distinct populations (west coast, east coast, south coast). Closest relative is Heaviside's dolphin.

Similar species This species is very distinctive—no other dolphin in New Zealand waters has a rounded dorsal fin

Birth weight 18–22 lb (8–10 kg)

Adult weight 103 lb (47 kg) female; 95 lb (43 kg) male

Diet Small fish, juvenile squid

Group size 2–10, very occasionally several dozen when following inshore trawlers

Major threats Entanglement in gill nets and trawls

IUCN status Endangered: Hector's dolphin; Critically Endangered: Maui's dolphin

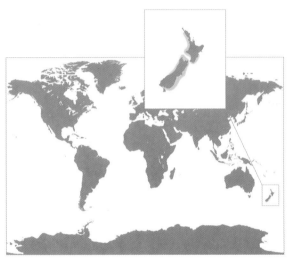

Range and habitat This species is found in inshore waters of New Zealand, at depths of less than 330 ft (100 m). Maui's dolphin is restricted to the central North Island west coast.

Identification checklist

- Small body—less than 5 ft (1.5 m) long
- Convex trailing edge of rounded dorsal fin
- Light gray body, black flippers, flukes, and dorsal fin, and dark crescent mark over blowhole

Anatomy

Hector's dolphins are among the smallest of all dolphins. Their small size and rounded dorsal fin make this species hard to misidentify. The two subspecies (Hector's dolphin and Maui's dolphin) look identical, though Maui's dolphin adults are marginally larger. Calves are born with a dark gray hue, which fades to the adults' light gray over about six months.

Behavior

Normally found in small groups of 2–10, which occasionally coalesce, often swapping members in the process. These dolphins are usually seen swimming quietly, making dives of one to two minutes followed by five to eight breaths before diving again. Social contexts can include play-chases, lobtails, spyhops, and jumps—some seemingly intended to make as much splash (and noise) as possible. Off open coasts Hector's dolphins are often seen surfing. They are inquisitive and attracted to slow-moving boats. Unlike most dolphins they do not have a rich repertoire of audible sounds. Almost all their sounds are high-frequency, narrow-band clicks. In both Hector's and Maui's dolphins, individuals have small home ranges that typically cover less than 30 miles (50 km) of coastline.

Food and foraging

Both subspecies are generalists, eating a wide variety of fish species, and juvenile squid. Most prey items are less than 1/3 ft (10 cm) long, mostly taken on the bottom, but also throughout the water column. Prey species differ by location, apparently reflecting what is available.

Life history

Hector's dolphins reach maturity between seven and nine years. Females produce one calf every two to four years thereafter. Maximum lifespan is almost 30 years, though most individuals do not reach 20. Hector's dolphin is the most intensively studied member of the genus *Cephalorhynchus*.

Conservation and management

Hector's and Maui's dolphins have serious problems with entanglement in fishing gear, particularly gill nets, but also trawls. To reduce impacts, extensive inshore areas are now closed to gill netting. Numbering around 50 adults, Maui's dolphin is on the verge of extinction.

MALE

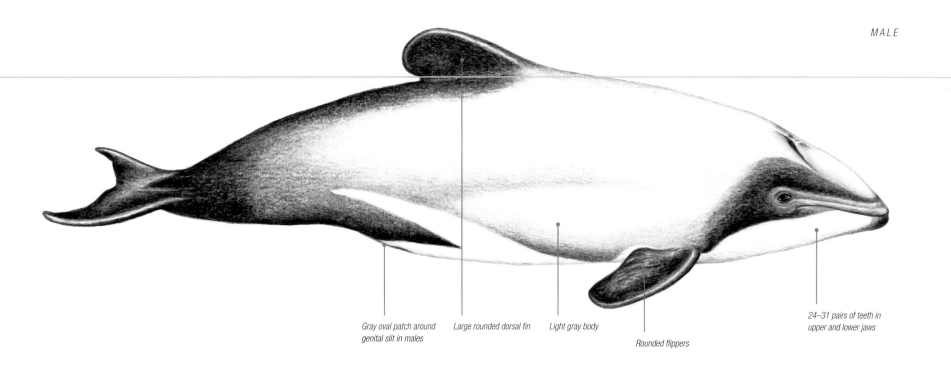

Gray oval patch around
genital slit in males

Large rounded dorsal fin

Light gray body

Rounded flippers

24–31 pairs of teeth in
upper and lower jaws

Dorsal fin
The dorsal fin has a distinctive
convex trailing edge.

Size
Newborn 2–2½ ft (60–75 cm)
Adult 4½–5 ft (1.35–1.5 m) female; 4–4½ ft (1.2–1.35 m) male

Dive sequence
1. Normally no blow is visible
as the dolphin surfaces.

2. Most surfacings are slow
rolls, without any splash.

3. The rounded dorsal
fin then shows.

4. Soon after, the head
submerges.

5. Lastly, the dorsal
fin disappears below
the surface.

Breaching
Most jumps are single vertical leaps
with headfirst re-entry. Side flops,
such as this one, may be repeated
more than 10 times in a row.

LONG-BEAKED COMMON DOLPHIN

Family Delphinidae

Species *Delphinus capensis*

Other common names None

Taxonomy Most closely related to the short-beaked common dolphin. There are two subspecies recognized: *D. c. capensis* and *D. c. tropicalis* (endemic to the Indian Ocean).

Similar species It can be easily mistaken for the short-beaked common dolphin, which has the same coloration pattern

Birth weight Not available

Adult weight 331–518 lb (150–235 kg)

Diet Small schooling fish (sardines, anchovies, hake) and squid

Group size 10–500 animals

Major threats Direct takes off Peru that are used for human food and shark bait

IUCN status Data Deficient

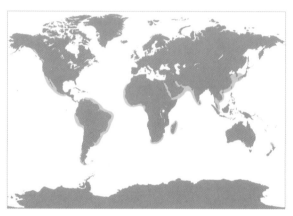

Range and habitat This species is found in shallow tropical and subtropical waters of the Atlantic, Pacific, and Indian Oceans.

Identification checklist

- Dark, lighter gray, and yellow crisscross coloration pattern on the side
- Rounded melon
- Sleek, robust body
- Long beak
- Falcate dorsal fin in the middle of the back

Anatomy

The long-beaked common dolphin is very similar to its sister species, the short-beaked common dolphin (*Delphinus delphis*), differing mainly in having a larger body size and a longer beak with more teeth (up to 60) than other delphinid. The complex coloration pattern it presents also differs from that of the short-beaked common dolphin by having a much darker and wider eye stripe that extends down from the eye toward the flippers.

Behavior

This species behaves very similarly to the short-beaked common dolphin. They can occur in small groups of 10–30 individuals, which may segregate by age and sex, or larger groups of hundreds or even thousands of individuals. Long-beaked common dolphins are energetic animals, often leaping out of the water and bow-riding boats.

Food and foraging

They feed mostly on small schooling fish—such as sardines, anchovies, and herring—as well as squid. Their diet varies geographically and with availability of prey. They tend to feed on prey in more shallow waters.

Life history

This species is sexually mature at 6½ ft (2 m) long. Gestation lasts 10–11 months, followed by a calving period of 1–3 years. Breeding is between spring and fall. The lifespan is around 40 years.

Conservation and management

These dolphins are not as well-studied as short-beaked common dolphins but they are likely not as abundant. They have been hunted off Peru and northern Venezuela and the current status of how many animals are taken is not known. They are also involved in incidental catches in gill nets off southern California.

MALE / FEMALE

Often has a flank darker stripe that is absent in the short-beaked common dolphin

Crisscross coloration pattern on the sides resembling a hourglass, formed by the interaction of a dark cape, a yellow thoracic patch, a light gray flank patch, and white belly

Dark eye stripe

Dorsal fin may present a gray or whitish area in the center

Coloration patterns

The long-beaked common dolphin has a long, slender beak and different facial coloration pattern—with a darker, wider eye stripe extending further toward the flipper—than the short-beaked common dolphin.

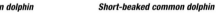

Long-beaked common dolphin   *Short-beaked common dolphin*

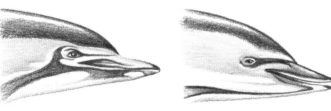

Size
Newborn 2½–3½ ft (0.8–1 m)
Adult 6¼–7¼ ft (1.9–2.2 m) female; 6½–8 ft (2–2.4 m) male

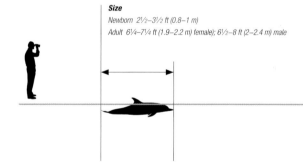

Dive sequence
1. The dorsal fin and the dark dorsal cape are the features most commonly seen on the surface when these dolphins are on the move.

2. When preparing to jump, these dolphins usually show their beak first. The coloration of their heads usually allows species identification.

3. When jumping out of the water, the full body is visible, allowing a correct species identification.

SHORT-BEAKED COMMON DOLPHIN

Family Delphinidae

Species *Delphinus delphis*

Other common names Saddleback dolphin, white-bellied porpoise

Taxonomy Most closely related to the long-beaked common dolphin—there are two subspecies recognized: *D. d. delphis*, which has a worldwide distribution, and *D. d. ponticus*, which is endemic to the Black Sea

Similar species It can be easily mistaken for the long-beaked common dolphin, which has the same crisscross coloration pattern on the sides of the body

Birth weight Not available

Adult weight 331–440 lb (150–200 kg)

Diet Small schooling fish (sardines, anchovies, hake) and squid

Group size 100–500 animals

Major threats Incidental catches in a number of fisheries around the world, including gill nets, purse seines, and trawls

IUCN status Least Concern

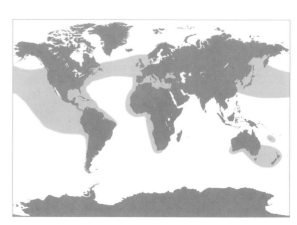

Range and habitat This species is found in oceanic and offshore warm-tropical to cool-temperate waters of the Atlantic and Pacific Oceans.

Identification checklist

- Dark, lighter gray, and yellow crisscross coloration pattern on the side
- Rounded melon
- Moderately long beak
- Sleek robust body
- Falcate dorsal fin in the middle of the back

Anatomy

The short-beaked common dolphin is very similar to its sister species, the long-beaked common dolphin, but differs in having a smaller body size and a shorter beak. Both these species can be distinguished from other dolphins by their complex coloration pattern, which forms a yellow and gray hourglass on the sides. These colors are usually stronger in this species than in the long-beaked form.

Behavior

This species' behavior is very similar to that of the long-beaked common dolphin. It occurs in large groups of hundreds and sometimes even thousands of individuals. Social subgroups usually consist of 10–30 individuals that may segregate by age and sex. These dolphins can associate with other species such as pilot whales, striped dolphins, and Risso's dolphins. They are often seen bow-riding boats.

Food and foraging

These dolphins feed mostly on small schooling fish such as anchovies, sardines, and mackerel, as well as squid. Their diet can vary according to the geographical region where they occur and whether they prefer offshore or more coastal waters.

Life history

Males become sexually mature from 3–12 years and females from 2–7 years. Gestation varies from 10–11 months, with a lactation period that lasts at least 10 months and a calving period of 1–3 years. The lifespan is approximately 35 years.

Conservation and management

In the Black and Mediterranean Seas populations have declined due to habitat degradation, lack of prey, and fisheries interactions. Elsewhere, this species is subject to incidental takes in fisheries, particularly in South Australia and the eastern tropical Pacific.

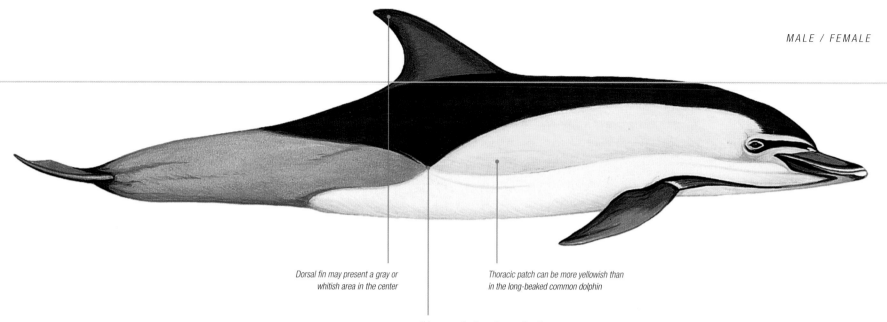

MALE / FEMALE

Dorsal fin may present a gray or
whitish area in the center

Thoracic patch can be more yellowish than
in the long-beaked common dolphin

Crisscross coloration pattern on the sides
resembling a hourglass, formed by the
interaction of a dark cape, a yellow thoracic
patch, a light gray flank patch, and white belly

Skull shape
The long beaked common dolphin has a long, slender skull with
more teeth (47–67 pairs) while the short-beaked common
dolphin has a short, broad skull with fewer teeth (41–57 pairs)
in the upper and lower jaws.

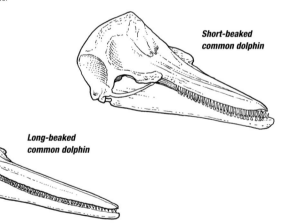

**Short-beaked
common dolphin**

**Long-beaked
common dolphin**

Size
Newborn 2½–3½ ft (0.8–1 m)
Adult 5¼–6¼ ft (1.6–1.9 m) female; 5¾–8 ft (1.7–2.4 m) male

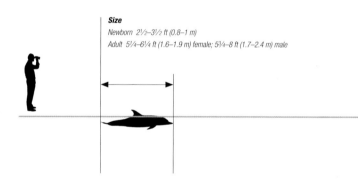

Dive sequence
1. Like the long-beaked common
dolphin, when this species dives
the dorsal fin and dorsal cape
are the features most commonly
seen on the surface.

2. In preparation for a jump
these dolphins tend to show
their beak first. Identification is
possible at this point from the
coloration of their heads.

3. The full body is visible
as the dolphin exits the
water, again allowing for
species identification.

PYGMY KILLER WHALE

Family Delphinidae

Species *Feresa attenuata*

Other common names Blackfish (although this is also used for several other species), sea wolf

Taxonomy Closest relatives are other oceanic delphinids

Similar species Most often confused with melon-headed whales and false killer whales

Birth weight Unknown

Adult weight 243–375 lb (110–170 kg)

Diet Primarily cephalopods and small fish

Group size Typically 10–20 individuals, although larger groups have been reported

Major threats Fisheries as bycatch and from directed hunts in Japan, Indonesia, and the Caribbean

IUCN status Data Deficient

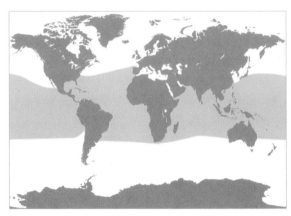

Range and habitat This species is found in tropical and subtropical oceanic waters, primarily in offshore and deep-water habitats with some populations found closer to shore near oceanic islands such as Hawaii.

Identification checklist

- Body is mostly dark gray to black in coloration with white or gray markings around the belly
- Lips of adults are white
- Rounded pectoral flippers
- Central fin position

Anatomy

Pygmy killer whales are often confused with melon-headed whales due to their similar size and coloration. Pygmy killer whales typically have a more rounded melon as well as broader, more rounded tips to their pectoral fins. In good light, the dorsal cape of the pygmy killer whale is very evident and only exhibits a shallow dip below the dorsal fin. In comparison, there is a steep angle by which the dorsal cape dips below the fin of the melon-headed whale.

Behavior

Pygmy killer whales are generally found in small groups, typically with fewer than 50 individuals, although groups of more than 100 animals have been encountered. They are less active at the surface than melon-headed whales. In many areas they avoid boats, and will rarely bow-ride. Spyhopping is common.

Food and foraging

Relatively little information exists on the prey of pygmy killer whales, but they are believed to feed at depth or during the night on squid and small fish. There is also some evidence that they may attack smaller dolphins.

Life history

Almost nothing is known of the life history of this species. What little is known primarily comes from stranded animals. The smallest known lactating female was a 6¾ ft (2.04 m) individual that stranded in the Caribbean, and the smallest sexually mature male was 6⅝ ft (2.07 m) individual that stranded in Florida.

Conservation and management

Harpoon and driftnet fisheries in the Caribbean, Indonesia, and Sri Lanka have targeted pygmy killer whales both for human consumption and to be used as bait for longline fisheries. They are also caught incidentally in other fisheries throughout their range. Pollutants and contaminants may build up in the species, but because they feed on lower-trophic level animals, these levels are not as high as in other species, such as false killer whales. The species remains relatively uncommon throughout most of its range.

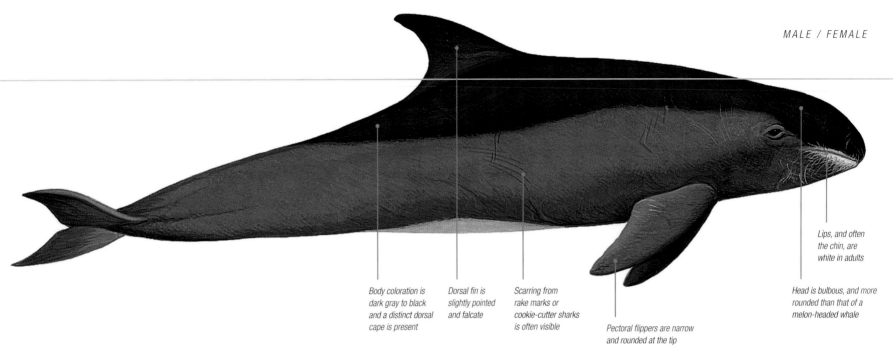

MALE / FEMALE

Body coloration is
dark gray to black
and a distinct dorsal
cape is present

Dorsal fin is
slightly pointed
and falcate

Scarring from
rake marks or
cookie-cutter sharks
is often visible

Pectoral flippers are narrow
and rounded at the tip

Lips, and often
the chin, are
white in adults

Head is bulbous, and more
rounded than that of a
melon-headed whale

Head shapes
The head shape of the pygmy killer whale is round and
slightly bulbous when viewed from above or from the
side. The head of the melon-headed whale appears
slightly more pointed when viewed from the side and
triangular when viewed from above.

Melon-headed whale

Pygmy killer whale

Size
Newborn 2½ ft (80 cm)
Adult 7–8½ ft (2.1–2.6 m)

Dive sequence
1. When surfacing, this
species does so discreetly
and low in the water.

2. The tip of the rounded
head becomes visible.

3. Next, the dorsal fin presents
fully from the water.

4. Then the dorsal
cape becomes visible.

5. Lastly, the body appears
to roll slowly while
re-entering the water.

Surface behavior
Pygmy killer whales often surface
synchronously and in pairs. Slight tail-flicking is
not uncommon. Breaching occurs occasionally,
although the tail flukes rarely leave the water
completely. While resting, pygmy killer whales
frequently roll on their sides, with their heads
partially or fully out of the water.

SHORT-FINNED PILOT WHALE

Family Delphinidae

Species *Globicephala macrorhynchus*

Other common names Blackfish (although this is also used for several other species), pilot whales, pilots, potheads

Taxonomy The closest relative is the long-finned pilot whale. In Japan, two distinct geographic forms exist, however they are not yet recognized taxonomically.

Similar species Most often confused with long-finned pilot whales in areas where their ranges overlap, and false killer whales

Birth weight 133 lbs (60 kg)

Adult weight 2,200–6,600 lbs (1,000–3,000 kg)

Diet Primarily squid as well as fish and other cephalopods

Group size Typical group size ranges from 15 to 50

Major threats Fisheries (directed takes and bycatch), pollution

IUCN status Data Deficient

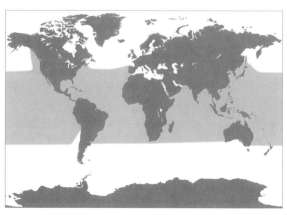

Range and habitat This species occurs in tropical and warm-temperate waters worldwide. It is primarily found in deep water and offshore habitats though sometimes seen close to shore around oceanic islands.

Identification checklist

- Large, robust body that is dark gray to black in coloration
- Bulbous square-shaped head with little to no rostrum evident
- Low, falcate dorsal fin with wide base
- Pectoral flippers are, on average, one-sixth the length of the body
- Fin position is about one-third of the way back from the head

Anatomy

Short-finned pilot whales are sexually dimorphic, with adult males being longer and heavier than their female counterparts. Males have a more pronounced melon with significantly larger dorsal fins than females. Size and coloration varies geographically.

Behavior

Photo-identification and genetic studies suggest that short-finned pilot whales live in relatively stable social groups comprised of both males and females. Females often remain with their natal group for life although males do not. While group size usually ranges from about 15–50 individuals, some groups may number many more than that. As with other social odontocetes, short-finned whales are prone to mass strandings. Daytime observations of the species often finds groups resting or logging at the surface. Dive-depth data from tagged animals suggests that this species actively feeds at depth during the night.

Food and foraging

Squid is thought to make up the majority of the diet for these whales, but other cephalopods and fish may be consumed.

Life history

Short-finned pilot whales reach sexual maturity between 8 and 9 (females) and 13 and 17 (males) years of age. Gestation is around 15 months with calves born year round. Calving peaks in the spring and fall in the southern hemisphere, and the fall and winter in the northern hemisphere. Calving occurs typically once every five to eight years and lactation lasts at least two years. These whales can live into their 60s, but females usually stop reproducing in their 40s.

Conservation and management

For centuries, short-finned pilot whales have been hunted in harpoon and drive fisheries in Japan. They are also hunted in the Lesser Antilles, Sri Lanka, and Indonesia and taken incidentally in fisheries throughout their range. Despite these takes, the species remains relatively abundant on a global scale.

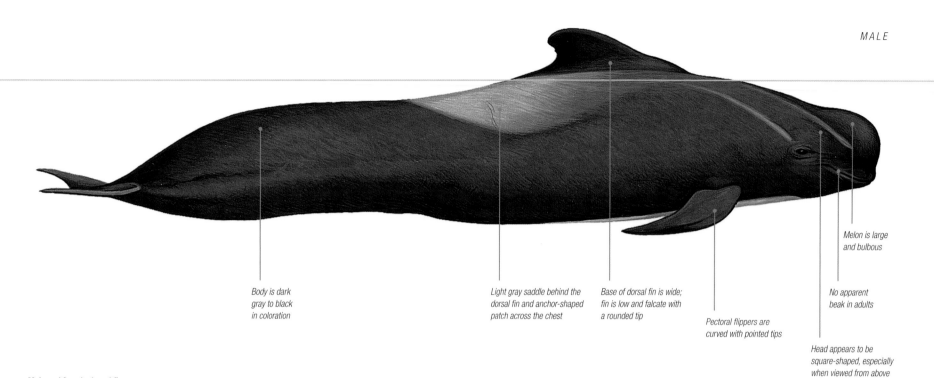

MALE

Melon is large
and bulbous

Body is dark
gray to black
in coloration

Light gray saddle behind the
dorsal fin and anchor-shaped
patch across the chest

Base of dorsal fin is wide;
fin is low and falcate with
a rounded tip

No apparent
beak in adults

Pectoral flippers are
curved with pointed tips

Head appears to be
square-shaped, especially
when viewed from above

Male and female dorsal fin
The dorsal fin of an adult male pilot whale is significantly
larger than that of a female. The base on the male is
wider, the fin height is taller, and the tip of the dorsal fin
is more hooked than that of the female.

Male dorsal fin

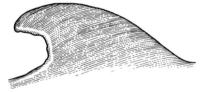

Female dorsal fin

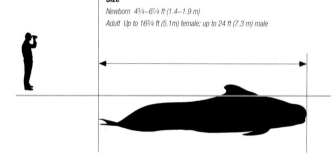

Size
Newborn 4¾–6¼ ft (1.4–1.9 m)
Adult Up to 16¾ ft (5.1m) female; up to 24 ft (7.3 m) male

Dive sequence
1. Initially, the head and melon
break high above the surface
of the water. The blow is
conspicuous, low, and bushy.

2. Soon the dorsal fin
becomes fully visible as the
body breaks high above the
surface of the water.

3. The dorsal fin becomes
more prominent as the
body rounds.

4. Next the peduncle arches
high before the dive. Fluking
may occur, but most frequently
does not.

Surface behavior
Surfacing often occurs synchronously with other
animals. Spyhopping by lone or multiple individuals
is common. Short-finned pilot whales rarely engage
in highly acrobatic behavior unlike many of their
smaller delphinid relatives. When breaching does
occur, it is often by juveniles who usually do not
fully leap above the water.

LONG-FINNED PILOT WHALE

Family Delphinidae

Species *Globicephala melas*

Other common names Blackfish (although this is also used for several other species), pilot whales, pilots, potheads

Taxonomy Closest relation is the short-finned pilot whale and three subspecies are recognized: southern hemisphere (*G. m. edwardii*), North Atlantic (*G. m. melas*) and North Pacific (*G. m.* un-named subsp.)

Similar species Most often confused with short-finned pilot whales in areas where their ranges overlap, false killer whales, and female killer whales

Birth weight 165–220 lbs (75–100 kg)

Adult weight Up to 2,900 lb (1,300 kg) female; 5,000 lb (2,300 kg) male

Diet Primarily squid and fish

Group size Typical group size of 10–20 individuals

Major threats Fisheries (directed takes and bycatch), pollution

IUCN status Least Concern

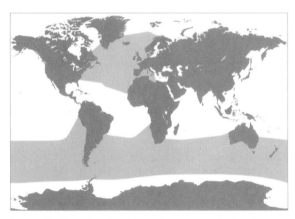

Range and habitat This species inhabits cool-temperate and subpolar waters. It is typically found in deep, pelagic waters but may occur in coastal waters in some parts of its range.

Identification checklist

- Large, robust body that is dark gray to black in coloration
- Bulbous head
- Low, falcate dorsal fin with wide base
- Pectoral flippers are, on average, one-fifth the length of the body
- Fin position is about one-third of the way back from the head

Anatomy

Long-finned pilot whales are among the largest of the delphinids. Like short-finned pilot whales, they are sexually dimorphic, with adult males being larger than females. Although both subspecies have a gray to white saddle and eye stripe, the southern hemisphere subspecies is considerably more pigmented than the North Atlantic subspecies, having a more conspicuous saddle and a more distinct eye stripe.

Behavior

Long-finned pilot whales usually occur in tight, social groups of fewer than 50 individuals although groups of several hundred animals have been reported. Their groups tend to be stable and maternally based. They are known to mass strand frequently, especially in certain parts of their range, such as Cape Cod, Massachusetts. Long-finned pilot whales are often slow-moving and found resting or logging at the surface. Spyhopping and lobtailing are both more common surface active behaviors.

Food and foraging

Cephalopods (primarily squid) make up the majority of the diet, especially in the southern hemisphere. North Atlantic whales take small- to medium-sized fish, but squid is still the main food source.

Life history

Females reach sexual maturity around 8 years of age and males at 12. Lifespan may be 35–45 years for males and more than 60 for females. Breeding occurs year-round with a peak in summer and a gestation of around 16 months. Calves are born every 3–5 years and females continue to reproduce into their late 40s or early 50s.

Conservation and management

Directed takes of these whales occur in several drive fisheries worldwide, most notably in the Faroe Islands. Incidental takes occur in longline, driftnet, and trawl fisheries. Pollutants pose a risk to the species where high levels of DDT, PCB, and other toxins can build up in their tissues.

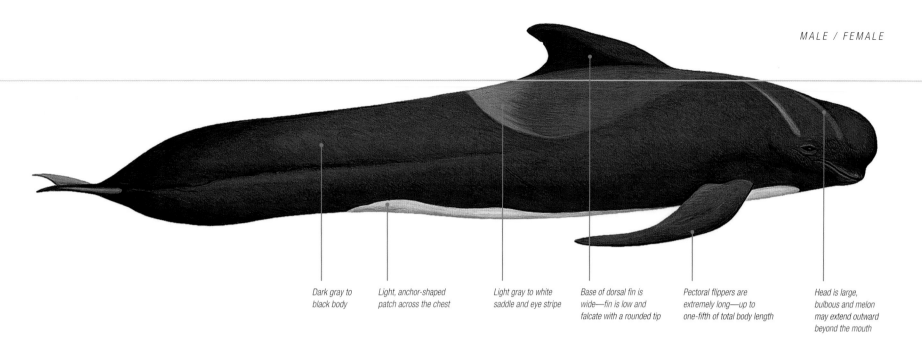

MALE / FEMALE

Dark gray to
black body

Light, anchor-shaped
patch across the chest

Light gray to white
saddle and eye stripe

Base of dorsal fin is
wide—fin is low and
falcate with a rounded tip

Pectoral flippers are
extremely long—up to
one-fifth of total body length

Head is large,
bulbous and melon
may extend outward
beyond the mouth

Long-finned and short-finned pectoral flippers
Both long-finned and short-finned pilot whales have sickle-shaped
pectoral flippers with strongly angled leading edges and pointed tips.
The overall length of the fin is longer in long-finned pilot whales—up
to one-fifth of total body length—than in short-finned pilot whales, in
which it is up to one-sixth of the body length.

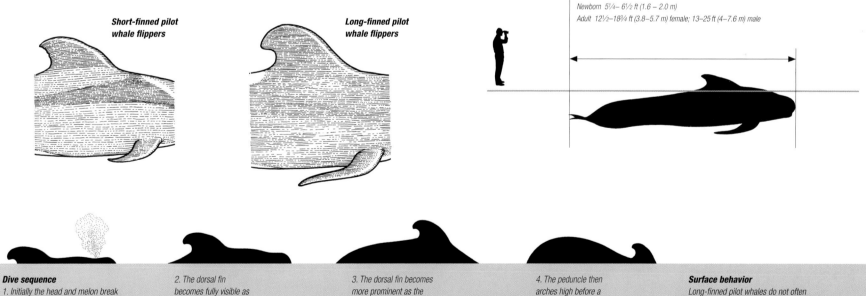

***Short-finned pilot
whale flippers***

***Long-finned pilot
whale flippers***

Size
Newborn 5¼– 6½ ft (1.6 – 2.0 m)
Adult 12½–18¾ ft (3.8–5.7 m) female; 13–25 ft (4–7.6 m) male

Dive sequence
1. Initially the head and melon break
high above the surface. Next there is
a strong low bushy blow, which can
often be seen from more than half a
mile away in clear conditions.

2. The dorsal fin
becomes fully visible as
the body breaks high
above the surface.

3. The dorsal fin becomes
more prominent as the
body rounds.

4. The peduncle then
arches high before a
dive. Sometimes fluking
may occur.

Surface behavior
Long-finned pilot whales do not often
engage in high-energy behaviors, such
as breaching or porpoising.

RISSO'S DOLPHIN

Family Delphinidae

Species *Grampus griseus*

Other common names Gray dolphin, grampus

Taxonomy Closest relatives are false killer whale and pilot whale

Similar species Bottlenose dolphins

Birth weight 44 lb (20 kg)

Adult weight 660–1,100 lb (300–500 kg)

Diet Squid (mainly), octopus, cuttlefish, anchovies, and krill

Group size 3–30 (average), occasional super-pods of 1,000

Major threats Ocean noise, fisheries bycatch, hunting, aquarium display

IUCN status Least Concern

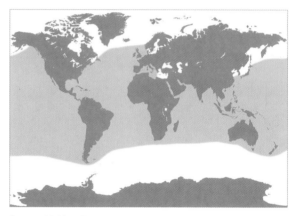

Range and habitat This species inhabits continental slopes and outer shelves with steep bottom topography in waters around 1,300–3,330 ft (400–1,000 m) deep. It is found in tropical and temperate waters between 60°N and 60°S. Not much is known about seasonal migrations, but they are likely influenced by oceanographic changes and prey shifts.

Identification checklist

- Vertical crease in bulbous head
- Dark gray with extensive white scarring, circular marks
- Flippers long, narrow, and pointed straight
- No beak

Anatomy

This is a robust-bodied dolphin with a narrow tail stock. Males and females are roughly the same size. The dorsal fin is tall and falcate, and located in the middle of the back. The body is heavily scarred with scratches, bite marks, splotches, and circular marks produced by squid, cookie-cutter sharks, lamprey, and other Risso's dolphins. Calves lack scarring but, as these dolphins age, their coloration changes from dark to light gray and pale white. They typically have no teeth in the upper jaw, but two to seven pairs in the lower jaw.

Behavior

Risso's are shy and rarely approach boats. They exhibit a range of surface behaviors such as leaping, breaching, and spyhopping. These are deep-diving dolphins with maximum reported depths of around 980 ft (300 m) and dive duration of 30 minutes. They can maintain stable, long-lasting bonds and form age- and sex-specific social groupings.

Food and foraging

Risso's prefer deep offshore waters but their distribution and movement patterns are driven by the behavior of their primary prey, market and jumbo squid. Acoustic studies suggest they forage at night, travel and socialize in the morning, and rest in the afternoon.

Life history

Calving season is summer to fall in the northwest Pacific Ocean, summer in South Africa, and fall along the California coast. Risso's reach sexual maturity between eight and ten years. Gestation is about 13–14 months and average lifespan is 30–34 years.

Conservation and management

Risso's are widely distributed and regionally abundant and are not considered to be at immediate risk. However, as deep divers they are vulnerable to military sonar and seismic surveys. Other threats, such as drive fisheries, intentional hunting, fisheries bycatch, and climate change, can potentially cause localized declines.

MALE / FEMALE

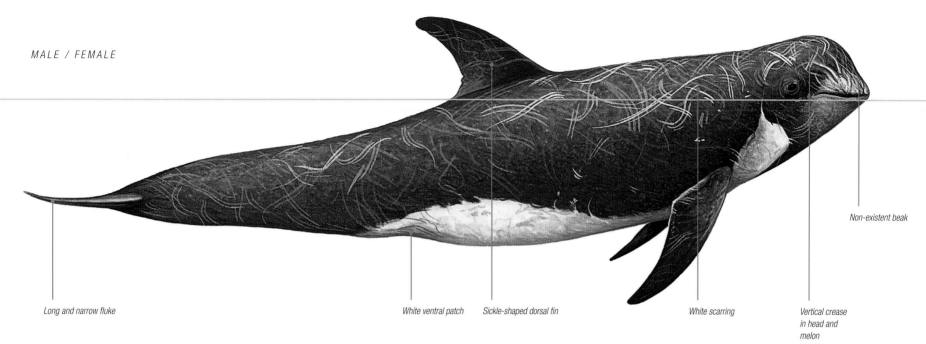

Non-existent beak

Long and narrow fluke

White ventral patch   Sickle-shaped dorsal fin

White scarring

Vertical crease
in head and
melon

Vertical crease
A Risso's dolphin's trademark features are
the unusual vertical cleft in the head and
melon, and extensive white scarring and
markings on the body that increase with age.

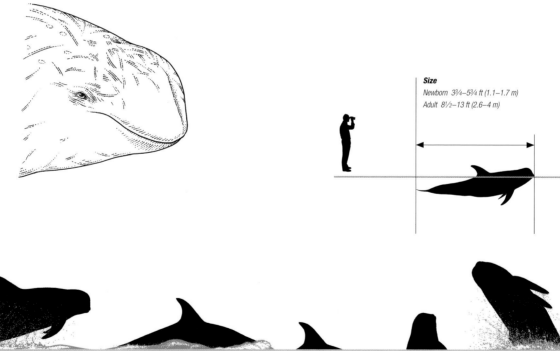

Size
Newborn 3¾–5¾ ft (1.1–1.7 m)
Adult 8½–13 ft (2.6–4 m)

Dive sequence
1. During slow
surfacing, just the head
is raised initially.

2. Next, the striking dorsal
and square head becomes
visible with the long and
narrow front flippers
pointed toward the water.

3. The body arches well above
the surface of the water, making
the dorsal fin more prominent,
as the head disappears back
below the surface.

4. Lastly, just the
dorsal fin is visible.

Surface behavior
Often, Risso's dolphins will raise their head at
a 45-degree angle when breaking the water
surface. Sometimes, they may slowly raise
their head straight out of the water.

These dolphins are also
capable of energetic
breaching, sometimes
leaping clear of the water.

FRASER'S DOLPHIN

Family Delphinidae

Species *Lagenodelphis hosei*

Other common names Sarawak dolphin

Taxonomy More closely related to *Stenella*, *Tursiops*, *Delphinus,* and *Sousa* species than to *Lagenorhynchus* species—populations in Japan and the Philippines differ morphologically

Similar species At a distance the stripe from eye to anus is similar to that of the striped dolphin, *Stenella coeruleoalba*

Birth weight 44 lb (20 kg)

Adult weight 353–463 lb (160–210 kg)

Diet Deep-dwelling fish, squids, and shrimps

Group size 100–500, but can be up to 1,000

Major threats Directed hunt (drive fishery in Japan, harpoon in Sri Lanka and the Lesser Antilles), bycatch in gill nets, ocean debris, boat strikes

IUCN status Least Concern

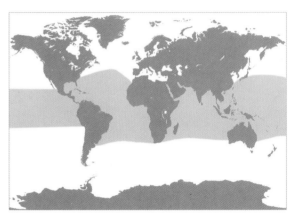

Range and habitat Found in tropical oceanic waters but sometimes seen close to shore in areas where deep water approaches the coast. High sighting rates observed in waters 2,300–11,500 ft (700–3,500 m) deep.

Identification checklist

- Stocky body
- Short but distinct beak
- Small, triangular, sometimes slightly falcate dorsal fin
- Small flippers and flukes
- Distinctive black band from eye to anus in adult males
- Black stripe from eye to flipper
- Back is dark to brownish-gray
- White belly that sometimes appears pink

Anatomy

Coloration varies with sex and age. Adult males have a distinct black band that runs from the face to the anus—this is less apparent or absent in females, young adults, and calves. The dark stripe that runs from the mid-lower jaw to the front of the flipper sometimes merges with the side stripe to form a "bandit mask" in adult males. A post-anal hump is well developed in adult males but is either absent or slight in females and young. This species has two grooves on the palate similar to those of the common dolphin.

Behavior

Fraser's dolphins are energetic swimmers, displaying frequent and rapid porpoising when swimming tightly in groups. Group size tends to be large, usually from about 100–500 and sometimes up to 1,000 animals. They are often seen mixed with other species, particularly melon-headed whales and short-finned pilot whales.

Food and foraging

Fraser's dolphins feed on mesopelagic fish, crustaceans, and cephalopods (squid). Physiological adaptations allow them to forage in deep waters. In the eastern tropical Pacific they feed in two depth zones, at 820 ft (250 m) and 1,640 ft (500 m). In the Philippines they appear to feed from near surface to as deep as 2,000 ft (600 m). In South Africa and in the Caribbean, however, they were observed to feed near the water surface.

Life history

Fraser's dolphin lives up to 19 years or more and grows to about 9 ft (2.7 m). Sexual maturity is reached at about 7–10 years in males and 5–8 years in females. Schools show mixed-age groups, with male to female ratio of 1:1. Gestation period is about 12½ months. Calving interval is approximately two years. Seasonal peak in calving varies with geography—in Japan it is in spring and fall and in South Africa, in summer.

Conservation and management

Threats include incidental catch in gill-net fisheries, drive fishery in Japan, and hunting in several localities for food or shark bait. For Southeast Asia it has been recommended that Fraser's dolphin would benefit from international cooperation.

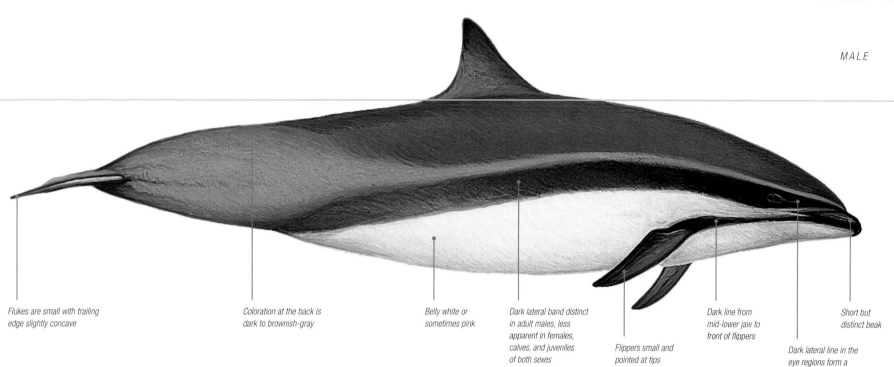

MALE

Flukes are small with trailing
edge slightly concave

Coloration at the back is
dark to brownish-gray

Belly white or
sometimes pink

Dark lateral band distinct
in adult males, less
apparent in females,
calves, and juveniles
of both sexes

Flippers small and
pointed at tips

Dark line from
mid-lower jaw to
front of flippers

Dark lateral line in the
eye regions form a
"bandit mask"

Short but
distinct beak

Dorsal fin variation
Some variation in the shape of the dorsal fin is present in this
species. In adult males the dorsal fin is more erect or triangular,
while in females and younger males, the dorsal fin is generally
more curved, or falcate.

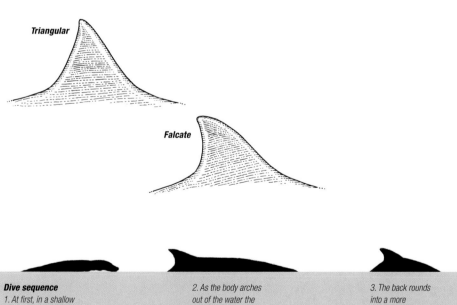

Triangular

Falcate

Size
Newborn 3½– 3¾ ft (1–1.1 m)
Adult 7–7¼ ft (2.1–2.2 m) female; 7¼ –8 ft (2.2–2.4 m) male

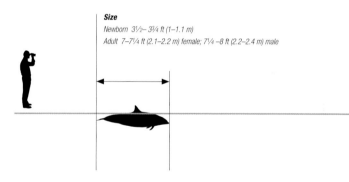

Dive sequence
1. At first, in a shallow
dive, just the head and
back are visible.

2. As the body arches
out of the water the
dorsal fin appears.

3. The back rounds
into a more
pronounced arch.

4. Finally, just the tip of
the dorsal fin remains
above the surface before
the shallow dive.

Breaching
This species sometimes
breaches, leaving the water
with a burst of spray.

ATLANTIC WHITE-SIDED DOLPHIN

Family Delphinidae

Species *Lagenorhynchus acutus*

Other common names Springer and jumper dolphin (names also applied to white-beaked dolphins)

Taxonomy The nomenclature and taxonomy of *Lagenorhynchus* species remain unresolved

Similar species Similar in size, shape, and distribution to white-beaked dolphins

Birth weight 53 lb (24 kg)

Adult weight 397 lb (180 kg) female; 507 lb (230 kg) male

Diet Herring, mackerel, cod, smelt, squid, sand lance, shrimp, and hake

Group size Small subgroups of 2–10, aggregations of 50–500, but occasionally thousands

Major threats Historically subject to large drive hunts, now susceptible to entanglement in nets and trawls, plus habitat disruption from climate change, and chemical pollution

IUCN status Least Concern

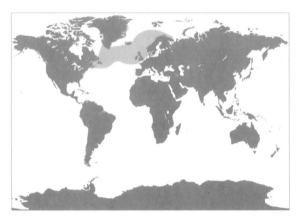

Range and habitat This species is typically found in cold-temperate to subpolar waters over continental shelf and slope, and is occasionally seen in shallow coastal waters or the central North Atlantic.

Identification checklist

- Back, dorsal fin, flippers, and flukes are dark gray or black
- Belly and lower jaw are white
- Stocky and deep-bodied with blunt beak
- Bright white blaze on the mid-lateral flanks
- Yellow-ocher blaze from white blaze to the tail stock

Anatomy

This is a relatively large, robust dolphin with striking coloration. Although these animals appear to have a short, blunt beak they have a long row of 29–40 pairs of small conical teeth on the upper and 31–38 pairs on the lower jaws. They have 15 ribs, and between 77 and 82 vertebrae (among the highest number of all dolphins).

Behavior

Group size seems to vary with location. In the eastern Atlantic and around Iceland, groups are usually formed of fewer than 10 individuals, whereas off the coast of New England around 40 individuals per group is more common. Segregation by age and sex has been suggested from analysis of mass strandings, with few large juveniles found in groups containing adults of both sexes and calves. These dolphins may feed with large baleen whales, and are known to associate with other dolphin species.

Food and foraging

This species does not appear to feed at depth, with most dives less than one minute in length. Prey species include short-finned squid, herring, smelt, silver hake, mackerel, and various species of shrimp and squid. Seasonal north–south shifts in distribution are likely associated with changing abundance and density of primary prey. In the 1970s a shift in habitat use may have been related to an increase in sand lance, or sand eels, over continental shelf waters.

Life history

The males of the species are larger and heavier than the females, but both sexes appear to attain sexual maturity at around five years of age. Lactation appears to last an average of 18 months, but some animals may breed annually. The average lifespan is between 22 and 27 years.

Conservation and management

Total abundance is thought to be in the hundreds of thousands. Once subject to large-scale directed catches, relatively few of these dolphins are currently taken purposely. Bycatch in gill nets and trawls causes some mortality, and mass-strandings involving up to 100 animals are common for this species. Accumulation of contaminants, including pesticides, may cause the immune systems of these dolphins to become suppressed and increase their susceptibility to disease.

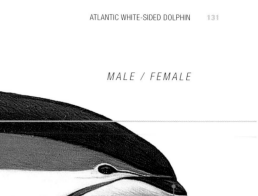

MALE / FEMALE

Complex and striking color pattern
with patches of gray, bright white,
and yellow-ocher along the flanks

Robust and stocky body
with girth up to 60
percent of total length

Tall, pointed dorsal fin

Blunt beak

Head, beak, and eye coloration
A black ring sets off the eye, connected
to the beak and upper jaw by a thin line
extending forward and an even thinner
line extending backward to the position
of the external ear. An oblique gray stripe
connects the eye to the flipper.

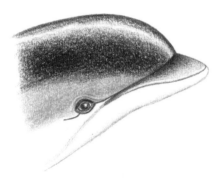

Size
Newborn 3½–4 ft (1–1.2 m)
Adult 8¼ ft (2.5 m) female; 9¼ ft (2.8 m) male

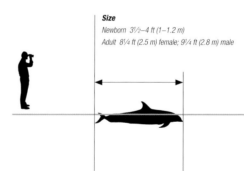

Blow
Usually initiated just before the
head reaches the surface. At slow
speed an underwater blow may
produce a single bubble or a
stream before the head emerges
to allow a new breath.

Dive sequence
1. The entire anterior portion of
the back, from beak to posterior
edge of the dorsal fin, breaks
the surface at the same time,
just after or during the blow.

2. At slow to medium
speeds, the body flexes
strongly after the blow so
that the body submerges
again without the flukes
breaking the surface.

3. At higher speeds, the back
just posterior to the dorsal fin
may emerge very briefly.

Breaching
Although not as acrobatic as
some other dolphin species,
these animals may leap
completely clear of the surface
and sometimes "belly-flop" to
produce a splash on re-entry.

Subsurface roll
While bow-riding or approaching a
stationary vessel, these dolphins may
swim just below the surface and
sometimes roll to the side and look
up at observers, providing a good
look at their distinctive side stripes.

Tail-slaps
While interacting with a boat (and
other dolphins) these animals may
deliberately splash water with their
flukes, either while rolled to the side
or from an upright position, or wriggle
wildly to produce "white water."

WHITE-BEAKED DOLPHIN

Family Delphinidae

Species *Lagenorhynchus albirostris*

Other common names White-nosed dolphin, squidhound, white-beaked porpoise

Taxonomy Molecular data indicates closest relatives are other *Lagenorhynchus* spp.

Similar species May be confused with the similar-sized white-sided dolphin

Birth weight 44–88 lb (20–40 kg)

Adult weight 397–640 lb (180–290 kg) females; 507–772 lb (230–350 kg) males

Diet Fish (mainly codfish) and squid

Group size 1–10

Major threats Ocean pollution, noise pollution, and global warming

IUCN status Least Concern

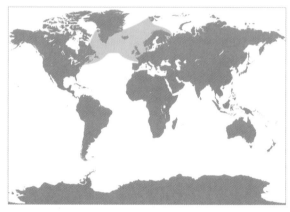

Range and habitat This species is found in temperate and subarctic waters of the North Atlantic. It occurs primarily over the continental shelves.

Identification checklist

- Black body with gray and white markings
- Stocky, robust body shape
- Tall, falcate dorsal fin—set in the center, mid-length
- Short, thick, white or whitish beak
- Long, pointed flippers

Anatomy

The coloration exhibits variable patterns of black and white on both flanks and belly. The most spectacular feature is the white beak, sometimes ranging to dark gray. Males grow somewhat larger than females. The white-beaked dolphin is a very stocky and muscular delphinid. It has up to 93 vertebrae—more than in any other cetacean, except Dall's porpoise. The numerous short vertebrae are believed to be an adaptation to fast, dynamic swimming. There are 25–28 pairs of teeth in the upper and lower jaws.

Behavior

This is an energetic and fast-swimming dolphin. The species exhibits aerial behavior typically seen among dolphin, with jumps, breaches, and even loops. White-beaked dolphins often approach vessels and perform bow-riding. They are occasionally observed with other cetacean species, such as fin and humpback whales and other dolphin species.

Food and foraging

The diet mainly consists of codfish, such as whiting, cod, and haddock. In addition, other pelagic and deep-water benthic fish species and squids. Foraging usually takes place close to the sea bottom and social feeding is known to occur.

Life history

Mating and giving birth occur mainly in midsummer. The gestation period is estimated to last 11 months. Sexual maturity is reached at lengths between 7½–8¼ ft (2.3–2.5 m) in males and 7½–8 ft (2.3–2.4 m) in females—around seven and nine years in both sexes. Females rest for several years between births. Lifespan estimated to at least 40 years.

Conservation and management

The white-beaked dolphin is at risk of being affected by climate change—for example, competing with subtropical and warm temperate species for declining fish stocks. Incidental catches of young animals are known. It is possible that pollution may impact this species.

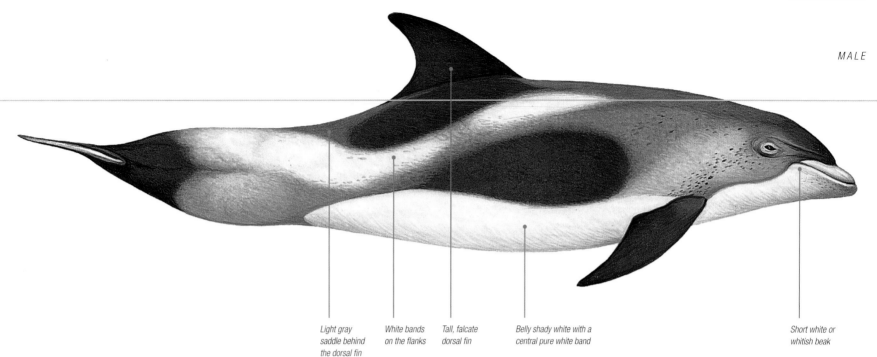

MALE

Light gray
saddle behind
the dorsal fin

White bands
on the flanks

Tall, falcate
dorsal fin

Belly shady white with a
central pure white band

Short white or
whitish beak

Variation of beak coloration
Despite the vernacular name, the color of the beak
of the white-beaked dolphin exhibits considerable
variation, ranging from white to an even ashy gray.
The beak color may also extend beyond the actual
beak, including the anterior onset of the melon.

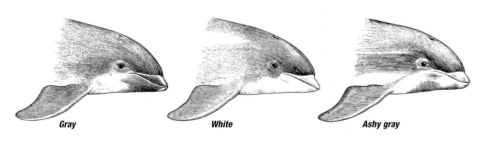

Gray   **White**   **Ashy gray**

Size
Newborn 3¾–4¼ ft (1.1–1.3 m)
Adult 7½–9¼ ft (2.3–2.8 m) female; 8–10 ft (2.4–3.1 m) male

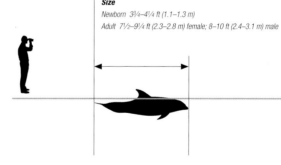

Dive sequence
*1. No visible blow. Initially, the
head and anterior dorsal part
of the body appears. Often,
the beak does not appear
above the surface.*

*2. The tall falcate
dorsal fin appears*

*3. The dorsal fin is clearly
visible, while the dolphin
rolls at the surface.*

*4. The white-beaked dolphin
is generally fast-moving and
this sequence occurs in
rapid succession.*

Breaching
*The white-beaked dolphin is an
acrobatic dolphin, which may display
a range of aerial behaviors, including
vertical breaches and various jumps.*

PEALE'S DOLPHIN

Family Delphinidae

Species *Lagenorhynchus australis*

Other common names Blackchin dolphin

Taxonomy Closely related to dusky and hourglass dolphins. The genus *Lagenorhynchus* is under revision.

Similar species Easily confused with dusky dolphin, possibly with the hourglass dolphin

Birth weight Unknown

Adult weight 220–250 lb (100–115 kg) females, males unknown

Diet Fish, cephalopods, crustaceans

Group size 1–15, occasionally up to 100

Major threats Bycatch, habitat degradation, global warming

IUCN status Data Deficient

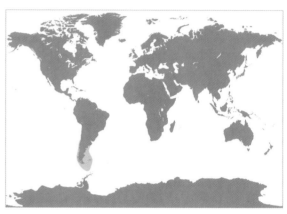

Range and habitat The Peale's dolphin is found in both coasts of South America, occupying deep, protected channels and fjords of the southwest and shallow continental shelves—usually less than 660 ft (200 m) depth—to the southeast.

Identification checklist

- Robust body
- Dark gray or black dorsally
- Two light areas on flanks
- Dark face
- Short beak
- White patch under the flipper
- Tall, falcate dorsal fin

Anatomy

Peale's dolphin is the most robust of the three southern *Lagenorhynchus* species. Their coloration pattern is complex and varies individually, mainly in the posterior border of the throat patch, along the sides, and the genital region. This species is unusual in having wide flat gums in the mouth, ideal for capturing small octopuses without smashing them. Flippers and dorsal fin are dark with lighter back edges. Young animals are lighter gray than adults.

Behavior

Peale's dolphins are often seen swimming slowly in or near coastal kelp beds. They can be very active, with head movement, rolling, breaching, spyhopping, and spinning. When traveling, their head makes a wide splash, earning them the name "plowshare" dolphins. They associate with other cetaceans, such as Commerson's dolphin; they often swim together along the north coast of Tierra del Fuego.

Food and foraging

In the southwestern South Atlantic, Peale's dolphin forages in the coastal ecosystem, feeding mainly on demersal and bottom fish such as southern cod and Patagonian grenadier; also octopus,

squid, and shrimps have also been found in stomachs examined. They feed in or near kelp beds and in open waters, with cooperative feeding, such as straight-line and large circle formations or star-burst feeding in which large groups encircle prey.

Life history

Peale's dolphin usually occur in small groups of 2–5 animals, but also in aggregations of up to 100. There is little information on reproduction and breeding season. Ovaries of only three females have been studied, suggesting that 6¼ ft (1.9 m) could be taken as a possible length at sexual maturity. There is no information on sexual maturity in males. Calves have been reported from spring through fall. The oldest animal studied was 13 years old, but the maximum age is probably greater.

Conservation and management

Peale's dolphins have been heavily exploited for southern king crab bait, especially in southern Chile, but this activity has declined in recent years. Although this species is very common within its range, no estimation of abundance is available, with the exception of a local population of some 200 animals off Chiloé, southern Chile.

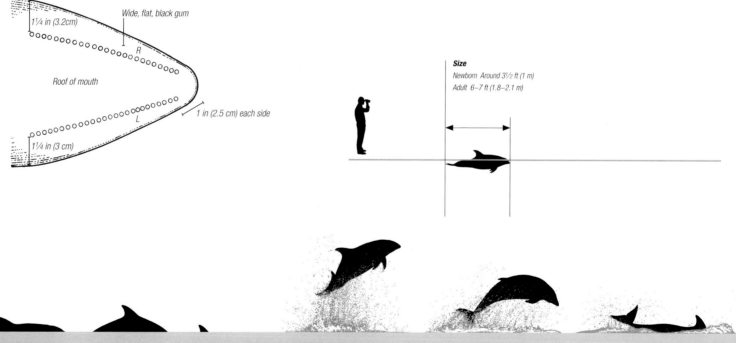

MALE / FEMALE

Wide lips and gums,
probably as feeding
adaptation to capture
small octopus

Tall dorsal fin

White axillary
(armpit) patches

Dark face

Flukes curved with rounded
tips, dark on both surfaces

Two light patches
on flanks

Dark gray or black
dorsal surface

Stocky body

Pointed,
recurved
flippers

White on ventral
surface behind
the throat patch

Short beak

Wide platform

A highly distinctive anatomical
feature of Peale's dolphin is the wide
"platform" between the teeth and the
edge of the mouth. The teeth are set
well in from the edge of the lip in both
the upper and lower jaw. This is very
unusual compared to the condition
in other dolphin species.

1¼ in (3.2cm)

Wide, flat, black gum

R

Roof of mouth

1 in (2.5 cm) each side

L

1¼ in (3 cm)

Size

Newborn Around 3½ ft (1 m)

Adult 6–7 ft (1.8–2.1 m)

Dive sequence

1. When rising to the surface, the
dolphin initially shows part of the
head. Due to the small size of
this species, the blow is usually
not visible.

2. The dorsal fin and
back appear on the
surface while the head
is still visible.

3. The head drops
below the surface while
the dorsal fin and back
remain on the surface.

4. Finally, the dolphin
disappears below the
surface. Sometimes the
flukes appear on diving,
or may slap the water.

Surface activity

Peale's dolphins carry out normal jumps
and other aerial behavior, including
breaching, spyhopping, head slaps, and
spins. They often leap and land on their
belly or side repeatedly.

HOURGLASS DOLPHIN

Family Delphinidae

Species *Lagenorhynchus cruciger*

Other common names None

Taxonomy Closely related to dusky and Peale's dolphins—the genus *Lagenorhynchus* is under revision

Similar species Could be confused with dusky or Peale's dolphin

Birth weight Unknown

Adult weight 194–207 lb (88–94 kg)

Diet Small fish, squids, crustaceans

Group size 1–10, occasionally up to 100

Major threats No major threats identified

IUCN status Least Concern

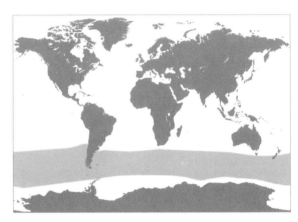

Range and habitat Distribution for this species is circumpolar, inhabiting sub-Antarctic and Antarctic waters from around 40–67°S. Primarily oceanic, these dolphins occur in deep waters but also are sighted near islands or banks.

Identification checklist

- Robust body
- Black dorsally, white ventrally
- Two elongated lateral white areas that form an hourglass pattern
- Beak short and black
- Dorsal fin and flukes totally black
- Tall, recurved dorsal fin set at the mid-back

Anatomy

The hourglass dolphin has a robust body, with a tall, recurved dorsal fin. The dorsal fin shape varies between sex and/or age classes, from erect to hooked. The tail stock is often keeled, especially in adult males. Coloration is mainly black or dark with two elongated white areas along the sides, connected by a white line. The tail flukes and dorsal fin are totally black, while part of the underside of the flippers is white. The ventral region is also generally white. Pigmentation of juveniles is unknown.

Behavior

Hourglass dolphins are active and fast swimmers with quick and energetic movements. They usually bow-ride, leaping out of the water while swimming in the wake of the boat. This species often associates with other cetaceans, especially fin and pilot whales, and seabirds such as albatross, giant petrels, cape petrels, and prions.

Food and foraging

Small fish, squid, and crustaceans were found in the stomachs of five specimens from different regions. They often join large aggregations of seabirds in feeding associations.

Life history

These dolphins usually occur in groups of 2–8, but concentrations of 60–100 dolphins have been reported. Although they are often observed, this is one of the least-known species of dolphin. The only data available on reproduction is based on four specimens—two females and two males—of unknown age. A 5¼ ft (1.6 m) female was sexually immature while another of 6 ft (1.8 m) was near maturity. Both males— 5¾ ft and 6¼ ft (1.7, 1.9 m)—were mature. Calves have been seen during January and February.

Conservation and management

With the exception of five dolphins taken for scientific research and a few incidental captures, no major threats due to human activities have been reported for this species. The only abundance estimate available is for south of the Antarctic Convergence, of about 144,000 hourglass dolphins during summer months.

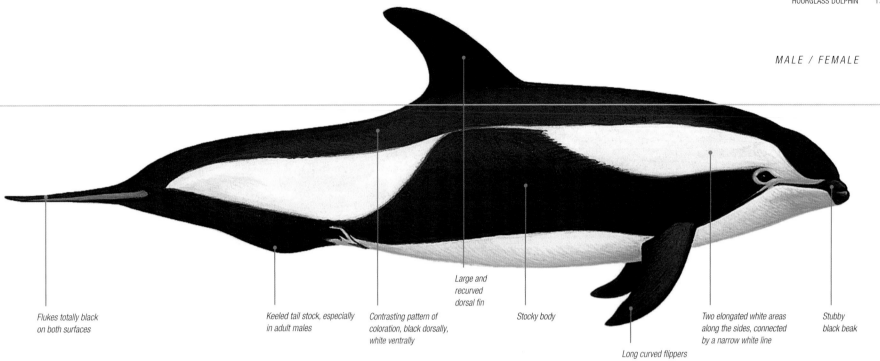

Flukes totally black on both surfaces

Keeled tail stock, especially in adult males

Contrasting pattern of coloration, black dorsally, white ventrally

Large and recurved dorsal fin

Stocky body

Two elongated white areas along the sides, connected by a narrow white line

Stubby black beak

Long curved flippers

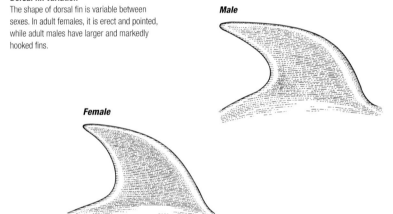

Dorsal fin variation
The shape of dorsal fin is variable between sexes. In adult females, it is erect and pointed, while adult males have larger and markedly hooked fins.

Male

Female

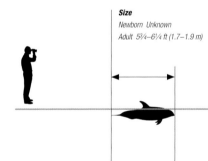

Size
Newborn Unknown
Adult 5¾–6¼ ft (1.7–1.9 m)

Dive sequence
1. When rising to the surface, the dolphin initially shows part of the head. The dorsal fin and back appear on the surface while the head is still visible. Due to the small size of these dolphins, the blow is usually not visible.

2. The head drops below the surface while the dorsal fin and back remain on the surface.

3. Lastly, the dolphin disappears below the surface. Sometimes the flukes appear on diving, or may slap the water.

Surface behavior
Hourglass dolphins are active bow-riders. They have often been observed surfing and leaping in the wake of vessels. They have also been seen playing around large whales.

PACIFIC WHITE-SIDED DOLPHIN

Family Delphinidae

Species *Lagenorhynchus obliquidens*

Other common names Lags, hookfin porpoise

Taxonomy Given uncertain relationships among *Lagenorhynchus* spp., a new genus, *Sagmatias*, has been proposed—the closest relative of this species is the dusky dolphin

Similar species Southern hemisphere dusky dolphin

Birth weight 33 lb (15 kg)

Adult weight 300–400 lb (135–180 kg)

Diet Cephalopods (such as squid), lantern fish, or myctophids, schooling fish (such as northern anchovy)

Group size 10–100 is common, but sometimes as much as 1,000

Major threats Gill net, mid- and bottom-water trawl fisheries bycatch, climate change

IUCN status Least Concern

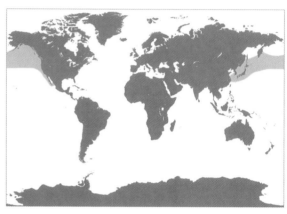

Range and habitat This species inhabits cool temperate waters of the North Pacific Ocean from California to the Bering Sea, and far south to Taiwan. It is found typically along continental shelf and slopes, and in open ocean waters.

Identification checklist

- Large, falcate dorsal fin located at mid-back
- Distinct light gray "suspender" body stripes
- White belly, bicolored gray and white dorsal fin, notched tail fluke
- Small beak, and black lips

Anatomy

The Pacific white-sided dolphin is larger then its sister species, the dusky dolphin. Two morphological (northern and southern) forms are recognized with dolphins north of Baja California having a smaller head, but the two are indistinguishable at sea. These dolphins have short beaks with 23–32 pairs of small, conical teeth in the upper and lower jaws. They have a bicolored, falcate dorsal fin and a fluke notched at the center.

Behavior

This is a gregarious, acrobatic dolphin species found in large pods and often in mixed species schools of Risso's dolphins and northern right whale dolphins. They are responsive to oceanic changes and follow seasonal distribution of their prey. The two forms have different acoustic call types. Unlike most delphinids, Pacific white-sided dolphins rarely whistle; instead they primarily use echolocation or sonar clicks for foraging and communication.

Food and foraging

These dolphins are flexible foragers, capable of feeding at night and during the day. They mainly consume a variety of schooling fish such as northern anchovies, sardines, herring, and Pacific whiting. At night, they feed on the rising layer of vertically migrating organisms composed of mostly mesopelagic fish (myctophids) and squid. They can feed cooperatively to corral fish or split into smaller feeding groups depending on prey size and density.

Life history

Females reach sexual maturity around 11 years of age—males between 10 and 11 years. Mating generally occurs in late summer to early fall. Gestations lasts about 11½ months and with a calving interval of 3 years. Average lifespan is believed to range between 36 and 40 years.

Conservation and management

Due to this species' abundant and widespread occurrence they are currently not at risk. These dolphins were historically targeted in the high seas driftnet and gill-net fisheries for salmon and squid in the western and central Pacific, but these fisheries were banned in 1993. Small numbers are now killed in thresher shark and swordfish driftnet, gill net, and groundfish trawl fisheries in the eastern Pacific but with limited impacts on their overall population.

MALE / FEMALE

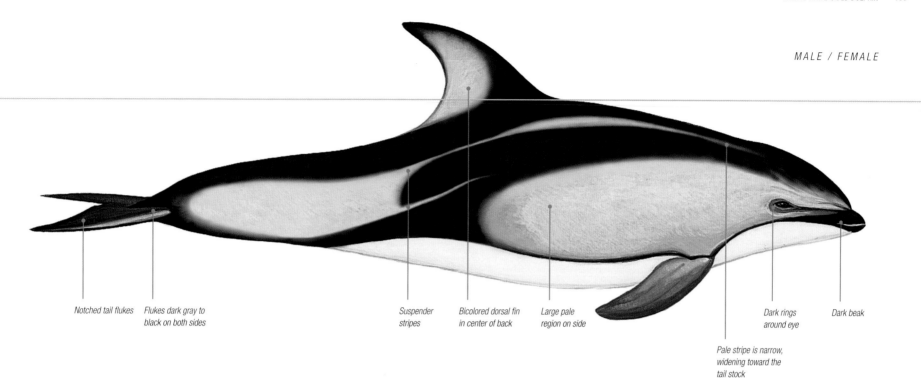

Notched tail flukes

Flukes dark gray to black on both sides

Suspender stripes

Bicolored dorsal fin in center of back

Large pale region on side

Dark rings around eye

Dark beak

Pale stripe is narrow, widening toward the tail stock

Bicolored dorsal fin
Pacific white-sided dolphins are distinguished easily at sea by their unique, bicolored, falcate dorsal fin, which is located in the middle of the back.

Size
Newborn 3½–4 ft (1–1.2 m)
Adult 5¾–8¼ ft (1.7–2.5 m) male; 7½ ft (2.3 m) female

Dive sequence
1. As the dolphin surfaces, only the top of the head is visible to start with.

2. Next, the bicolored dorsal fin breaks the surface of the water. Note how large it appears in relation to the body size.

3. The body arches, making the dorsal fin more prominent, as the dolphin dives.

Surface behavior
Frequently, they somersault or do backward flips with their belly fully exposed.

These dolphins are fast and agile swimmers, often exhibiting porpoising surface behaviors in which they arch their body entirely out of the water but parallel to the water surface.

These dolphins are energetic and exhibit a variety of surface active behaviors.

DUSKY DOLPHIN

Family Delphinidae

Species *Lagenorhynchus obscurus*

Other common names Dusky, duskies

Taxonomy *Sagmatias* is a proposed genus for *Lagenorhynchus spp.*—the closest relative is the Pacific white-sided dolphin. Four subspecies are recognized: *L. o. fitzroyi* (Fitzroy's dolphin), *L. o. obscurus* (African dusky dolphin), *L. o. posidonia* (Peruvian/Chilean dusky dolphin) and *L. o.* unnamed subsp. (New Zealand dusky dolphin).

Similar species Northern hemisphere Pacific white-sided dolphins

Birth weight 22 lb (10 kg)

Adult weight 150–190 lb (69–85 kg)

Diet Cephalopods (such as squid), lantern fish, or myctophids, schooling fish (such as sardines), and anchovy

Group size From less than 20 to 1,000

Major threats Gill nets and trawl fisheries, climate change, aquaculture, and tourism

IUCN status Data Deficient

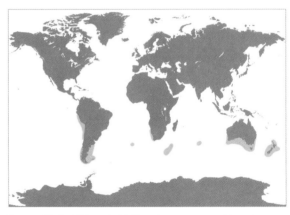

Range and habitat This species inhabits cool temperate waters of the southern hemisphere. It is generally found close to shore and along the continental shelf and slope in waters up to around 6,600 ft (2,000 m) deep. They have a discontinuous distribution across the coastal shallow waters off New Zealand, Australia, South America, Southern Africa, and neighboring oceanic islands.

Identification checklist

- Bicolored, falcate dorsal fin with a darker leading edge and grayish-white trailing edge
- Distinct white blaze or stripes from the tail stock to just below the dorsal fin
- White throat and belly
- Dark beak and jaw

Anatomy

Dusky dolphins are smaller in size relative to most other dolphin species. Those from South Africa and New Zealand are ¼–⅓ ft (8–10 cm) shorter than duskies from Peru. Males tend to have larger dorsal fins than females but they are difficult to distinguish at sea. Duskies have a narrower and longer rostrum than their sister species, the Pacific white-sided dolphin.

Behavior

This is one of the most acrobatic species of dolphins with an extensive repertoire of leaps that may influence group cohesion and behavior. Duskies are often found in large pods near Kaikoura, New Zealand, where they are sometimes joined by common dolphins and southern right whale dolphins.

Food and foraging

Group size and structure vary by activity, prey consumed, and geographic location. In some areas small groups of dolphins feed cooperatively on schooling fish that form prey balls during the day.

Elsewhere, dusky dolphins can form larger pods of more than 500 animals, which split into smaller subgroups to feed nocturnally on a rising and rich layer of squid and mesopelagic fish in offshore waters more than 16,000 ft (5,000 m) deep.

Life history

Like most dolphins, reproduction and calving is seasonal. Females and males reach sexual maturity at around seven to eight years in New Zealand. In Peru, males and females reach sexual maturity at between three and five years. Mating generally occurs in late summer to early fall. Gestation is roughly about 11 months. Average lifespan is believed to range between 36 and 40 years.

Conservation and management

There is insufficient data available about their abundance and distribution globally. Primary threats continue to be hunting for consumption, bycatch in gill nets, and midwater trawl-net fisheries. Marine shellfish farms in New Zealand also pose a growing problem that may affect foraging behavior and limit habitat.

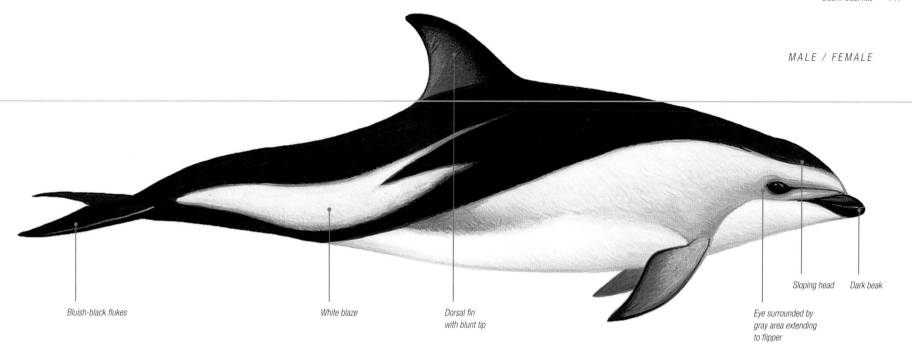

MALE / FEMALE

Bluish-black flukes

White blaze

Dorsal fin
with blunt tip

Sloping head

Dark beak

Eye surrounded by
gray area extending
to flipper

Blaze
An evenly sloping head from blowhole to beak
and the whitish streaks or blaze running from
the tail to the base of the dorsal are the
quintessential features of the dusky dolphin.

Size
Newborn 2½–3 ft (80–92 cm)
Adult 6–7 ft (1.8–2.1 m)

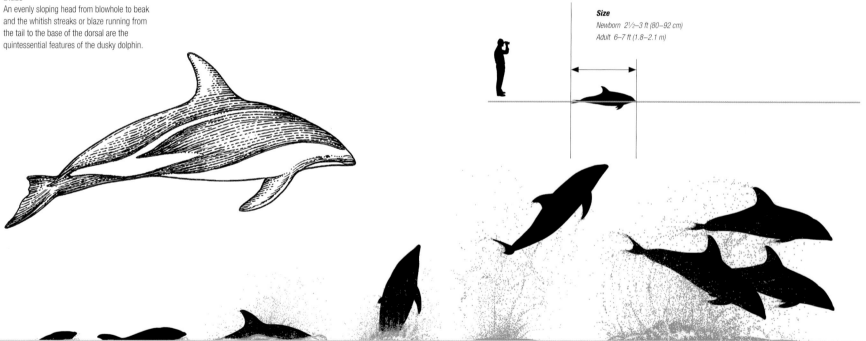

Dive sequence
*1. Their sloping head and
white blaze are clearly
visible when they surface
or travel at swift speeds.*

*2. They are fast swimmers,
slowing down surface
behaviors only when resting.*

Leaping
*Dusky dolphins are the gymnasts of the
sea, exhibiting a mindboggling array of
leaps, many of which are synchronized to
perfection and produce noisy splashes.*

*Their characteristic leaps vary from
acrobatic ones to highly coordinated
aerial maneuvers depending on
behavioral activity such as feeding,
mating, traveling, or socializing.*

NORTHERN RIGHT WHALE DOLPHIN

Family Delphinidae

Species *Lissodelphis borealis*

Other common names None

Taxonomy No subspecies, population structure uncertain, closest relative is the southern right whale dolphin

Similar species At a distance, porpoising sea lions or fur seals with similarly low-surfacing profile and no dorsal fin

Birth weight Not available

Adult weight Up to 253 lb (115 kg)

Diet Squid, deep-water schooling fish

Group size From a few individuals to several hundred, but usually about 10–50 in a group

Major threats Driftnet fishing in deep water seaward of shelf edge

IUCN status Least Concern

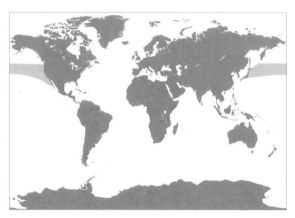

Range and habitat This species is found in the temperate North Pacific, from about 30°N to 51°N, and occasionally to 55°N. It primarily inhabits deep oceanic waters along or seaward of the continental shelf edge.

Identification checklist

- All black except for a white band on the ventral side from throat to flukes and a small white patch on the chin
- Unusually slender, elongated body with a small head and very slim tail stock
- Short beak, demarcated from the melon by a shallow crease
- Dorsal fin absent
- Small, narrow flippers and flukes

Anatomy

The lack of a dorsal fin, or a dorsal hump or ridge, makes right whale dolphins unique among the toothed cetaceans. Their streamlined shape, with tapered extremities and small appendages, has been described as eel-like. The mouthline is straight and there are 37–56 pairs of small, slender, pointed teeth in the upper and lower jaws—usually more in the upper than the lower.

Behavior

These dolphins are gregarious, occurring in schools of more than 1,000 animals. They are often seen with Pacific white-sided dolphins, whose presence may embolden them to bow-ride (see page 56). Otherwise northern right whale dolphins flee from powered vessels, attaining speeds of more than 19 mph (30 kmph). Their surfacing mode can be slow or fast. In slow mode they make little disturbance and expose only the head and blowhole. In fast mode they swim rapidly beneath the surface and catch their breath while making graceful low-angle leaps. Fast-swimming animals sometimes engage in belly-flops and side- or fluke-slaps.

Food and foraging

Although no regular pattern of north–south migration has been documented, these dolphins appear to move southward and inshore during the winter and northward and offshore during the summer. The diet is dominated by squid and deep-water fish, particularly lanternfish (family Myctophidae). Foraging dives can be to 656 ft (200 m) or deeper.

Life history

Males and females attain sexual maturity at about 10 years. Gestation lasts about one year, and females calve, at most, every other year. They can live for at least 42 years.

Conservation and management

The most serious threat comes from bycatch in driftnets. Concern over the huge toll of northern right whale dolphins—as many as 15,000–24,000 died each year during the 1980s—in part prompted the United Nations to ban large-scale driftnetting on the high seas in 1994, by which time numbers in the central North Pacific had been reduced by up to 30 percent. It was recently estimated that there are around 68,000 in the entire North Pacific.

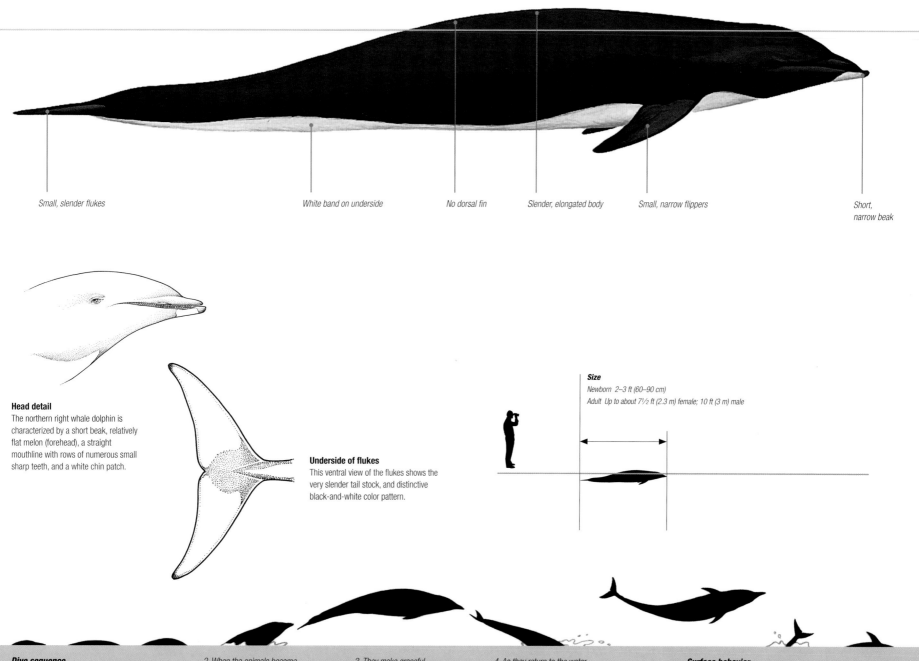

MALE / FEMALE

Small, slender flukes

White band on underside

No dorsal fin

Slender, elongated body

Small, narrow flippers

Short,
narrow beak

Head detail
The northern right whale dolphin is
characterized by a short beak, relatively
flat melon (forehead), a straight
mouthline with rows of numerous small
sharp teeth, and a white chin patch.

Underside of flukes
This ventral view of the flukes shows the
very slender tail stock, and distinctive
black-and-white color pattern.

Size
Newborn 2–3 ft (60–90 cm)
Adult Up to about 7½ ft (2.3 m) female; 10 ft (3 m) male

Dive sequence
1. In slow surfacing mode, the
unremarkable black backs of these
finless dolphins are easy to miss
even if sea conditions are fairly calm.

2. When the animals become
active, they swim rapidly just
below the surface.

3. They make graceful
low-angle leaps as they
flee from boats.

4. As they return to the water,
the overall impression is of an
all-dark, slender, elongated,
almost eel-like form.

Surface behavior
Fast-swimming animals sometimes make
repeated belly-flops, side-slaps or fluke-slaps.

SOUTHERN RIGHT WHALE DOLPHIN

Family Delphinidae

Species *Lissodelphis peronii*

Other common names None

Taxonomy No subspecies, population structure uncertain, closest relative is the northern right whale dolphin

Similar species At a distance, "porpoising" sea lions or fur seals with similarly low-surfacing profile and no dorsal fin

Birth weight Not available

Adult weight Up to 253 lb (115 kg)

Diet Squid, deep-water schooling fish

Group size From a few individuals to several hundred or a thousand

Major threats Driftnet fishing (especially for squid or swordfish) in deep water seaward of shelf edge

IUCN status Data Deficient

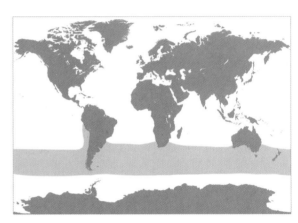

Range and habitat This species is circumpolar in cool-temperate and sub-Antarctic waters of the Southern Ocean between 40°S and 55°S. Its range extends northward with cold boundary currents such as the Benguela Current off West Africa and the Humboldt Current off western South America. The southward limit is around 63–64°S.

Identification checklist

- All-black upper side except for the face and throat, which along with the flippers, belly, and flanks are white
- Unusually slender, elongated, and slightly flattened body that tapers to small head and very slim tail stock
- Short beak, demarcated from melon by shallow crease
- Dorsal fin completely absent
- Small, narrow flippers and flukes

Anatomy

The lack of a dorsal fin, or even a dorsal hump or ridge, makes right whale dolphins unique among toothed cetaceans. Their streamlined body shape, with tapered extremities and proportionally small appendages, has been described as almost eel-like. The mouthline is straight and there are 39–50 pairs of small, slender, sharply pointed teeth in the upper and lower jaws.

Behavior

These dolphins are gregarious, sometimes occurring in schools of up to 1,000 animals. Four types of school configurations have been described: tightly packed with no identifiable subgroups, composed of distinct subgroups, "V"-formation, and arranged in a "chorus-line." These dolphins often associate with long-finned pilot whales and dusky dolphins. Observations off Argentina suggest occasional hybridization with dusky dolphins. Like their northern relatives, southern right whale dolphins are fast swimmers and engage in graceful, "bouncing," low-angled leaps when on the move. Fast-swimming animals sometimes engage in repeated belly-flops and side- or fluke-slaps. When in a less energetic mode, they can make little disturbance and expose only the head and blowhole.

Food and foraging

The diet of southern right whale dolphins has been less well studied than that of their northern counterparts but they are known to prey on squid and small schooling fish, probably diving to depths of at least 660 ft (200 m). Observations indicate that southern right whale dolphins associate with areas of strong upwelling, making only local seasonal north–south and perhaps offshore–inshore movements in response to changing foraging conditions.

Life history

Most characteristics are probably similar to those of northern right whale dolphins, with both males and females maturing at about 10 years of age, gestation lasting about a year, and females giving birth at intervals of two or more years.

Conservation and management

There are no good estimates of abundance of this species but it is considered common in some areas, for example southwestern South America. They are susceptible to entanglement in offshore driftnets. Substantial numbers were killed in a rapidly developing driftnet fishery for swordfish off northern Chile in the early 1990s.

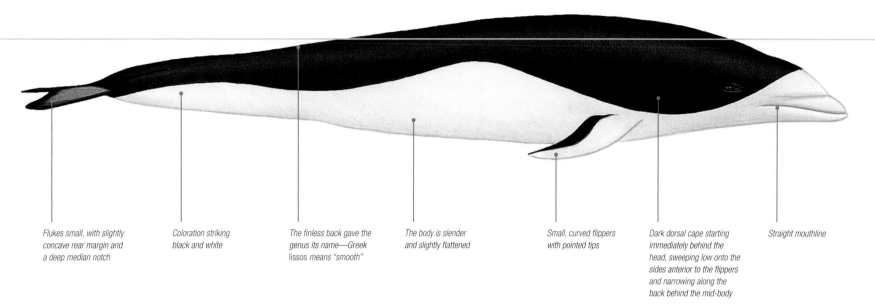

Flukes small, with slightly concave rear margin and a deep median notch

Coloration striking black and white

The finless back gave the genus its name—Greek *lissos means "smooth"*

The body is slender and slightly flattened

Small, curved flippers with pointed tips

Dark dorsal cape starting immediately behind the head, sweeping low onto the sides anterior to the flippers and narrowing along the back behind the mid-body

Straight mouthline

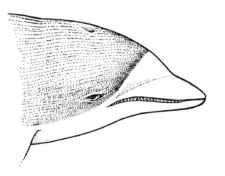

Head and mouthline detail
The southern right whale dolphin has a short beak, relatively flat melon (forehead), straight mouthline, and rows of numerous small sharp teeth. Herman Melville, author of *Moby Dick*, called this the "mealy-mouthed porpoise" because its white face reminded him of an animal that had "just escaped from a felonious visit to a meal bag."

Size
Newborn 2–3 ft (60–90 cm)
Adult 7½ ft (2.3 m) female; 10 ft (3 m) male

Dive sequence
1. In slow surfacing mode, these slender, streamlined finless dolphins may be easy to miss even if sea conditions are calm.

2. When active, such as when approached by a boat, they swim rapidly just below the surface and make graceful low-angle leaps as they flee. The bold, sharply defined black-and-white color pattern is especially striking.

Surface behavior
Fast-swimming animals sometimes make repeated belly-flops, side-slaps, or fluke-slaps.

IRRAWADDY DOLPHIN

Family Delphinidae

Species *Orcaella brevirostris*

Other common names None

Taxonomy Their closest relative is the Australian snubfin dolphin

Similar species Similar to the snubfin dolphin but their range does not overlap—can be confused with finless porpoises but they lack a dorsal fin

Birth weight 22–26 lb (10–12 kg)

Adult weight Maximum 287 lb (130 kg)

Diet Small fish, crustaceans, and cephalopods

Group size Generally 2–6 individuals but feeding aggregations of up to 25 dolphins have been observed in some areas

Major threats Entanglement in fishing gear, particularly gill nets, and threatened in rivers and estuaries by construction of dams

IUCN status Vulnerable but with five populations in the Malampaya Sound, Songkhla Lake or Lagoon, and the Mekong, Irrawaddy, and Mahakam rivers classified as Critically Endangered

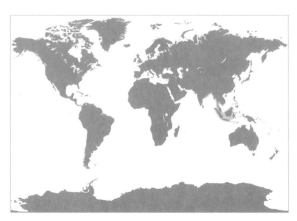

Range and habitat This species occurs in nearshore waters of Southeast and South Asia. They also occur in three large rivers: the Irrawaddy, Mekong, and Mahakam, and two brackish lagoons: Chilika and Songkhla.

Identification checklist

- Moderately robust body with a bulbous head, slight neck crease, and no beak
- Small triangular dorsal fin positioned slightly behind the mid-back with a falcate trailing edge and rounded tip
- Large, wide flippers with a rounded leading edge
- Broad flukes with a distinct median notch
- Gray overall but becoming slightly lighter on the underside

Anatomy
The neck of Irrawaddy dolphins is flexible with only the first two vertebrae fused. Irrawaddy dolphins share this feature with the true river dolphins, narwhals, and belugas. Additional vertebrae are fused in other cetaceans to stiffen their neck for fast swimming. The length of rostrum differentiates Irrawaddy from snubfin dolphins.

Behavior
In rivers, Irrawaddy dolphins prefer habitat associated with channel confluences. In the Sundarbans mangrove forest the species shifts its distribution in response to changes in freshwater input. The affinity of this species for low-salinity waters is probably due to ecological preferences related to prey. Irrawaddy dolphins are social animals exhibiting frequent physical interactions.

Food and foraging
Irrawaddy dolphins are generalist feeders. They spit water in a directed stream in order to herd fish. In places, the dolphins cooperate with cast-net fishermen. They benefit by easily preying on fish whose movements are confused by the falling net, and the fishermen's catch is two to three times greater when the dolphins herd fish toward them.

Life history
These dolphins are born throughout the year but there is a peak during the pre-monsoon season. Gestation is about 14 months and weaning about 2 years. Irrawaddy dolphins live for about 30 years.

Conservation and management
Protected areas for Irrawaddy dolphins have been established especially in large rivers and mangrove channels but enforcing fishery regulations, restricting vessel traffic, and protecting critical habitat has proved challenging. Irrawaddy dolphins have been promoted as flagships for aquatic conservation and are an important species for monitoring climate change impacts.

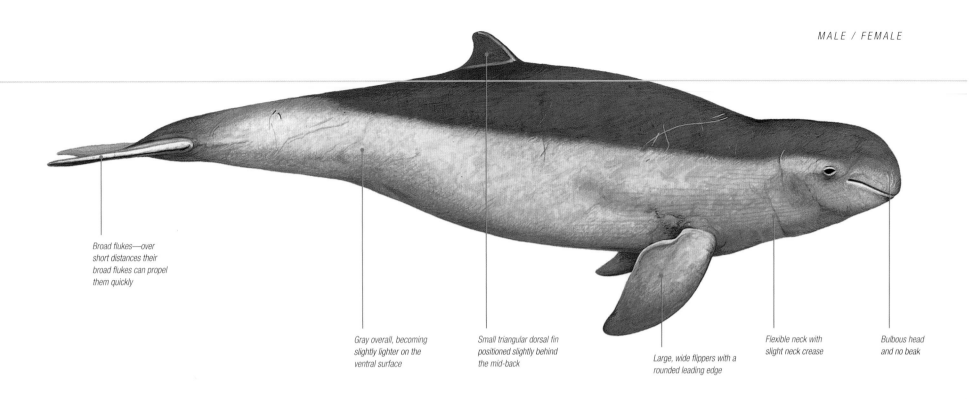

MALE / FEMALE

Broad flukes—over short distances their broad flukes can propel them quickly

Gray overall, becoming slightly lighter on the ventral surface

Small triangular dorsal fin positioned slightly behind the mid-back

Large, wide flippers with a rounded leading edge

Flexible neck with slight neck crease

Bulbous head and no beak

Adapted for prey

The body shape of the Irrawaddy dolphin is well-adapted to foraging in large rivers and estuary habitats. They are less streamlined than marine dolphins that must swim long distances in search of schooling prey. However, with a flexible neck and large paddle-like flippers they can maneuver easily to catch diverse prey in complex habitats.

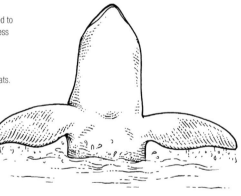

Size
Newborn About 3½ ft (1 m)
Adult Up to 7¼ ft (2.2 m) females; 9¼ ft (2.8 m) males

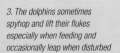

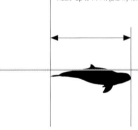

Dive sequence
1. Irrawaddy dolphins generally surface low in the water. They first show the top of their head.

2. This is followed by a short roll and the appearance of their dorsal fin before the animal submerges below the surface.

3. The dolphins sometimes spyhop and lift their flukes especially when feeding and occasionally leap when disturbed

Surface behavior
Activity increases during feeding and socializing when the dolphins surface more vigorously and show more of their body. They also often lift their flukes and spyhop holding their head high out of the water. Only very occasionally do Irrawaddy dolphins leap out of the water, generally when they are disturbed.

AUSTRALIAN SNUBFIN DOLPHIN

Family Delphinidae

Species *Orcaella heinsohni*

Other common names Australian snubnose dolphin

Taxonomy Dolphins of the genus *Orcaella* were recently split into two species: the Australian snubfin dolphin, *O. heinsohni*, and the Irrawaddy dolphin, *Orcaella brevirostris*

Similar species Can be confused with dugongs in most parts of their range

Birth weight 22–26 lb (10–12 kg)

Adult weight 250–420 lb (114–190 kg)

Diet Fish, cuttlefish, and squid

Group size 1–10, up to 15 while socializing

Major threats Incidental capture in gill nets, habitat degradation and loss, coastal-zone development; pollution, vessel traffic, and global warming

IUCN status Near Threatened

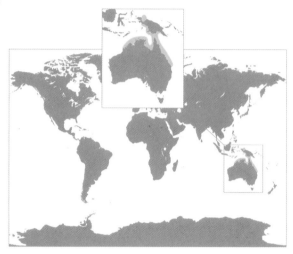

Range and habitat This species inhabits primarily shallow coastal and estuarine waters off northern Australia. It may also occur in southwest Papua New Guinea, West Papua, and Sulawesi.

Identification checklist

- Light to brownish-gray body
- Blunt rostrum with a round head
- Small triangular and slightly falcate dorsal fin set in the posterior portion of the body
- Flippers are broad and paddle-like

Anatomy

Snubfin dolphins are medium-sized and are characterized by a rounded head and blunt rostrum. They have a small dorsal fin situated in posterior portion of the body. A subtle three-tone body coloration pattern consists of a dark brown dorsal cape, light to brownish-gray side, and white abdomen. The flippers are broad, highly mobile, and paddle-like in shape.

Behavior

Snubfin dolphins are generally shy and elusive and do not ride the bow of boats. Surfacing can be inconspicuous and unpredictable, with a low profile roll showing only a small portion of their head, back, and dorsal fin. However, while traveling fast, socializing, and foraging individuals will show their dorsal fin and their flukes while diving. They are most active while socializing and individuals can be seen with much splashing doing low side and front leaps, tail-slapping, and spyhopping.

Food and foraging

Analysis of stomach contents from stranded dead animals suggests that snubfin dolphins are opportunistic, generalist feeders, preying on a wide variety of fish and cephalopods associated with shallow coastal-estuarine environments. Similar to their closest relative, the Irrawaddy dolphin, snubfins have occasionally been observed spitting water, which appears to be associated with feeding.

Life history

Australian snubfin dolphins are generally found in small groups of 2–6 animals, however groups of up to 15 animals have been observed. The social system of snubfin dolphins is relatively stable with individuals forming strong and long-lasting associations. Snubfin dolphins reach adult size 7 ft (2.1 m) in 4–6 years and may live for at least 30 years.

Conservation and management

There is no global population estimate—small populations of 50–200 individuals are found in local areas of Australia. Habitat loss and degradation due to the rapid increase in coastal-zone urbanization, ports and shipping infrastructure, and an increase in vessel activities throughout their range is considered one of the most important threats for Australian snubfin dolphins.

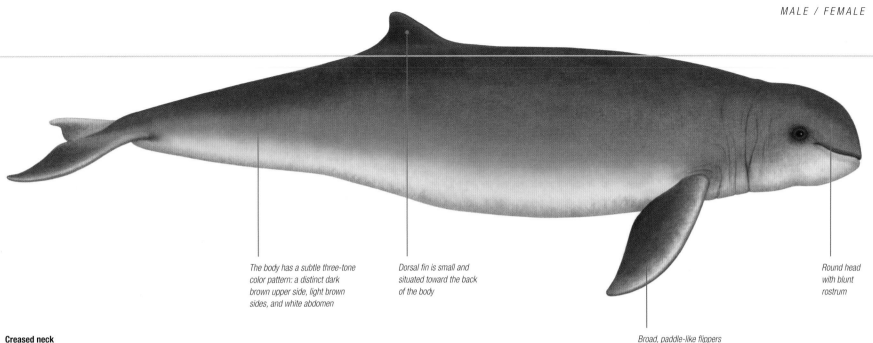

MALE / FEMALE

The body has a subtle three-tone color pattern: a distinct dark brown upper side, light brown sides, and white abdomen

Dorsal fin is small and situated toward the back of the body

Round head with blunt rostrum

Broad, paddle-like flippers

Creased neck
Distinctive neck crease about halfway between the eye and front insertion of the flipper.

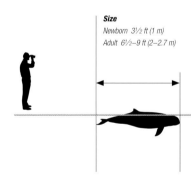

Size
Newborn 3½ ft (1 m)
Adult 6½–9 ft (2–2.7 m)

Dive sequence
1. Initially, the dorsal portion of the head is visible on the surface and sometimes the entire head breaks the surface

2. The body then arches slightly, slowly showing the back and small dorsal fin. Often animals do not arch their body much while at the surface, only showing the front part of the back and tip of the dorsal fin.

3. As the dorsal fin drops below the surface the tail is arched, showing only the dorsal region of the flukes. During steep dives, the flukes come clear of the surface.

Surface behavior
Snubfin dolphins rarely lift their entire bodies out of the water; breaches are usually characterized by shallow, arching leaps with the dolphin coming partially out of the water up to their flippers and falling back into the water either ventrally or laterally. Animals are also seen occasionally spyhopping.

KILLER WHALE

Family Delphinidae

Species *Orcinus orca*

Other common name Orca

Taxonomy At least six distinct ecotypes within the species that may in fact represent different species or subspecies

Similar species None—unique and distinctive

Birth weight 440 lb (200 kg)

Adult weight 14,550 lb (6,600 kg)

Diet Mainly fish, marine mammals, or both, depending on ecotype

Group size 1–100 or more, normally 5–20

Major threats Overfishing and shooting (among fish-eaters); pollutants (among mammal-eaters)

IUCN status Data Deficient; at least one population Critically Endangered

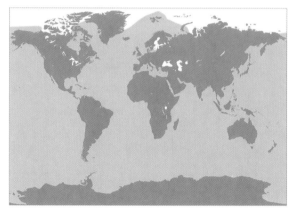

Range and habitat The most widespread cetacean, occurring in coastal and offshore waters throughout the world, and most common in higher latitudes and near-shore areas.

Identification checklist

- Large size—adults average 20–29½ ft (6–9 m) in length
- Bold black-and-white coloration
- Very large dorsal fin—3½–6½ ft (1–2 m); the male's is larger than the female's
- White eye patch
- Little or no beak evident

Anatomy

The killer whale or orca is the largest of the dolphins. There is a marked difference in the sexes with adult males being around 3½ ft (1 m) longer and weighing twice as much as females. The male dorsal fin is unmistakable—erect and triangular, and twice as tall as the female's—and male pectoral flippers and flukes are also proportionately much larger. Killer whales have 10–12 pairs of teeth in the upper and lower jaws; they are large at 4–5 in (10–12 cm) long, and conical.

Behavior

Individuals in groups travel fairly close together unless they are hunting, then they can be scattered over several miles. Their entire life is spent in family groups comprising a mother, her offspring, and her daughter's offspring. All breeding occurs outside the group. Group cohesion is maintained through underwater vocalizations. Fish-eaters are constantly vocalizing and use echolocation to find prey, as fish can't hear well. In contrast, mammal-eaters are normally quiet when they are foraging—marine mammals hear extremely well underwater—but they become vocal during and after an attack.

Food and foraging

Some populations are generalists, while others specialize in hunting either fish or mammals. Larger prey is hunted cooperatively and shared among the group. Prey includes seabirds, sea turtles, seals, sea lions, sharks, large whales, dolphins, and walruses. Killer whales are not often observed attacking, but mammal-eaters in particular can be vigorous when attacking large, fast prey.

Life history

Females first breed at age 12–14 years; males at 15 years or more. Males can live for 50–60 years and females for 80–90 years. The gestation period is 15–18 months. Females have one calf every five years, producing about five during her lifetime. Weaning probably occurs between one to two years.

Conservation and management

Mammal-eating killer whales are mostly doing well because of increased prey due to marine mammal protection measures. Fish-eaters are declining in some cases due to competition with humans. Because they live long and feed high on the food chain, accumulating pollutants is a potential problem in some populations.

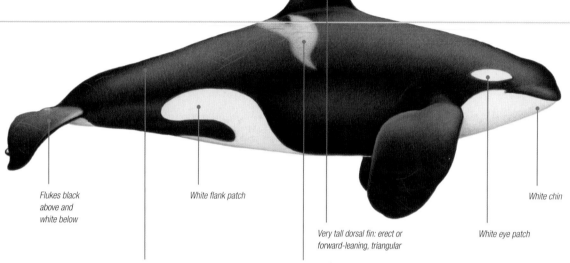

MALE

Flukes black
above and
white below

White flank patch

Very tall dorsal fin: erect or
forward-leaning, triangular

White chin

White eye patch

Shiny black in most populations;
two-tone gray in others

Gray saddle patch

Dorsal fin

The dorsal fin is strongly sexually dimorphic in adults: that of
the female (and younger animals) is shorter and falcate, while
the male's is nearly twice as tall and erect, sometimes even
forward-canted.

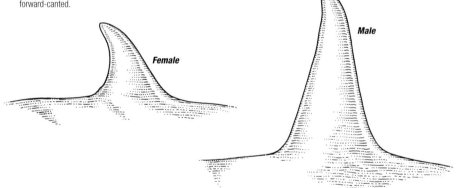

Female

Male

Size
Newborn 6½–8¼ ft (2–2.5 m)
Adult 18½–25 ft (5.6–7.7 m) female; 20–29½ ft (6–9 m) male

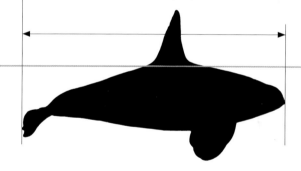

Diving sequence
1. Slow travel is the most
commonly seen behavior.

2. When feeding, the
body arches over hard
and shows tail stock. The
blow is low and bushy.

3. The killer whale
normally does not
fluke when diving but
will wave flukes in the
air when socializing.

Surface behavior
It breaches occasionally,
coming most of the way
out of the water—
especially juveniles.

It occasionally spyhops,
usually to look at things
above the surface.

In pursuit of fast prey, such as
dolphins or porpoise, individuals
may porpoise—launching out
of the water at high speed and
re-entering headfirst.

Mammal-eating killer
whales often head-butt
fast-moving prey out of
the water.

Seals and seal lions
are often swatted up in
the air by the whale's
tail. Fish-eating killer
whales sometimes bring
fish (here a salmon) to
the surface.

MELON-HEADED WHALE

Family Delphinidae

Species *Peponocephala electra*

Other common names Blackfish (although this is also used for several other species), electra dolphin, many-toothed blackfish, Hawaiian blackfish, Hawaiian porpoise, and the Indian broad-beaked dolphin

Taxonomy Closest relatives are other oceanic delphinids

Similar species Most often confused with pygmy killer whales

Birth weight 33 lb (15 kg)

Adult weight 353–496 lb (160–225 kg)

Diet Squid, small fish, crustaceans

Group size Ranges from single animals to thousands, but typically found in groups of 100–500

Major threats Fisheries (both directed hunts and incidental takes), anthropogenic ocean noise

IUCN status Least Concern (although most populations are not well known)

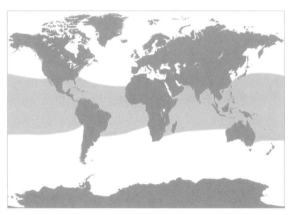

Range and habitat This species inhabits tropical and subtropical oceanic waters. It is primarily found in offshore and deep-water habitats with some populations found closer to shore near oceanic islands such as Hawaii and French Polynesia.

Identification checklist

- Dark gray body coloration with lighter gray ventral regions
- White pigmentation common around the mouth in adults
- Slightly bulbous head lacking a distinct beak
- Pointed pectoral flippers
- Central fin position

Anatomy

Melon-headed whales most closely resemble pygmy killer whales, and, to a lesser extent, false killer whales and pilot whales—all four species are collectively referred to as "blackfish." Compared to other oceanic delphinids, their bodies are darker and lack a distinct beak. White may be present around the lips and is most distinct on older individuals. Males tend to be slightly longer and more robust than females, with taller dorsal fins and rounder heads. A pronounced ventral keel posterior to the anus is present in adult males.

Behavior

Melon-headed whales are typically found in large groups of several hundred animals, but occasionally groups may number in the thousands. They are often seen associating with other cetacean species, most frequently with Fraser's dolphins or rough-toothed dolphins. The species is prone to mass-stranding events, with a minimum of 35 events known to have occurred worldwide.

Food and foraging

Stomach-content analysis suggests that mesopelagic fish—those that live at a depth of around 660–3,300 ft (200–1,000 m)—and squid are the most abundant prey items in the melon-headed whale diet. Crustaceans may also be consumed on occasion.

Life history

Much of what is known of the life history of melon-headed whales comes from the examination of stranded individuals. The age of sexually mature females from a mass stranding of 119 animals in Japan in 1982 was found to be between 11½ and 44½ years of age. Sexually mature males were between 16½ and 38½ years.

Conservation and management

Several conservation threats exist. Individuals are taken in small numbers in subsistence, drive, and harpoon fisheries in locations worldwide, including Japan, Taiwan, St. Vincent, and Indonesia. Dorsal fin disfigurements of melon-headed whale populations in Hawaii suggest that interactions with fisheries occur. This species has been recorded as bycatch in driftnet, gill-net, longline, and purse seine fisheries in tropical waters around the world. Melon-headed whales have shown susceptibility to anthropogenic noise. Strandings in Kauai in 2004 and Madagascar in 2008 have been linked to the use of mid- and high-frequency sonar.

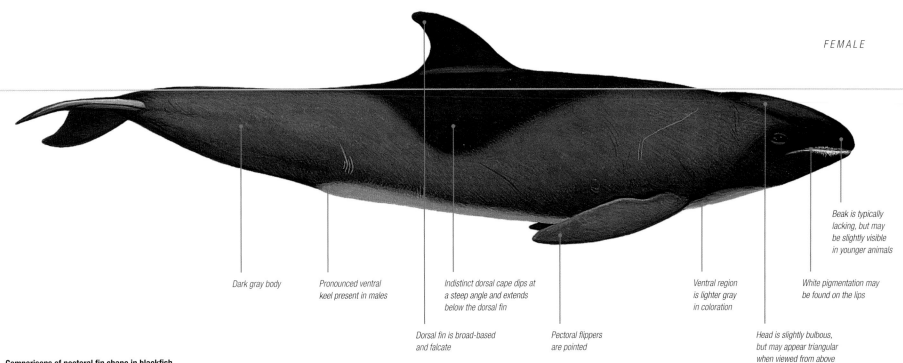

FEMALE

Dark gray body

Pronounced ventral keel present in males

Indistinct dorsal cape dips at a steep angle and extends below the dorsal fin

Dorsal fin is broad-based and falcate

Pectoral flippers are pointed

Ventral region is lighter gray in coloration

Head is slightly bulbous, but may appear triangular when viewed from above

White pigmentation may be found on the lips

Beak is typically lacking, but may be slightly visible in younger animals

Comparisons of pectoral fin shape in blackfish
The shape of the pectoral fin in closely related blackfish species varies. The flipper of the melon-headed whale is long and pointed, while that of the pygmy killer whale is short with a rounded tip, and the flipper of the false killer whale is "S"-shaped.

Size
Newborn 3.3 ft (1 m)
Adult 7–9¼ ft (2.1–2.8 m)

Melon-headed whale   *Pygmy killer whale*   *False killer whale*

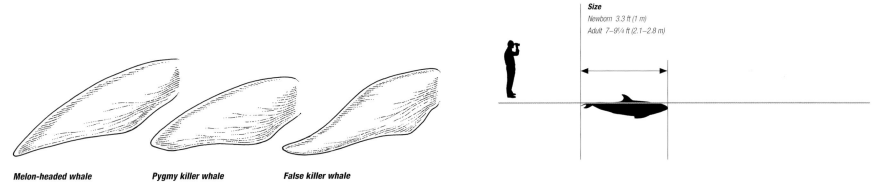

Dive sequence
1. The head and melon break the surface of the water in a torpedo-like fashion; blow is usually not visible.

2. The dorsal fin becomes visible as the upper two-thirds of the animal's body breaks the surface of the water.

3. The typical surface profile is low and duration is brief.

4. The caudal peduncle, or tail stock, is only visible during high-speed travel.

5. Fluking is rare.

Surface behavior
Melon-headed whales rarely engage in full breach behavior. During high-speed travel they will often porpoise with all to most of their body clearing the water. Small aggregations of animals within the larger group typically travel together.

FALSE KILLER WHALE

Family Delphinidae

Species *Pseudorca crassidens*

Other common names Blackfish (although this is also used for several other species)

Taxonomy Closest relatives are other oceanic dolphins

Similar species Most often confused with pilot whales, pygmy killer whales, and melon-headed whales

Birth weight 176 lb (80 kg)

Adult weight 2,200–4,400 lb (1,000–2,000 kg)

Diet Fish (including large species such as tuna) and squid

Group size Typical group sizes range from 10–40 individuals although groups of more than 300 animals have been recorded

Major threats Bycatch in longline fisheries, directed takes by fishermen, and pollution

IUCN status Globally listed as Least Concern; genetically distinct population around Hawaii listed as Endangered

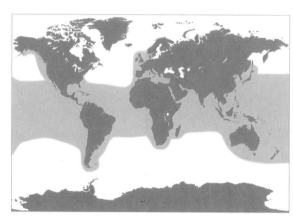

Range and habitat This species is found in tropical and temperate waters worldwide. It is primarily found in deep-water habitats although it is sometimes seen close to shore around oceanic islands.

<table>
<tr><td>

Identification checklist

</td><td>

- Slender, elongated body that is uniform black in color
- Narrow, tapered head with bulbous melon
- Falcate dorsal fin with a rounded tip
- "S"-shaped pectoral flippers
- Central fin position

</td></tr>
</table>

Anatomy

False killer whales are often confused with pilot, melon-headed, and pygmy killer whales. The base of the dorsal fin on false killer whales is narrower than on pilot whales, and the body and head shape is comparatively more slender with a less bulbous head.

Behavior

False killer whales are a highly social, energetic species, often seen traveling fast and low through the water. The species frequently approaches boats and engages in bow-riding behavior. Large groups of false killer whales are often comprised of several small clusters of individuals spread out over many miles. As with other highly social species, false killer whales are known to mass strand, with the largest event involving more than 800 individuals.

Food and foraging

The false killer whale feeds on a variety of fish and squid, including large game fish, such as mahi mahi and tuna. They hunt cooperatively, often spread out over many miles, and come together to share prey. In some parts of the world, they have been observed attacking other dolphin species.

Life history

Much of what is known on the life history of false killer whales comes from the examination of stranded and captive specimens. From dental studies, the oldest male was estimated to be 58 years of age and the oldest female 63 years. Females reach sexual maturity at 8–11 years and males at about 18 years. Gestation is around 14 months.

Conservation and management

False killer whales frequently take prey from fishing gear in tropical waters. This behavior threatens the species with entanglement in gear. Pollution and toxins also pose a threat because false killer whales consume large prey items—heavy metals and toxins may build up in their bodies making them susceptible to disease or other health problems.

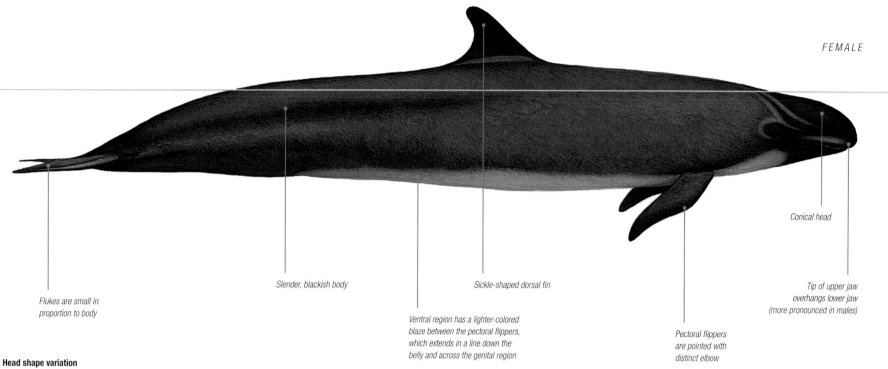

FEMALE

Flukes are small in
proportion to body

Slender, blackish body

Sickle-shaped dorsal fin

Ventral region has a lighter-colored
blaze between the pectoral flippers,
which extends in a line down the
belly and across the genital region

Pectoral flippers
are pointed with
distinct elbow

Conical head

Tip of upper jaw
overhangs lower jaw
(more pronounced in males)

Head shape variation
There are slight variations in head shape between the
sexes. The melon of the male is more bulbous than that
of the female and the tip of the jaw overhangs the lower
jaw more prominently in males.

Male

Female

Size
Newborn 5–6¼ ft (1.5–1.9 m)
Adult 11½–16½ ft (3.5–5m) female; 12¼–20 ft (3.7–6.1 m) male

Dive sequence
1. The head and melon
break the surface of the
water, and the whale's eyes
are visible.

2. There is a
conspicuous,
bushy blow.

3. Dorsal fin is fully visible
as the body quickly breaks
the surface.

4. Fluking is rare.

Surface behavior
False killer whales frequently breach and their
acrobatic behavior is especially evident while
pursuing prey. They have been observed knocking
their prey out of the water and breaching with
prey items in their mouth. Small aggregations of
animals frequently travel together scattered over
several miles. Bow-riding is common.

TUCUXI

Family Delphinidae

Species *Sotalia fluviatilis*

Other common names None

Taxonomy The marine counterpart, the Guiana dolphin, was recently recognized as a distinct species

Similar species The Guiana dolphin is very similar (although somewhat larger), but the two species do not occur together

Birth weight Estimated 18 lb (8 kg)

Adult weight 77–99 lb (35–45 kg)

Diet Fish up to 14 in (35 cm) long

Group size Commonly 2–5, up to 20

Major threats Entanglement in fishing nets and, in places, hunting for use as fish bait

IUCN status Data Deficient

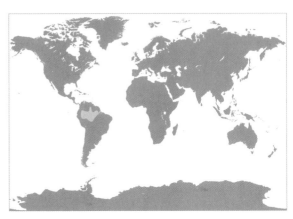

Range and habitat This species occurs in freshwater habitats only. It is found in the Amazon basin in Brazil, Colombia, Peru, and Ecuador. It may also occur in the Orinoco Basin of Colombia and Venezuela.

Identification checklist

- Small size
- Gray above, pink below
- Triangular dorsal fin on the center back
- Fast swimming, often breaches
- Blow not visible

Anatomy

The stiff, streamlined body of this species, with its distinct dorsal fin, is in strong contrast to the Amazon River dolphin with which it shares much of its range. Dark gray above and pinkish white below, this active small dolphin has 26–35 pairs of small teeth in the upper and lower jaws.

Behavior

Flamboyant swimmers, tucuxis' surfacing is normally announced by a noisy exhalation, and full breaches are frequent. As with other delphinid dolphins, clicks and whistles allow members of a group to communicate beyond the range of visibility, which is very limited in turbid Amazonian waterways. Tucuxis do not associate either with the Amazon River dolphins they commonly encounter, or with boats. Unlike many of their marine cousins, they do not bow-ride.

Food and foraging

Tucuxis forage in tight groups, often chasing fish in rapid dashes just below the water surface, with fish jumping out of their way. Thirty species of fish are known to be prey, some living in protected lakes and channels, while others occur in fast-flowing rivers.

Life history

Males have large testes for their body size, are no larger than females, and show no evidence of fighting, all of which suggests that male–male rivalry occurs in the form of sperm competition rather than aggression between animals. Gestation is about 11 months, and breeding is seasonal. The oldest known animal was 36 years of age.

Conservation and management

Threats to tucuxis are all human-related. Having adapted to live in freshwater over hundreds of thousands of years, the tucuxis now rely on a habitat that has been severely damaged—life for these dolphins has become very dangerous in less than one percent of that time. The most important cause of mortality in this species is entanglement in gill nets, which are used by every human riverine community where this dolphin occurs. Recently, direct hunting in pursuit of fish bait has become a significant problem.

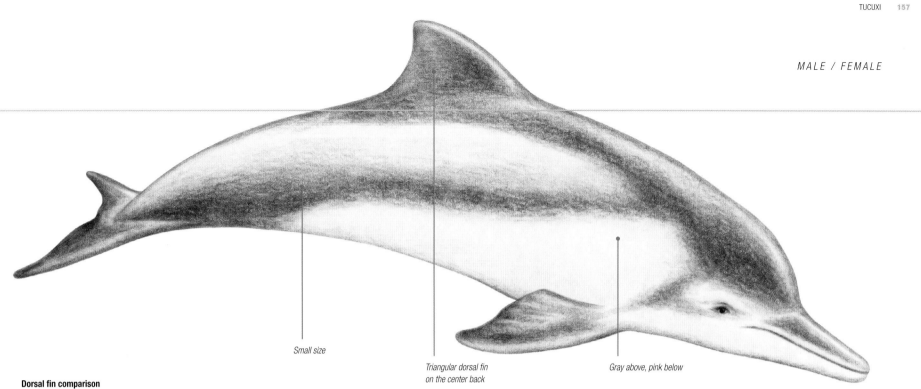

Small size

Triangular dorsal fin
on the center back

Gray above, pink below

Dorsal fin comparison
The dorsal fins of the two freshwater dolphins of South
America are strikingly different. The tucuxi's is upright
and falcate, whereas the Amazon river dolphin has little
more than a long dorsal ridge.

Tucuxi dorsal fin

Amazon River dolphin dorsal fin

Size
Newborn 2¼–2½ ft (70–80 cm)
Adult Around 5 ft (1.5 m)

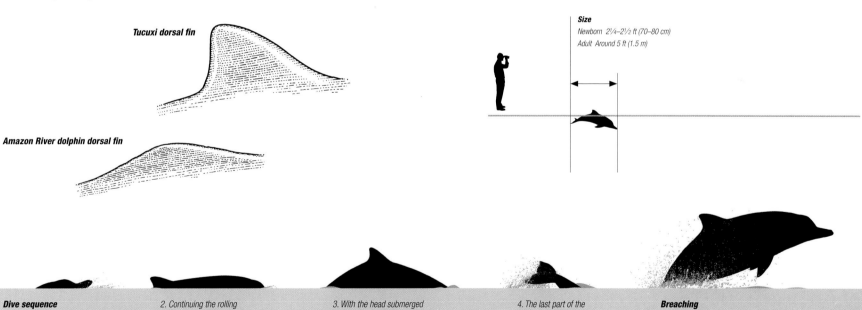

Dive sequence
*1. This species' surfacing sequence
lasts less than a second, and starts
with an exhalation.*

*2. Continuing the rolling
movement, the dolphin
inhales and shows its back.*

*3. With the head submerged
again, the back is arched as
the dolphin begins its next
dive at a steep angle.*

*4. The last part of the
tucuxi to disappear is
the tail stock.*

Breaching
*Tucuxis often leap clear of the water,
returning to the surface with a splash.
But blink and you miss it. This small
dolphin is very rapid in everything it does.*

GUIANA DOLPHIN

Family Delphinidae

Species *Sotalia guianensis*

Other common names Costero, marine tucuxi, estuarine dolphin

Taxonomy Recently separated from its freshwater counterpart, the tucuxi

Similar species Tucuxi, but the two species do not co-occur

Birth weight Around 26–33 lb (12–15 kg)

Adult weight Less than 220 lb (100 kg)

Diet Mostly fish, but also squid and shrimps

Group size 2–10, up to 60

Major threats Entanglement in nets and hunting for fish bait

IUCN status Data Deficient

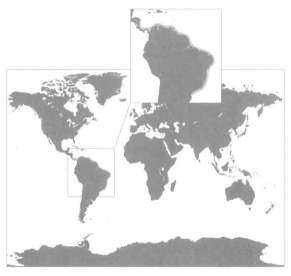

Range and habitat This species is found in nearshore habitats from Honduras to South Brazil. It is often found in bays and estuaries. Distribution is discontinuous.

Identification checklist

- Small, dark gray dolphin
- Central, triangular dorsal fin
- Aerially active, performing frequent leaps

Anatomy

This is a small, robust dolphin with a short beak, triangular dorsal fin, and slim flippers. The upper parts are dark gray, fading to light gray or pinkish gray underneath. There are 26–30 pairs of small teeth in the upper and lower jaws. Vibrissal pits on rostrum (see opposite) indicate the presence of electroreceptor sensitivity, likely used for prey detection in turbid water.

Behavior

A social species, Guiana dolphins are normally found in groups in estuaries and bays. Like their freshwater counterpart, the tucuxi, this dolphin is very active at the surface, often leaping clear of the water, but it does not bow-ride boats. Tooth rakes are often seen on the skin, but fighting is rare. Communities show high site fidelity, and home ranges are small compared to most marine dolphins. They may dive for up to two minutes.

Food and foraging

More than 60 species of demersal and pelagic schooling fish have been reported as prey. Small fish of 8 in (20 cm) or less are preferred. Foraging may be carried out individually or in groups.

Different dolphin communities may adopt their own foraging strategies based on local circumstances. One of the best-studied groups herds fish onto beaches and half-strand themselves for a few seconds while grabbing their prey.

Life history

Gestation is 11–12 months; the inter-birth interval is 2–3 years. There is no obvious seasonality in breeding. Females give birth for the first time at five to eight years of age, and males mature at around seven years. The oldest animals are about 30 years of age. Relatively large testes in males indicate the likelihood of sperm competition and promiscuous mating behavior.

Conservation and management

As with all coastal cetaceans, the Guiana dolphin suffers from negative interactions with humans. Entanglement in gill nets, seine nets, and shrimp traps is responsible for the death of many animals each year. There is very limited gene flow between concentrations of this dolphin, and large stretches of coast contain no animals at all, so recovery from depletion of a local population may take a long time.

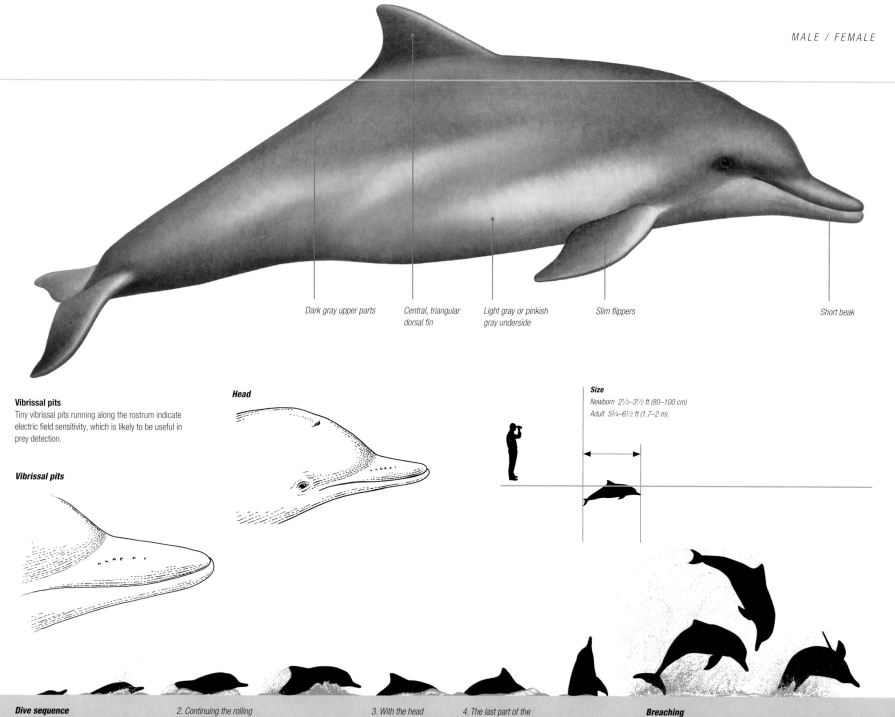

MALE / FEMALE

Dark gray upper parts

Central, triangular dorsal fin

Light gray or pinkish gray underside

Slim flippers

Short beak

Vibrissal pits
Tiny vibrissal pits running along the rostrum indicate electric field sensitivity, which is likely to be useful in prey detection.

Vibrissal pits

Head

Size
Newborn 2½–3½ ft (80–100 cm)
Adult 5¾–6½ ft (1.7–2 m).

Dive sequence
1. The surfacing sequence lasts less than a second, and starts with an exhalation.

2. Continuing the rolling movement, the dolphin inhales and shows its back.

3. With the head submerged again, the back is arched as the dolphin begins its next dive at a steep angle.

4. The last part of the dolphin to disappear is the tail stock.

Breaching
Guiana dolphins often leap clear of the water, returning to the surface with a splash. This small dolphin is very rapid in everything it does.

INDO-PACIFIC HUMPBACK DOLPHIN

Family Delphinidae

Species *Sousa chinensis*

Other common names Chinese white dolphin

Taxonomy Four *Sousa* species have been recently proposed: *S. teuszii* (Atlantic humpback dolphin) in the Atlantic Ocean, *S. plumbea* (Indian Ocean humpback dolphin) in the Indian Ocean, *S. chinensis* (Indo-Pacific humpback dolphin) in the eastern Indian and West Pacific Oceans, and *S. sahulensis* (Australian humpback dolphin) in the waters of the Sahul Shelf from northern Australia to southern New Guinea

Similar species Can be confused with bottlenose dolphins in most parts of their range

Birth weight 88–110 lb (40–50 kg)

Adult weight 510–550 lb (230–250 kg)

Diet Fish and cephalopods

Group size 1–10, although large aggregations of 20–30 animals have been seen off Hong Kong

Major threats Incidental capture in gill nets, habitat degradation and loss, coastal-zone development; pollution, vessel traffic, and global warming

IUCN status Near Threatened; the Eastern Taiwan Strait subpopulation is Critically Endangered

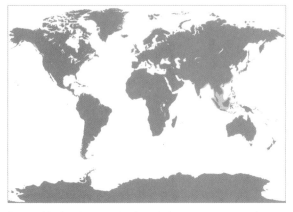

Range and habitat This species inhabits tropical to warm-temperate coastal waters from eastern India to central China and throughout Southeast Asia. It is found in shallow coastal waters, estuaries, and inshore reefs, and often enters rivers.

<table>
<tr><td>

Identification checklist

</td><td>

- Robust body with long well-defined rostrum
- Calves and juveniles are dark gray
- Adults may become mostly white
- The dorsal fin is larger than in other *Sousa* species, triangular in shape, and slightly falcate; it has a wide base without a basal hump

</td></tr>
</table>

Anatomy

Indo-Pacific humpback dolphins are characterized by a robust and medium-sized body. The coloration is generally gray but can vary greatly with age and geographical area. Most adults from Southern China are pure white—while others throughout the range retain some of their dark gray pigmentation and develop dark blotches or spotting. The hump characteristic of *S. teuszii* and *S. plumbea* is absent in Indo-Pacific humpback dolphins and their dorsal fin tends to be larger and more triangular in shape.

Behavior

Indo-Pacific humpback dolphins are generally shy and elusive and do not bow-ride. However, they display a variety of aerial acrobatics including vertical leaps, side leaps, and forward-backward somersaults. Socializing, including mating, in humpback dolphins is characterized by individuals in close proximity showing high levels of physical interaction including body contact—animals touching and biting each other and rubbing their bodies—and frequent aerial behavior such as leaps and somersaults. Fins and flukes often break the surface of the water.

Food and foraging

These dolphins are thought to be opportunist, generalist feeders, eating a variety of coastal-estuarine and inshore reef fish. Some cephalopods and crustaceans are also taken. This species feeds in association with fishing trawlers in Hong Kong.

Life history

Mating and calving appears to occur year round. The gestation period lasts 10–12 months, lactation may last more than 2 years, sexual maturity is reached at 9–10 years of age for females and 12–13 year for males, and a 3-year calving interval has been suggested. Individuals may live for at least 30 years.

Conservation and management

There is no global population estimate—available subpopulation estimates are in low tens to low hundreds, but at least 1,200 animals in the Pearl River estuary of Southern China. Because of their coastal distribution, Indo-Pacific humpback dolphins are vulnerable to a variety of threats including incidental captures in gill nets and shark nets set for bather protection, habitat loss and degradation, vessel strikes, pollution, and climate change.

MALE / FEMALE

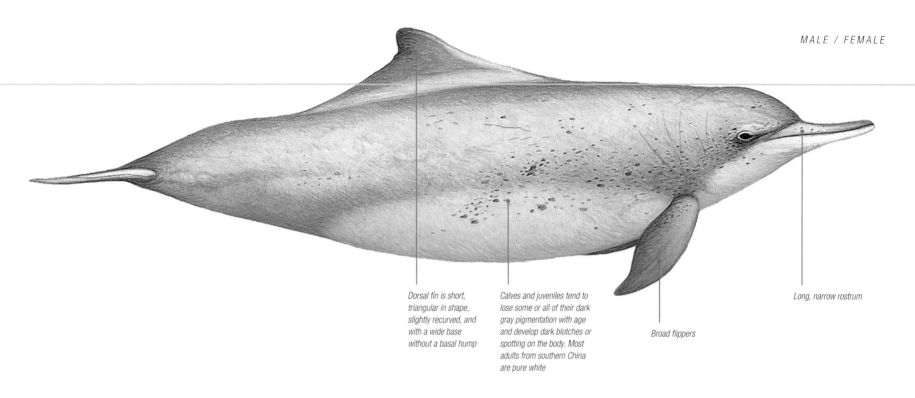

Dorsal fin is short,
triangular in shape,
slightly recurved, and
with a wide base
without a basal hump

Calves and juveniles tend to
lose some or all of their dark
gray pigmentation with age
and develop dark blotches or
spotting on the body. Most
adults from southern China
are pure white

Broad flippers

Long, narrow rostrum

Coloration

Calves and juveniles are dark gray. Adults tend to lose
some or all of their dark gray pigmentation and may
become mostly white, particularly in southern China.

Size
Newborn 3½ ft (1 m)
Adult 6½–8½ ft (2–2.6 m)

Dive sequence

1. Initially, the long, narrow rostrum
breaks water, followed by the melon.
Sometimes the whole head is lifted
completely clear of the water.

2. When the blowhole is exposed
above the surface, most of the body
remains submerged. The body then
arches, showing the dorsal fin.

3. Finally, the head drops below the
surface, the back is arched a little higher,
and the animal dives down, showing only
dorsal region of the flukes. While diving,
animals arch their back steeply and
show their fluke as they dive.

Breaching

Aerial behavior is generally
infrequent but humpback dolphins
can be seen doing vertical leaps,
side leaps, and somersaults.

INDIAN HUMPBACK DOLPHIN

Family Delphinidae

Species *Sousa plumbea*

Other common names Plumbeous dolphin

Taxonomy Four *Sousa* species are proposed: *S. teuszii* (Atlantic humpback dolphin), *S. plumbea* (Indian Ocean humpback dolphin), *S. chinensis* (Indo-Pacific humpback dolphin), and *S. sahulensis* (Australian humpback dolphin)

Similar species Most likely to be confused with bottlenose dolphins in most parts of their range

Birth weight 31 lb (14 kg)

Adult weight 550–573 lb (250–260 kg)

Diet Fish and cephalopods

Group size 1–20 individuals, although large groups of up to 100 have been observed in Arabian waters

Habitat Generally found in shallow coastal waters along sandy and rocky shorelines and in estuarine areas

Population There is no estimate of total population size—subpopulation estimates are low tens to low hundreds

Major threats Incidental capture in gill nets, habitat degradation and loss, coastal-zone development, pollution, vessel traffic, and global warming

IUCN status Has not been assessed

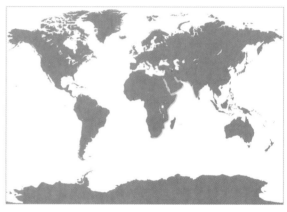

Range and habitat This species is found in coastal waters of the Indian Ocean from southwestern tip of South Africa eastward to at least Myanmar.

Identification checklist

- Robust body with long well-defined rostrum
- Coloration is generally gray with subtle white-pinkish underbody
- Prominent and well-developed dorsal hump
- Dorsal fin is smaller, slightly falcate, and less triangular in shape than in *S. chinensis*

Anatomy

Indian humpback dolphins are the largest of all humpback dolphins reaching lengths of up to 9¼ ft (2.8 m). A prominent and well-developed dorsal hump can be seen in both adults and young animals. The dorsal fin is smaller, slightly falcate, and less triangular than in *S. chinensis*. Calves are light gray with no spotting—while adults are dark gray with a lighter ventral surface. Scarring on the upper surface and head are often present and white in coloration.

Behavior

As with other humpback dolphins, Indian humpback dolphins are shy of boats and do not bow-ride. During daylight hours, Indian humpback dolphins are often observed feeding and traveling in shallow waters along sandy, rocky, and mangrove coastlines, and river mouths. Social behavior—appearing to be courtship and mating—is usually characterized by two individuals or more displaying prolonged physical contact and swimming vigorously side by side, rolling onto their sides, and exposing half of their bodies above the water. This would be followed by two individuals interlocking ventrally for 20–40 seconds, while swimming slowly and rolling beneath the surface.

Food and foraging

In the Arabian Gulf and in Bazaruto, Mozambique, Indian humpback dolphins have been observed herding fish onto exposed sand banks and beaching themselves in shallow waters to capture their prey. Both fish and cephalopods associated with estuarine waters have been found in stomach contents of stranded animals.

Life history

Female sexual maturity is attained at 10 years and male maturity at 12–13. The gestation period is 10–12 months, and the calving interval is around 3 years. Calves are fully weaned at two years of age and older. Individuals may live for at least 30 years.

Conservation and management

The conservation status of *S. plumbea* as a separate species from *S. chinensis* has not been assessed. Threats to Indian humpback dolphins throughout their range include: incidental capture in fishing nets; habitat loss and degradation through coastal and offshore development (such as land reclamation, dredging, port and harbor construction, oil and gas exploration); pollution; boat traffic; and climate change.

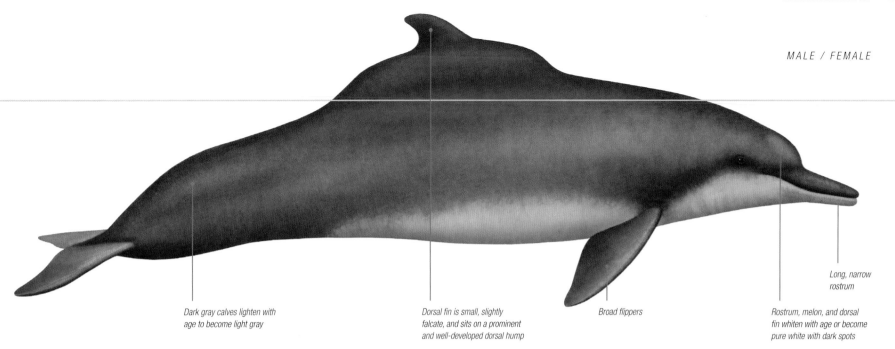

MALE / FEMALE

Dark gray calves lighten with age to become light gray

Dorsal fin is small, slightly falcate, and sits on a prominent and well-developed dorsal hump

Broad flippers

Long, narrow rostrum

Rostrum, melon, and dorsal fin whiten with age or become pure white with dark spots

Dorsal hump
The prominent dorsal hump is characteristic of specimens of *S. plumbea*.

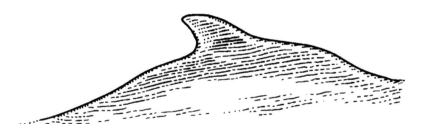

Size
Newborn 3½ ft (1 m)
Adult 6½–9¼ ft (2–2.8 m)

Dive sequence
1. Initially, the long, narrow rostrum breaks water, followed by the melon. Sometimes the whole head is lifted completely clear of the water.

2. When the blowhole is exposed above the surface, most of the body remains submerged. The body then arches, showing the dorsal hump and dorsal fin.

3. Finally the head drops below the surface, the back is arched a little higher, and the animal dives down, showing only dorsal region of the flukes. While diving animals arch their back steeply and show their fluke as they dive.

Breaching
Aerial behavior is generally infrequent but humpback dolphins can be seen doing vertical leaps, side leaps, and somersaults.

AUSTRALIAN HUMPBACK DOLPHIN

Family Delphinidae

Species *Sousa sahulensis*

Other common names Sahul dolphin

Taxonomy One of four species of *Sousa* that is found in Australian waters

Similar species Can be confused with bottlenose dolphins in most parts of their range

Birth weight 88–110 lb (40–50 kg)

Adult weight 507–550 lb (230–250 kg)

Diet Fish and cephalopods

Group size 1–5 individuals, larger schools of up to 30–35 individuals have been observed off Queensland while animals are feeding behind trawlers

Major threats Incidental capture in gill nets, habitat degradation and loss, coastal-zone development, pollution, vessel traffic, and global warming

IUCN status Not yet assessed

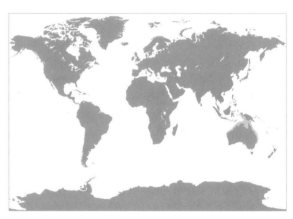

Range and habitat This species is found in tropical and subtropical waters of the Sahul Shelf from northern Australia to southern New Guinea. It inhabits shallow coastal waters, estuaries, and inshore reefs, and often enters rivers.

Identification checklist

- Robust body with long well-defined rostrum.
- Coloration is uniformly gray, with flanks shading to off-white. Adults develop variable amounts of white scarring and dark flecking throughout their body.
- Dorsal fin is low, triangular in shape, and has a wide base without a visible dorsal hump.

Anatomy

Known sizes for Australian humpback dolphins range from 3½–9 ft (1–2.7 m). The dorsal fin is short, triangular in shape, and lacks the dorsal "hump" typical of Atlantic and Indian humpback dolphins. The body is mainly dark gray in color, with flanks shading to light gray. A diagonal cape line extending from just above the eye and neck down to the urogenital area separates the dark back and lighter belly. White scarring and dark flecking on the head, back, dorsal fin, and tail stock are common in adult animals.

Behavior

Australian humpback dolphins are generally found in small groups of 2–5 animals—but groups of up to 30 animals have been observed while animals forage behind trawlers. Groups often change in size and composition and individual associations are mainly short-term. They are generally shy and elusive and tend to keep their distance from boats. Social interactions with snubfin and bottlenose dolphins have been observed. A case of hybridization between a male humpback dolphin and female snubfin dolphin was reported in northwestern Australia.

Food and foraging

These dolphins are thought to be opportunistic, generalist feeders, eating a wide variety of coastal-estuarine and inshore reef fish. Feeding may occur in a variety of habitats—mangroves, sandy-bottom estuaries, seagrass meadows, and inshore coral reefs—and involve animals dispersed over wide areas or tight groups targeting localized prey. Animals are occasionally seen chasing fish into shallows and beaching themselves to catch their prey.

Life history

Mating and calving occurs year round. The gestation period lasts 10–12 months, lactation may last more than 2 years and a 3-year calving interval is likely. Individuals may live for at least 30 years.

Conservation and management

There is no global population estimate—available subpopulation estimates are in low hundreds. Because of their coastal distribution, Australian humpback dolphins are vulnerable to a variety of threats including incidental captures in gill nets and shark nets set for bather protection, habitat loss and degradation, vessel strikes, pollution, and climate change.

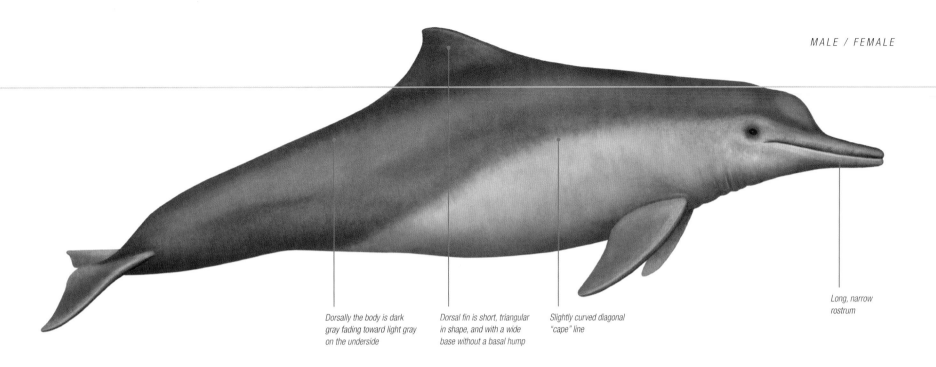

MALE / FEMALE

Long, narrow
rostrum

Dorsally the body is dark
gray fading toward light gray
on the underside

Dorsal fin is short, triangular
in shape, and with a wide
base without a basal hump

Slightly curved diagonal
"cape" line

Dorsal cape
Indistinct diagonal separation of the dark upper body
and lighter flanks and abdomen, which is not
characteristic of any other humpback dolphins.

Size
Newborn 3½ ft (1 m)
Adult 6½–9 ft (2–2.7 m)

Dive sequence
*1. Initially, the long, narrow rostrum
breaks water, followed by the melon.
Sometimes the whole head is lifted
completely clear of the water.*

*2. When the blowhole is exposed
above the surface, most of the body
remains submerged. The body then
arches, showing the dorsal fin.*

*3. Finally, the head drops below the surface,
the back is arched a little higher, and the
animal dives down, showing only the dorsal
region of the flukes. While diving animals
arch their back steeply and show their
fluke as they dive.*

Breaching
*Aerial behavior is generally infrequent
but humpback dolphins can be seen
doing vertical leaps, side leaps, and
somersaults.*

ATLANTIC HUMPBACK DOLPHIN

Family Delphinidae

Species *Sousa teuszii*

Other common names Cameroon dolphin, Teusz's dolphin

Taxonomy The validity of this species has been questioned in the past, but recent genetic and morphological analysis provided convincing evidence that this species is indeed valid and separate from other humpback dolphins in the Indian and Pacific oceans

Similar species Most likely to be confused with bottlenose dolphins, which also inhabit inshore areas

Birth weight At least 22 lb (10 kg)

Adult weight 550–630 lb (250–285 kg)

Diet Fish

Group size Ranges from 1–40 individuals, typically groups of 3–8 individuals

Major threats Incidental capture in gill nets, directed captures, habitat degradation, and reduction of prey through overfishing and climate change

IUCN status Vulnerable

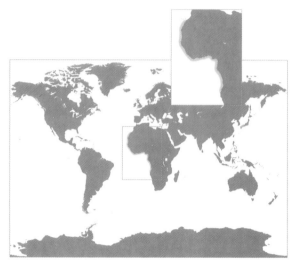

Range and habitat This species is endemic to the tropical to subtropical west coast of Africa in the eastern Atlantic Ocean, from the Western Sahara to Angola. It is mainly found in shallow coastal and estuarine waters, usually less than 66 ft (20 m) deep.

Identification checklist

- Robust body with a well-defined rostrum
- Typically slate gray on the back and sides, fading to light gray ventrally.
- Dorsal fin is small, slightly falcate, and triangular, and sits on a distinctive and well-developed dorsal hump

Anatomy

The physical appearance of Atlantic humpback dolphins is similar to that of Indo-Pacific humpback dolphins. Atlantic humpback dolphins have a robust body, broad flippers with rounded tips, a well-defined rostrum that is shorter than that of other humpback dolphins, and a small, falcate dorsal fin sitting on a distinctive and well-developed dorsal hump. Coloration is generally slate gray on the back and sides, fading to a lighter gray on the belly. The rostrum and dorsal fin tend to get lighter with age. Tooth counts are lower than in other *Sousa* species (27–32 upper teeth and 31–39 lower teeth).

Behavior

This species is generally shy, does not bow-ride, and aerial displays are rarely seen. Groups usually range from 1–8 animals, but gatherings of up to 20–40 animals have been observed. In Angola, some individuals appear to exhibit high site fidelity and strong association patterns. Animals tend to feed in small bays, sheltered waters behind reef-breaks and areas off dry river mouths, while traveling occurs mainly along exposed coastlines.

Food and foraging

Groups generally forage close to shore in shallow waters and often within the surf zone. They appear to feed mainly on inshore schooling fish such as mullet (*Mugil* species).

Life history

No life history studies have been conducted. Based on a related species—*Sousa plumbea*—it is suspected that males are larger than females.

Conservation and management

The total population size is unknown, but is thought to be only a few thousand. Atlantic humpback dolphins are listed as Vulnerable by the IUCN based on their restricted geographic range, low abundance, and apparent decline in recent decades. Incidental capture in gill nets is considered their greatest threat followed by directed takes, habitat loss and degradation, overfishing, marine pollution, anthropogenic sound, and climate change.

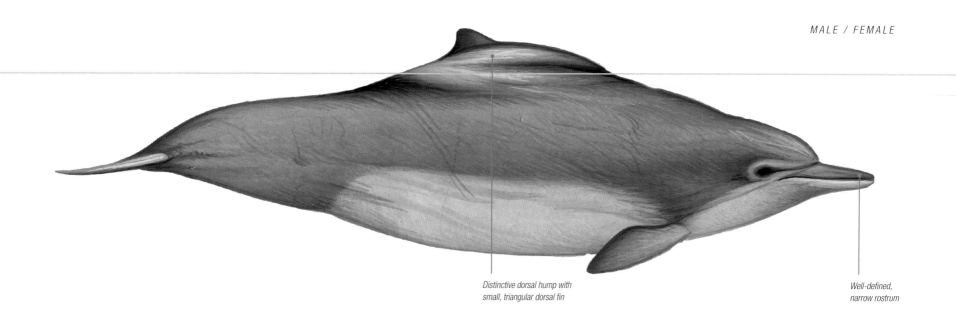

MALE / FEMALE

Distinctive dorsal hump with
small, triangular dorsal fin

Well-defined,
narrow rostrum

Dorsal hump
A prominent dorsal hump located
in front of the dorsal fin is
characteristic of specimens of
S. teuszii and *S. plumbea*.

Size
Newborn 3½ ft (1 m)
Adult 6–9¼ ft (1.8–2.8 m)

Dive sequence
1. Initially, the, narrow rostrum
breaks water, followed by the
melon. Sometimes the whole
head is lifted completely clear
of the water.

2. The blowhole is exposed
above the surface, while most
of the body remains
submerged. The body then
arches, showing the dorsal
hump and dorsal fin.

3. Finally the head drops below
the surface, the back is arched
a little higher, and the animal
dives down showing only the
dorsal region.

Breaching
Aerial behavior is rare but Atlantic
humpback dolphins have been
observed doing front and back
vertical leaps. While diving, animals
may show their flukes.

PANTROPICAL SPOTTED DOLPHIN

Family Delphinidae

Species *Stenella attenuata*

Other common names Spotted dolphin, spotter, white-spotted dolphin, slender-beaked dolphin, spotted porpoise, Graffman's dolphin

Taxonomy Two subspecies are recognized: an offshore, oceanic species (*S. a. attenuata*) and a coastal species found in the eastern tropical Pacific (*S. a. graffmani*)

Similar species Most often confused with other members of the genus *Stenella*, but can also be confused with common dolphins and common bottlenose dolphins

Birth weight Unknown

Adult weight 198–265 lb (90–120 kg)

Diet Epipelagic and mesopelagic fish, squid, and crustaceans

Group size Mean group size of 70–170 individuals

Major threats Bycatch in fisheries, predators (including killer whales, sharks, and possibly other blackfish), small directed drives for this species in Japan, West Africa, the Caribbean, and Indonesia

IUCN status Least Concern

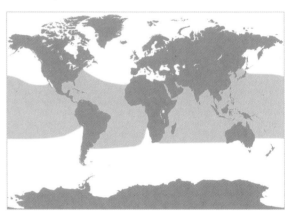

Range and habitat This species is cosmopolitan in tropical, subtropical, and warm-temperate waters worldwide, primarily in offshore waters. *S. a. graffmani* is primarily a coastal form, inhabiting waters inshore of the shelf break in the eastern tropical Pacific Ocean.

Identification checklist

- Slender, elongate build with long, narrow beak
- Distinct, dark cape
- Spotting (develops with age) and is usually restricted along the dorsal cape—in *S. a. graffmani*, adults are completely spotted
- In adults, the tip of the rostrum may be completely white
- Central fin position

Anatomy

Pantropical spotted dolphins are a relatively small delphinid with a slender build, elongate rostrum, and spotting on adults. The degree of spotting varies with geographic location. Coastal dolphins tend to exhibit heavier spotting than offshore dolphins. Lack of spotting on juveniles may contribute to species misidentification, but in comparison to common bottlenose dolphins, and even Atlantic spotted dolphins, the build of the pantropical spotted dolphin is more petite.

Behavior

The species is highly social, and group size ranges from a few to thousands of individuals. Coastal populations tend to occur in smaller groups than offshore populations. They are considered gregarious, and may associate with other species of dolphins. In the eastern tropical Pacific (ETP) the species is frequently associated with yellowfin tuna and seabirds. They are extremely acrobatic, frequently leaping high out of the water.

Food and foraging

Pantropical spotted dolphins feed on a variety of epipelagic and mesopelagic fish, squid, and crustaceans. In the ETP, offshore pantropical spotted dolphins are frequently found with large schools of yellowfin tuna—the nature of such behavior is unknown.

Life history

These dolphins reach sexual maturity between 9 and 11 years (females) and around 12–15 (males) years of age. Gestation lasts approximately 11 months and calves are born year round. The calving interval is typically once every two to three years and lactation lasts between nine months and two years of age.

Conservation and management

Numbers of this species in the ETP were reduced to about 25 percent of their previous population due to takes in the tuna purse seine fishery during the 1960–70s. While this no longer poses a serious threat, these dolphins are still taken in large numbers by drive fisheries in Japan and the Solomon Islands. Individuals are hunted for human consumption and bait in Sri Lanka, Indonesia, the Lesser Antilles, and the Philippines.

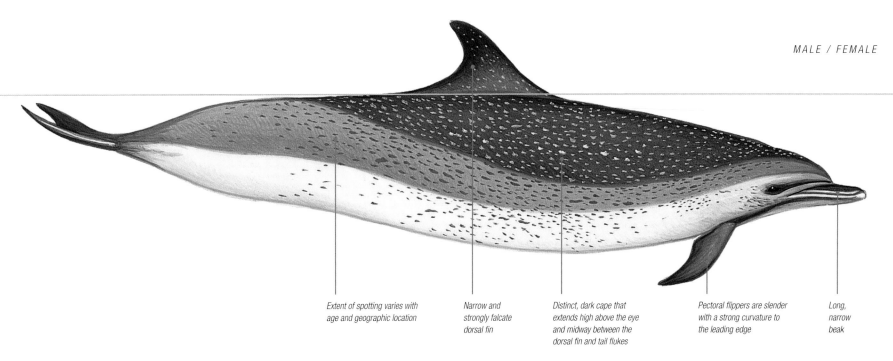

MALE / FEMALE

Extent of spotting varies with age and geographic location

Narrow and strongly falcate dorsal fin

Distinct, dark cape that extends high above the eye and midway between the dorsal fin and tail flukes

Pectoral flippers are slender with a strong curvature to the leading edge

Long, narrow beak

Calf coloration

Calves are born without spots. They are two-tone in coloration—dark gray dorsally and light gray to white ventrally. As they mature, spotting first appears on their ventral side and later on their dorsal side.

Calf

Size
Newborn 2½–2¾ ft (80–85 cm)
Adult 5¼–8 ft (1.6–2.4 m) female; 5¼–8½ ft (1.6–2.6 m) male

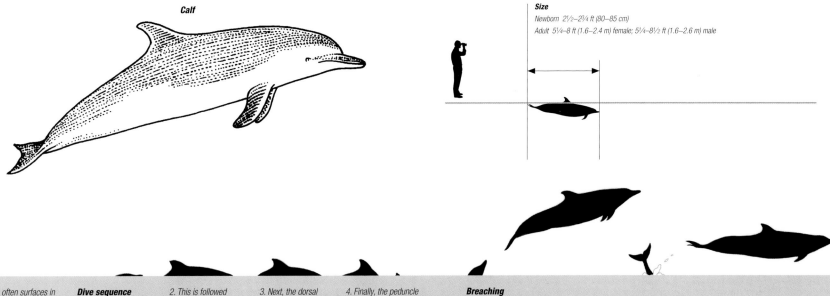

This species often surfaces in groups and engages in high-speed travel, jumping completely out of the water. Bow-riding occurs frequently in areas where the animals are not hunted or pursued by purse seine fishers.

Dive sequence
1. At first only the beak is visible as it breaks the surface of the water.

2. This is followed by the head as it becomes fully exposed.

3. Next, the dorsal fin and body stock quickly follow.

4. Finally, the peduncle and tail may be visible before the animal fully submerges.

Breaching
Breaching occurs frequently and individuals may leap high out of the water. Juveniles are particularly prone to such acrobatics.

CLYMENE DOLPHIN

Family Delphinidae

Species *Stenella clymene*

Other common names Short-snouted spinner dolphin, helmet dolphin

Taxonomy Closely related to the spinner dolphin and the striped dolphin

Similar species It can be easily mistaken for the spinner dolphin and for the common dolphin

Birth weight 22 lb (10 kg)

Adult weight 165 lb (75 kg) female; 176 lb (80 kg) male

Diet Small fish that occur in the water column, and squid

Group size 60–80 animals

Major threats Incidental takes in fisheries in Venezuela and off the coast of West Africa

IUCN status Data Deficient

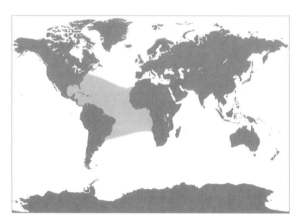

Range and habitat This species is endemic to deep tropical and subtropical waters of the Atlantic Ocean.

Identification checklist

- Tricolored pattern on the side
- Marked "moustache" on the top of the beak
- Falcate dorsal fin in the middle of the back

Anatomy

The clymene dolphin is shorter and smaller than most dolphins. Its external characteristics are very similar to those of the spinner dolphin, which is why the clymene dolphin was only recently recognized as a separate species. It can be distinguished from the other species by a dark gray dorsal cape, light gray sides, and a white belly. The lips and the tip of the beak are black, forming a marked "moustache."

Behavior

The clymene dolphin has been seen in association with short-beaked common dolphins and with spinner dolphins. It usually displays an acrobatic swimming behavior, and can spin like spinner dolphins. Schools may segregate by age and sex.

Food and foraging

The clymene dolphin feeds primarily on small fish occurring in the water column and squids. It is sometimes known to hunt during the night, when its prey migrate vertically in the water column. It tends to forage on more offshore waters.

Life history

Very little is known about this species' reproduction. It becomes sexually mature at about 6 ft (1.8 m) of length. Recent genetic studies suggest that the clymene dolphin may be a species that originated through natural hybridization between the spinner and the striped dolphins. There is also evidence for hybridization between the clymene and the spinner dolphin.

Conservation and management

This is probably one of the least-known dolphin species. No major threats for this species are known. However, the number of dolphins that are directly or indirectly taken in fisheries off the Caribbean and West Africa may be a reason for concern.

MALE / FEMALE

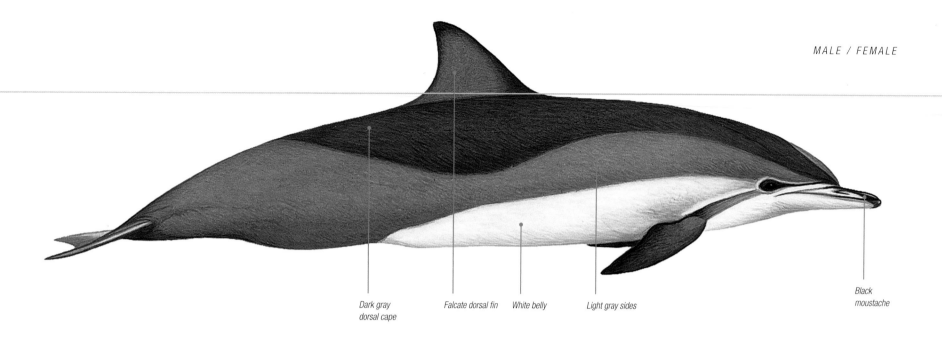

Dark gray
dorsal cape

Falcate dorsal fin

White belly

Light gray sides

Black
moustache

Beak with "moustache"
One of the features that distinguishes clymene
dolphins from other closely related species
such as spinner dolphins, is a dark line on top
of the snout, near the apex of the melon, often
called a "moustache."

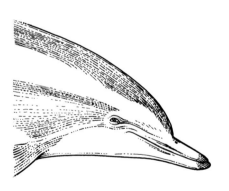

Size
Newborn Measurements unavailable
Adult 6¼ ft (1.9 m) female; 6½ ft (2 m) male

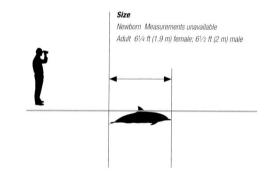

Dive sequence
*1. At first the rostrum breaks
the surface of the water.*

*2. Next, the head and
dorsal region surface.
There is no evident blow.*

*3. Lastly, the tail stock
disappears below the
surface with no fluking.*

Breaching
*Like spinner dolphins, clymene
dolphins engage in acrobatic behavior,
frequently leaping and spinning
completely out of the water.*

STRIPED DOLPHIN

Family Delphinidae

Species *Stenella coeruleoalba*

Other common names Euphrosyne dolphin (antiquated), streaker porpoise (in tropical Pacific tuna fishery)

Taxonomy Part of a large family and a diverse, controversial, genus. No subspecies are recognized but population structure is evident from morphological and genetic differences. The closest relative is the clymene dolphin.

Similar species Can be confused with other "white-bellied" pelagic dolphins including Fraser's, clymene, common, and spinner dolphins

Birth weight 15–24 lb (7–11 kg)

Adult weight Up to 343 lb (156 kg)

Diet Various small pelagic and benthopelagic fishes and squids

Group size Regionally variable from 10–30 in some regions, hundreds in others—occasionally in schools of up to 500

Major threats Hunting in Japan and entanglement in driftnets, as well as high contaminant levels, which may have contributed to large die-offs from viral infections in the Mediterranean in the early 1990s

IUCN status Least Concern (Mediterranean subpopulation assessed as Vulnerable)

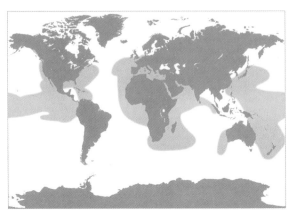

Range and habitat This species is common in tropical and warm-temperate marine regions around the world, mainly in slope and oceanic waters seaward of the continental shelf.

Identification checklist

- Robust, typical oceanic dolphin body shape
- Well-defined, moderately long beak
- Tall, falcate dorsal fin close to mid-back
- Bold dark-and-light color pattern: beak, flippers, and flukes are dark gray or bluish-black, as are the narrow eye-to-anus and eye-to-flipper stripes; the dark dorsal cape is invaded by a white or light gray spinal blaze

Anatomy

The robust body and distinctive color pattern are this oceanic dolphin's most striking features. It has numerous small, pointed teeth—39–53 pairs in the upper jaws, 39–55 in the lower jaws.

Behavior

These gregarious dolphins are active, energetic, and fast-swimming. They generally travel in dense schools, often consisting of more than 100 animals in areas where the species is abundant. Their aerial behavior can include breaches, chin slaps, and a maneuver called "rototailing" in which they perform a high, arcing jump while rotating the tail rapidly before re-entering the water. They don't often bow-ride. Based on observations in Japan, where striped dolphins have been killed in large numbers by driving entire schools to shore, the schooling system is complex. Schools can consist entirely of juveniles, adults, or a mix of the two, with the adult and mixed schools further divided into breeding and non-breeding schools. Young dolphins remain in mixed schools for a year or two after weaning, then join juvenile schools, and eventually adult or mixed schools.

Food and foraging

The diet consists of a large variety of small schooling fish and cephalopods. Striped dolphins may dive to depths of 660–2,300 ft (200–700 m) to reach their prey and by foraging in the evening and at night they probably take advantage of vertical migrations by deep-water species. In much of their range, striped dolphins appear to follow the fronts of warm currents, which move seasonally.

Life history

Females become sexually mature at 5–13 years and males at 7–15 years. Gestation is estimated at 12–13 months. Both males and females are believed to be capable of living for 57–58 years.

Conservation and management

These dolphins have a wide distribution and are globally abundant. Estimates include over half a million in the western North Pacific, 1.5 million in the eastern north and tropical Pacific, and close to 120,000 in the western Mediterranean Sea. Huge catches in Japan seriously depleted stocks there. In some regions, large numbers are killed incidentally in driftnet, purse seine, and other fisheries.

MALE / FEMALE

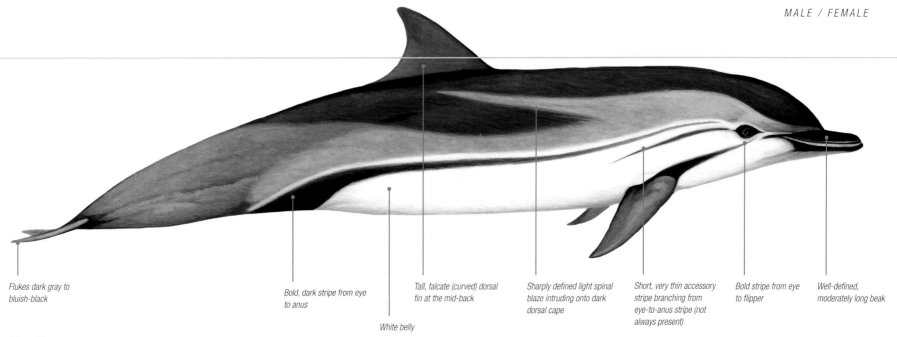

Flukes dark gray to
bluish-black

Bold, dark stripe from eye
to anus

White belly

Tall, falcate (curved) dorsal
fin at the mid-back

Sharply defined light spinal
blaze intruding onto dark
dorsal cape

Short, very thin accessory
stripe branching from
eye-to-anus stripe (not
always present)

Bold stripe from eye
to flipper

Well-defined,
moderately long beak

Color patterns

Though easily confused at a distance with other white-bellied
oceanic dolphins, the striped dolphin can be distinguished by its
bold eye-to-anus and eye-to-flipper stripes low on the side,
together with its light blaze intruding onto the dark dorsal
coloration mid-body.

Size
Newborn 3–3½ ft (0.9–1 m)
Adult 7¼ ft (2.2m) female; 8 ft (2.4 m) male

Not all surfacings by these energetic
dolphins are as revealing as this
one, but some individuals in a group
can often reveal enough of the color
pattern to allow identification.

Dive sequence
1. The moderate-length,
well-defined beak emerges
from the water.

2. The prominent dorsal fin surfaces
next. The sharp contrast between
the white underside and the rest of
the body may be evident.

3. As the animal leaps, more of the back is
revealed. It may be possible to see the
striped sides or the light spinal blaze that
intrudes onto the dark dorsal cape.

4. The dolphin splashes as it
clears the water. A fast-moving
school of striped dolphins can
create an abundance of froth.

ATLANTIC SPOTTED DOLPHIN

Family Delphinidae

Species *Stenella frontalis*

Other common names Bridled dolphin

Taxonomy No subspecies recognized but there are two forms—a larger, heavy-bodied coastal form and a slender offshore form

Similar species Easily confused with pantropical spotted dolphins and with common bottlenose dolphins

Birth weight Not known

Adult weight Up to 315 lb (143 kg)

Diet Extremely varied according to region and habitat type but encompasses fish, squid, and invertebrates from the seabed

Group size 1–15, sometimes up to 50, in coastal waters; traveling groups may number 100

Major threats None known to be serious but incidental mortality in fisheries and direct hunting may be a problem in local areas

IUCN status Data Deficient

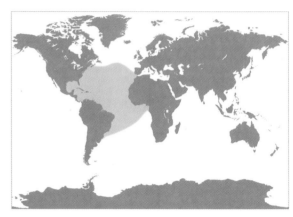

Range and habitat This species is endemic to the tropical and warm-temperate Atlantic Ocean including the Gulf of Mexico but not the Mediterranean Sea. The coastal form occurs mainly on the continental shelf and along the shelf edge. The offshore form occurs in slope waters and around oceanic islands.

Identification checklist

- Typical robust oceanic dolphin body shape
- Well-defined and moderately long beak
- Tall, falcate dorsal fin close to mid-back
- Generally, a dark upper side, gray along the flanks, and white underside but with highly variable spotting or mottling as the individual ages
- Underlying pattern includes a gray eye-to-flipper stripe and a light spinal (shoulder) blaze intruding onto the dark cape below the dorsal fin
- Fully grown adults have a white-tipped beak (rostrum)

Anatomy

These dolphins pass through a number of color stages, from a two-tone gray-white pattern at birth to speckled as juveniles (black spots underneath, white spots on top), mottled as young adults, and then as fully grown adults the black-and-white spots merge. Adults have a conspicuously white-tipped rostrum or beak. There are 32–42 pairs of teeth in the upper jaws and 30–40 in the lower.

Behavior

This species is only mildly gregarious, and is usually observed in fairly small groups of up to 15, sometimes 50–100. They are willing bow-riders and can be acrobatic. Several communities of these dolphins that are resident on the Bahama banks have become accustomed to divers and therefore are subject to both tourism and long-term research. They share much of their habitat and sometimes closely associate with bottlenose dolphins.

Food and foraging

The foraging behavior of the species is diverse, varying with the habitat. For example, in the Bahamas much foraging occurs in deep water, possibly at night when many organisms migrate toward the surface. During daytime, however, the dolphins not only rest and socialize in the shallow, clear waters but also forage there on both bottom-dwelling and schooling fish. In the Gulf of Mexico the dolphins sometimes follow trawlers to feed on discards. They also may corral fish schools and feed cooperatively.

Life history

Females become sexually mature at 8–15 years of age and give birth at intervals of 1–5 years. The average inter-birth interval for mothers whose calves survive is close to 3.5 years. Calves can be nursed for as long as five years. The maximum known longevity for this species is 23 years.

Conservation and management

Although endemic to a single ocean basin, the species is fairly abundant. There are few good estimates of numbers except in the United States where there are about 27,000 off the Atlantic coast and 37,000 in the northern Gulf of Mexico. Given their coastal distribution, these dolphins are vulnerable to entanglement in fishing gear.

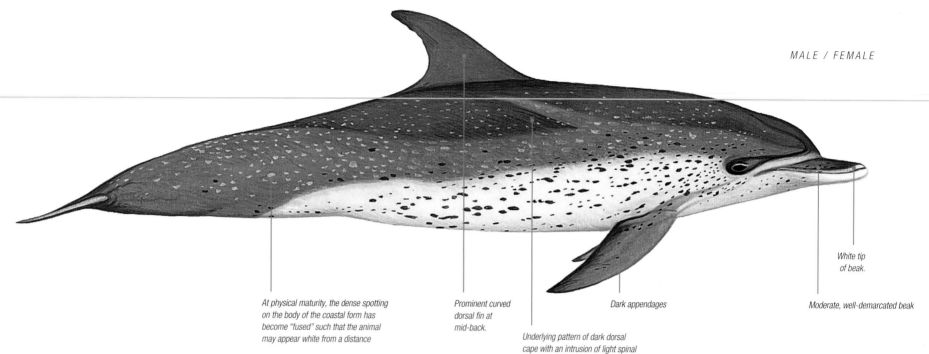

At physical maturity, the dense spotting on the body of the coastal form has become "fused" such that the animal may appear white from a distance

Prominent curved dorsal fin at mid-back.

Underlying pattern of dark dorsal cape with an intrusion of light spinal (shoulder) blaze just below dorsal fin

Dark appendages

White tip of beak.

Moderate, well-demarcated beak

Spotting and patterns

The basic color pattern—dark back, lighter sides, white belly—is present but muted on calves. As they age, they become speckled, then mottled, and finally heavily spotted as adults, with dark and light spots "fused."

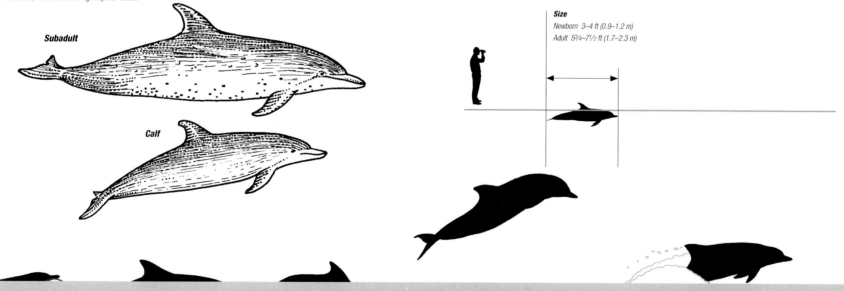

Subadult

Calf

Size
Newborn 3–4 ft (0.9–1.2 m)
Adult 5¾–7½ ft (1.7–2.3 m)

Dive sequence
1. These active and high-energy dolphins have a well-defined beak that emerges from the water along with the melon (forehead).

2. Next, the relatively large, erect dorsal fin and the middle of the back become visible.

3. Lastly, the tail stock (the area behind the dorsal fin and in front of the flukes) curves out of the water.

Surface behavior
Aerial activity can be expected when observing these dolphins.

These dolphins are fast swimmers and willing bow-riders, known for their acrobatic behavior.

SPINNER DOLPHIN

Family Delphinidae

Species *Stenella longirostris*

Other common names Long-snouted dolphin, long-beaked dolphin, spinner porpoise, spinning dolphin, spinner, Hawaiian spinner dolphin

Taxonomy Four subspecies are recognized: Gray's spinner dolphins (*S. l. longirostris*), Eastern spinner dolphins (*S. l. orientalis*) Central American spinner dolphins (*S. l. centroamericana*), and the dwarf spinner dolphin (*S. l. roseiventris*)

Similar species Clymene dolphins, pantropical spotted dolphins, striped dolphins, and common dolphins

Birth weight 22 lb (10 kg)

Adult weight 165 lb (75 kg)

Diet Primarily small mesopelagic fish

Group size 10–50 is typical

Major threats Fisheries (directed takes and bycatch), pollution

IUCN status Data Deficient

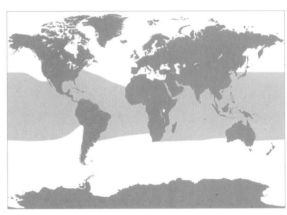

Range and habitat These dolphins are found in tropical and subtropical oceanic regions; they rest in shallow bays during the day and forage further offshore at night.

Identification checklist

- Small delphinid with long, slender body
- Narrow, elongate beak
- Dark stripe running from the eye to the flipper
- Color variation dependent on region
- Central fin position

Anatomy

Physical characteristics of spinner dolphins vary considerably throughout their geographic range. Gray's and dwarf spinner dolphins are tricolored with a dark gray cape, lighter gray coloration along their sides, and a whitish underside. Both Gray's and dwarf spinner dolphins have a slightly falcate dorsal fin, but Gray's may be more triangular in shape. Eastern spinner dolphins are almost uniformly dark gray with a triangular dorsal fin that slopes forward. Adult males have a strongly forward-sloping dorsal fin and a well-defined post-anal hump. Central American spinner dolphins more closely resemble the Eastern spinner dolphins with less-defined post-anal humps in the adult males.

Behavior

Spinner dolphins are highly gregarious and can occur in groups of 1,000 or more individuals. Typically, group size ranges from 10–50 animals. Spinner dolphins around oceanic islands spend most of the day resting in shallow, sandy-bottomed bays. They venture offshore late in the day in order to feed on a layer of vertically migrating prey.

Food and foraging

Most spinner dolphins feed primarily on mesopelagic fish that ascend to shallower water in the evening as part of their vertical migrations. Dwarf spinner dolphins predominantly feed on benthic fish, reef fish, and invertebrates.

Life history

Spinner dolphins breed year round although peak birthing varies by subspecies and region. Females reach sexual maturity between four and seven years of age and males between seven and ten years. Gestation is approximately 10½ months and calves are born once every 3 years, on average. Their lifespan is at least 20 years.

Conservation and management

Spinner dolphins in the eastern tropical Pacific were taken in large numbers by the tuna purse-seine fishery during the 1960–70s. In the Gulf of Thailand, dwarf spinner dolphins are taken incidentally as bycatch in shrimp trawl nets. In Australia they are caught in shark gill nets. They may also be taken directly for use as bait or for human consumption in places such as Sri Lanka, the Caribbean, Indonesia, the Philippines, and occasionally Japan and West Africa.

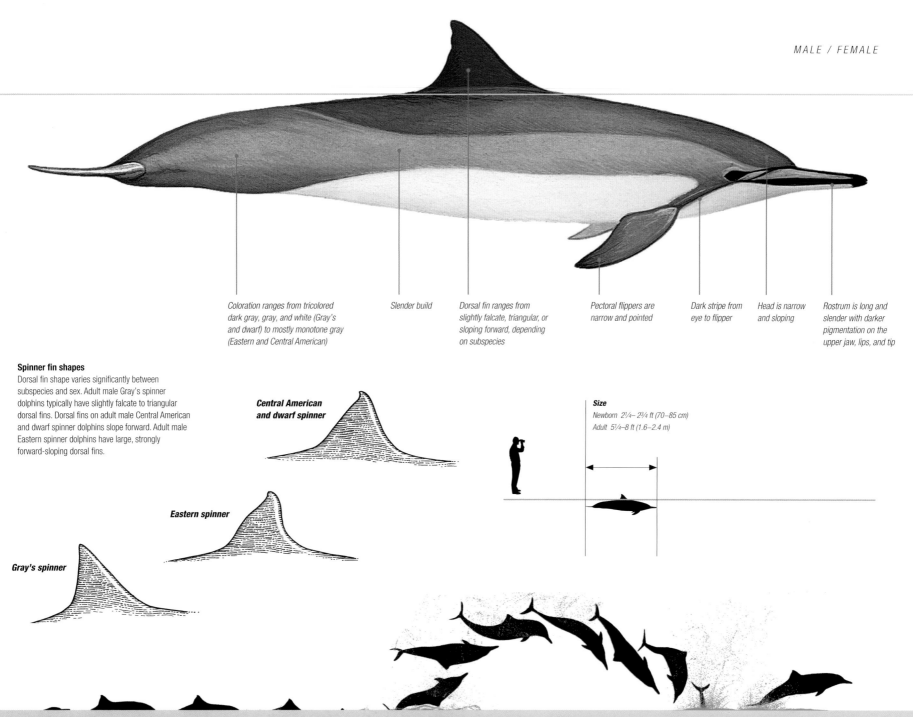

MALE / FEMALE

Coloration ranges from tricolored dark gray, gray, and white (Gray's and dwarf) to mostly monotone gray (Eastern and Central American)

Slender build

Dorsal fin ranges from slightly falcate, triangular, or sloping forward, depending on subspecies

Pectoral flippers are narrow and pointed

Dark stripe from eye to flipper

Head is narrow and sloping

Rostrum is long and slender with darker pigmentation on the upper jaw, lips, and tip

Spinner fin shapes
Dorsal fin shape varies significantly between subspecies and sex. Adult male Gray's spinner dolphins typically have slightly falcate to triangular dorsal fins. Dorsal fins on adult male Central American and dwarf spinner dolphins slope forward. Adult male Eastern spinner dolphins have large, strongly forward-sloping dorsal fins.

Central American and dwarf spinner

Eastern spinner

Gray's spinner

Size
Newborn 2¼– 2¾ ft (70–85 cm)
Adult 5¼–8 ft (1.6–2.4 m)

Dive sequence
1. The rostrum breaks the surface of the water first.

2. There is no evident blow as the head and dorsal region surface.

3. The stock submerges with no fluking.

Bow-riding is common in nearshore areas.

Breaching
Spinner dolphins are so-named for their acrobatic behavior, frequently leaping and spinning completely out of the water. An individual may spin up to seven times in a single leap. Repeated leaping and spinning is common. All age classes and sexes perform this behavior and there is no other species of cetacean that acts in this way. Spinner dolphins frequently carry remoras, and spinning may be one mechanism by which the species tries to rid themselves of these parasitic fish.

ROUGH-TOOTHED DOLPHIN

Family Delphinidae

Species *Steno bredanensis*

Other common names Slopehead

Taxonomy Closely related to the genera *Sotalia* and *Orcaella*

Similar species Difficult to distinguish from common bottlenose dolphins when viewed from above

Birth weight Unknown

Adult weight Up to 342 lb (155 kg)

Diet Small, surface fish, large predatory fish such as mahi-mahi, squid, and other cephalopods

Group size Average 10–30

Major threats Fishing depredation interactions with sport fishing around the island of Hawaii, and Moorea and Tahiti in the Society Islands—mass-stranding events have occurred in Maui and more often off the east coast of Florida

IUCN status Least Concern

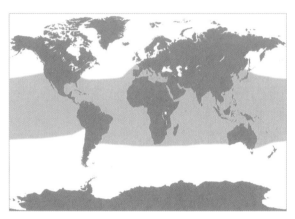

Range and habitat This species has a worldwide tropical-to-temperate distribution. It is most often observed in deep water around oceanic islands, but is sometimes seen in shallow waters along the continental shelf of Brazil.

Identification checklist

- Sloped, narrow melon
- Countershaded body with a white belly and dark gray back
- Visible white splotches on lower sides and underside of adults
- Dorsal fin slightly falcate (curved)
- Vertical ridges on teeth

Anatomy

This species is the only member of the genus *Steno*. The skull is indistinguishable from those of the humpback dolphins except for the number of teeth. Rough-toothed dolphins have 19–26 pairs of teeth in the upper jaws and 19–28 pairs of teeth in the lower jaws. The front flippers are set further back on the body than in most cetaceans, and are 17–19 percent of the body length.

Behavior

Evidence from genetic studies and photo-identification suggests that this species stays in a particular home range. This behavior, together with observed association patterns and a limited gene flow, indicates that the rough-toothed dolphin has an insular population structure. Behavior such as caring for sick or injured individuals, synchronized swimming, and cooperative foraging on large prey also suggest a level of social organization. These dolphins are curious and have been observed playing with flotsam floating on the water. They are known to approach boats and bow-ride. However boat-avoidance behavior has been observed around the islands of Hawaii and Tahiti. They are often encountered with other dolphin species and humpback whales.

Food and foraging

Their diet includes small fish and squid. They have been observed foraging on surface species such as needlefish and flying fish, occasionally in mixed species groups. Cooperative foraging on large predatory fish such as mahi-mahi has also been observed.

Life history

Life expectancy is up to 36 years. Males reach sexual maturity at 14 years, and females at 10 years. Females give birth to one calf and the gestation period is unknown, but could be up to 12 months, as in bottlenose dolphins. Group sizes average around 10–30, although singles and pairs are encountered. The largest groups recorded include around 90 in the main Hawaiian islands, 150 in French Polynesia, and 300 in the eastern tropical Pacific.

Conservation and management

Interactions with sport fishermen (depredation) around some islands are a threat to this species. Entanglements have been recorded in American Samoa and the eastern tropical Pacific. Several mass-stranding events have been recorded off the east coast of Florida and one around Maui.

MALE / FEMALE

Post-anal keel is sometimes (but not always) observed in males and not as prominent as in some species, such as spinner dolphins

White splotches on lower sides and underside of adults

Slightly falcate dorsal fin

Cookie-cutter scars in some populations.

Flippers set farther back than most dolphin species

Narrow melon with minimal rise and no crease visible

Patterns and flippers
Adult dolphins display unique color patterns of white splotches on the lower sides and underside. Their large front flippers are set farther back than other dolphins and are about 17–19 percent of the body length.

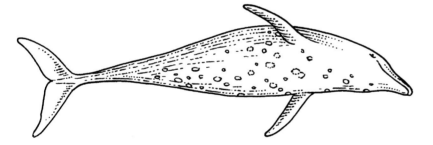

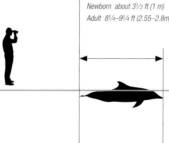

Size
Newborn about 3½ ft (1 m)
Adult 8¼–9¼ ft (2.55–2.8m) maximum; average: 8½ ft (2.6 m) female and 9 ft (2.7 m) male

Dive sequence
1. The initial rise out of the water for a breath includes only the top of the head and the rostrum (unique to this species).

2. The body then rises out of the water, exposing the dorsal fin and midway down the lateral part of the body.

3. As the rostrum returns to the water, after the breath is taken, the back is rounded slightly.

4. The tail may or may not rise out of the water.

5. In a typical rough-toothed dolphin surfacing, water often accompanies the dolphin along its body as it surfaces.

INDO-PACIFIC BOTTLENOSE DOLPHIN

Family Delphinidae

Species *Tursiops aduncus*

Other common names Indian Ocean bottlenose dolphin, inshore bottlenose dolphin

Taxonomy Closely related to its sister species, the common bottlenose dolphin

Similar species Common bottlenose dolphin

Birth weight 20–40 lb (9–18 kg)

Adult weight 385–440 lb (175–200 kg)

Diet Demersal, reef, epi- and mesopelagic fish, and cephalopods

Group size 1–15, although groups of several hundred have been observed on rare occasions

Major threats Habitat degradation, fisheries interactions, eco-tourism

IUCN Status Data Deficient

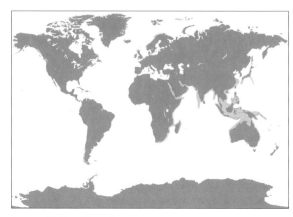

Range and habitat This species inhabits warm-temperate to tropical nearshore waters on the continental shelf or around oceanic islands of the Indian and western Pacific oceans.

Identification checklist

- Slender body, with a long and thin rostrum
- Dark to medium gray dorsal surface
- Ventral side is whitish with a pink hue
- Potential speckled ventral side (age-dependent)
- Triangular dorsal fin, in comparison to the more falcate version of the common bottlenose dolphin

Anatomy

This Indo-Pacific bottlenose dolphin has a slender body and long, thin rostrum in comparison to its close relative, the common bottlenose dolphin. The dorsal side is darker gray with a triangular or falcate dorsal fin. The ventral side is lighter gray with dark speckles developing on mature individuals of some populations.

Behavior

This species lives in fission–fusion societies where group membership changes on a daily or hourly basis. Associations between males and females are linked to female reproductive state. Adult males form alliances to optimize access to reproductively active females. Adult females range from solitary to large networks of affiliates. Juveniles form behaviorally specific associations and spend considerable time engaging in play to cultivate relationships.

Food and foraging

Diet, including mesodermal and reef fish as well as cephalopods, and foraging strategies vary between and within populations. Foraging can occur in solitude or cooperatively. One foraging strategy, referred to as "sponging," is the first documented tool use in cetaceans. It involves dolphins carrying a conical sponge on their rostrum as protection while searching for prey on the seafloor. Behavioral and genetic studies confirm that this behavior is passed from mother to offspring, particularly daughters. Other examples include beaching to catch fish, manipulating shells, and interacting with fisheries by feeding on discarded bycatch and begging.

Life history

The lifespan of this species is around 40 years. Sexual maturity is reached at 12–15 and 10–15 years for females and males, respectively. The peak birthing and mating seasons coincide with the months of highest water temperature with gestation lasting around 12 months. Nursing lasts for three to five years, although calves supplement their intake with solids by the age of six months. The inter-birth interval is typically between three and six years.

Conservation and management

The main threat to this species is habitat degradation due to coastal development. Other threats include fisheries interactions, such as bycatch and incidental entanglements, eco-tourism, noise, and chemical pollution.

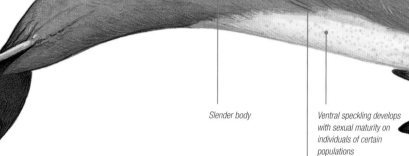

Slender body

Ventral speckling develops with sexual maturity on individuals of certain populations

Dark to medium gray dorsal surface often appearing as a cape

Pectoral fins, dorsal fin, and fluke larger and broader relative to body size compared to T. truncatus

Rostrum long and thin

Dorsal fin shape

The majority of individuals have unique marks, nicks, and notches, on the trailing edge of their dorsal fin. These marks evolve over time, usually during interactions with other adults of the same species, and make it possible to distinguish between individual animals.

Size

Newborn 3–4 ft (0.9–1.25 m)
Adult 6– 8¼ ft (1.8– 2.5 m)

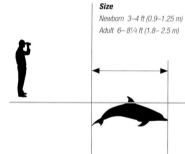

Dive sequence

1. A typical diving sequence for the Indo-Pacific bottlenose dolphin starts with the rostrum protruding through the surface. At this point the dolphin usually exhales.

2. The head and back break the surface next and the dolphin inhales when the blowhole is visible.

3. The dorsal fin and part of the body become fully visible during the surfacing. The dolphin arches its back while diving down and sometimes also exposes a proportion of its peduncle before becoming fully submerged.

Foraging dives

A typical foraging dive can be identified by a peduncle dive, which can be described as left apart from that upon diving more of the dolphin's peduncle is visible on the surface.

Another typical dive during foraging is a tail-out dive, which can also be described as left but upon diving the whole tail comes out of the water.

Although these two dive types are typical, they are not only associated with foraging behavior but may also occur during resting or socializing.

COMMON BOTTLENOSE DOLPHIN

Family Delphinidae

Species *Tursiops truncatus*

Other common names Bottlenose dolphin

Taxonomy Two subspecies: *T. t. ponticus* and *T.t. truncatus* are recognized. Also, coastal and offshore ecotypes recognized in some areas.

Similar species Indo-Pacific bottlenose dolphin, Pantropical spotted dolphin, Atlantic spotted dolphin, Risso's dolphin, rough-toothed dolphin

Birth weight 31–44 lb (14–20 kg)

Adult weight 573 lb (260 kg) maximum female; 1,433 lb (650 kg) maximum male

Diet A wide variety of schooling, and non-schooling fish; invertebrates such as shrimp, octopus, and squid

Group size 2–15 is most common, but may reach several hundred in offshore areas; highly variable depending on reproductive status, behavior, and habitat

Major threats Habitat degradation and loss, hunting, live-capture fisheries (for display, research, military use), drive fisheries, fisheries bycatch, anthropogenic contaminants

IUCN status Least Concern

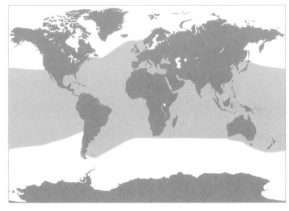

Range and habitat This species is found in temperate to tropical waters of approximately 50–90° F (10–32° C); primarily coastal areas (including bays and estuaries) over the continental shelf but may extend offshore to deep pelagic waters.

Identification checklist

- Body light to dark gray on upper side with pinkish-white coloration on the underside
- Stocky rostrum separated from the melon by a prominent crease
- Tall, falcate dorsal fin
- Fin located in the mid-back

Anatomy

The bottlenose dolphin is one of the most well-known and best recognized cetaceans, due to its prominence in marine parks and aquaria. Living up to its name, this medium-sized dolphin has a stocky, bottle-shaped rostrum. Individuals are often identifiable by prominent nicks and notches occurring along the trailing edge of the falcate dorsal fin. Sexual dimorphism is present and males may be one-third as large as females. Individuals in offshore populations are often larger than those in inshore (coastal) populations.

Behavior

Bottlenose dolphins are easily visible at the surface due to relatively short dive times and performance of aerial displays such as leaping, tail-slapping, and spyhopping. Many surface displays occur during foraging and socializing. Bottlenose dolphins will commonly bow-ride in the presence of boats.

Food and foraging

Bottlenose dolphins feed on a variety of schooling and non-schooling fish. Individuals may feed individually or by cooperating with other group members to herd and corral fish.

Life history

These dolphins occur in dynamic fission–fusion societies, where individuals frequently join and split from groups according to foraging, anti-predation, mating, and calf-rearing strategies. As such, group size varies widely but typically ranges from 2 to 15 individuals. Males may form tight alliance bonds with other males for mating purposes while females may form strong bonds with other females in nursery groups. Females typically give birth to one calf every 2–6 years after a gestation of 12 months. Wild bottlenose dolphins may live for 50 or more years, with females typically living slightly longer than males.

Conservation and management

Major threats include habitat degradation and loss; directed hunting for food, bait, and fishery conflicts; bycatch; drive fisheries; and environmental contaminants. However, the conservation status of Least Concern indicates the bottlenose dolphin as a species is doing well throughout its range. Inshore populations are more vulnerable to hunting and habitat degradation, particularly those in the Mediterranean, Black Sea, Taiwan, Japan, and coastal regions, of Peru, Ecuador, and Chile.

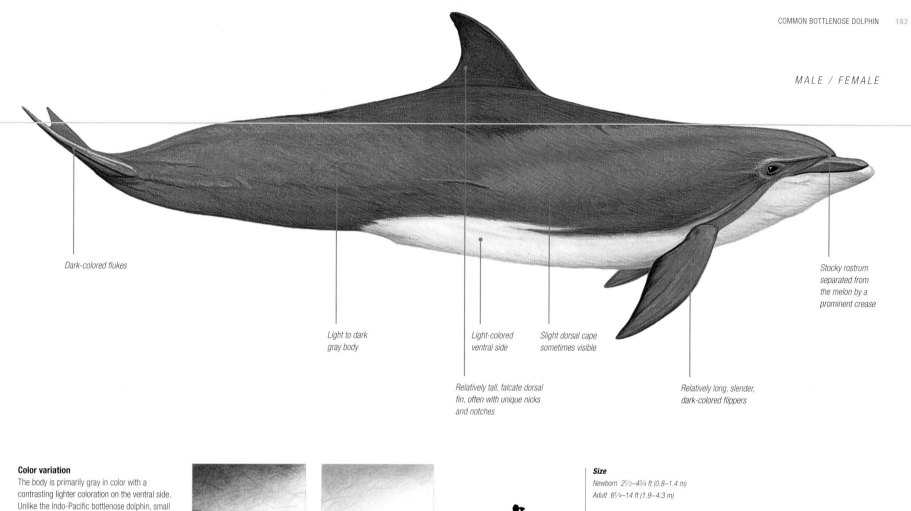

MALE / FEMALE

Dark-colored flukes

Stocky rostrum separated from the melon by a prominent crease

Light to dark gray body

Light-colored ventral side

Slight dorsal cape sometimes visible

Relatively tall, falcate dorsal fin, often with unique nicks and notches

Relatively long, slender, dark-colored flippers

Color variation

The body is primarily gray in color with a contrasting lighter coloration on the ventral side. Unlike the Indo-Pacific bottlenose dolphin, small spots are rarely present on the ventral surface.

Light to dark gray on dorsal side

White to pink ventral side

Size

Newborn 2½–4¾ ft (0.8–1.4 m)
Adult 6¼–14 ft (1.9–4.3 m)

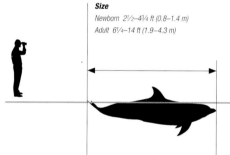

Dive sequence

A typical dive consists of a smooth arc at the surface, with the blowhole, dorsal fin, and dorsal ridge breaking the surface of the water.

Leap

The rostrum breaks the surface first and the body straightens vertically, then shifts horizontally with a slight flex to the tail stock before finally arcing headfirst back to the water.

TOOTHED WHALES
Sperm Whales

The sperm whales include the sole member of the Family Physeteridae, the sperm whale, as well as the much smaller and less well-known pygmy sperm whale and dwarf sperm whale that together comprise the Family Kogiidae. All sperm whales get their name from a spermaceti organ in the head that contains a waxy fluid important in sound production. The spermaceti organ is much larger and more pronounced in the sperm whale as compared to dwarf and pygmy sperm whales.

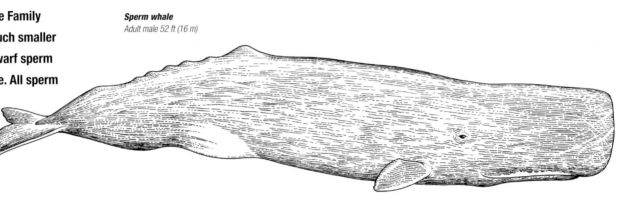

Sperm whale
Adult male 52 ft (16 m)

- The sperm whale is distributed throughout the world's ocean basins but with different distributional patterns between males and females. Adult males have a wider range, migrate long distances, and may be found in high latitudes near both the north and south poles. Females and juveniles have much smaller home ranges and inhabit deep waters of the tropics and subtropics.

- Dwarf and pygmy sperm whales are found worldwide in tropical and temperate ocean waters.

- The sperm whale is the largest of the toothed whales, with adults ranging in size from 36–52 ft (11–16 m) in length and weighing between 33,000–99,200 lb (15,000–45,000 kg). Adult males are much larger and heavier than adult female sperm whales.

- The dwarf and pygmy sperm whales are much smaller than the sperm whale. Adult dwarf and pygmy sperm whales range between 6½–11 ft (2–3.3 m) in length.

- Sperm whales are brown to gray in coloration. Dwarf and pygmy sperm whales are dark gray in color.

- Sperm whales are often identified at sea by their large body size, box-car shaped heads, wrinkled skin appearance, and blowhole positioning that is far left and forward on the head.

- Dwarf and pygmy sperm whales are more difficult to identify at sea because of their much smaller body size and their elusive behavior. The blunted head shape can be diagnostic and the size and positioning of the dorsal fin can be used to distinguish between dwarf and pygmy sperm whales.

- All of the sperm whales are deep divers with small, underslung lower jaws and are believed to rely on suction-feeding to forage at depth.

Dwarf sperm whale
Adult male 6¼–8½– ft (1.9 m–2.6 m)

Sperm whale size
The sperm whale is by far the largest of the different sperm whales with adult males reaching up to 99,200 lb (45,000 kg) in weight. Dwarf and pygmy sperm whales are much smaller, with only pygmy sperm whales rarely exceeding 1,000 lb (454 kg) in weight.

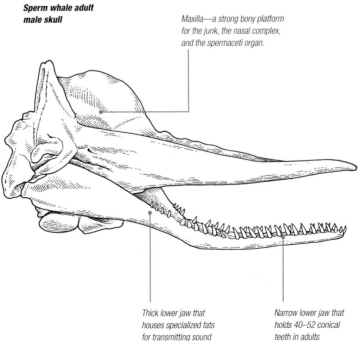

Sperm whale adult male skull

Maxilla—a strong bony platform for the junk, the nasal complex, and the spermaceti organ.

Thick lower jaw that houses specialized fats for transmitting sound

Narrow lower jaw that holds 40–52 conical teeth in adults

Skull
The lower jaw is very narrow and underslung in the sperm whale, and holds 40–52 conical teeth in adults. The upper jaw serves as a strong bony platform for the junk—analogous to the melon in other toothed whales—the nasal complex, and the spermaceti organ of sperm whales. The inside of the thick lower jaw houses specialized fats that are used to transmit sound to the inner ear of toothed whales. The largest brain of any mammal is found directly behind the sperm whale's protective cranium.

SPERM WHALE

Family Physeteridae

Species *Physeter macrocephalus*

Other common names Cachalot

Taxonomy The sperm whale is the only species in this family and its closest relatives are the much smaller dwarf and pygmy sperm whales

Similar species None but at sea sperm whales are most likely to be confused with baleen whales, such as the humpback whale and gray whale, from a distance because of their large sizes

Birth weight 1,100–2,200 lb (500–1,000 kg)

Adult weight 33,000 lb (15,000 kg) female; 99,200 lb (45,000 kg) male

Diet Cephalopods and fish

Group size 20–30 (females and immature groups); adult males are usually solitary but younger males form fluid groups of around 20

Major threats Pollution, ingestion of oceanic debris, ship strikes, and interaction with fisheries

IUCN status Vulnerable

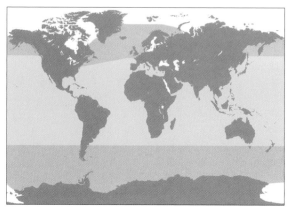

Range and habitat Sperm whales have a truly global distribution as they are found in all of the world's oceans. However, males and females have different distributions. Only adult males are found near the North and South Poles whereas females and immature animals are typically found in the deeper waters of tropical and subtropical ocean basins (lighter shaded area). Adult males migrate to warmer waters for breeding.

Identification checklist

- No teeth in the upper jaw
- Wrinkled skin appearance
- Large square head in adults
- 40–52 conical teeth in the lower jaw of adults
- Blowhole is far forward on the left side of the head and produces a low and angled blow
- Large body size

Anatomy

The sperm whale is the largest of the toothed whales and exhibits extreme sexual dimorphism. An adult male sperm whale weighs three times that of an adult female, is longer in length, and has a more pronounced and larger square head. The sperm whale name comes from the large spermaceti organ in the head that is now known to contain waxy fluids that are used in sound production. The nasal air sac complex includes the spermaceti organ and is almost one-third the length of the whale. Sperm whales have an asymmetrical skull, which is reflected in the blowhole positioned to the left and far forward on the head. Sperm whales have the largest brain of any mammal. Teeth are found only in the lower jaw after sexual maturity is reached.

Behavior

Male and female sperm whales have very different distributions and behaviors. Sperm whales are very social with a structure that is built around matriarchal groups of 20–30 adult females and young whales. These female-based groups are found primarily in deeper waters of the tropics and subtropics. Young whales are unable to dive as deep as adults and tend to stay near the surface where adult females in the group stagger diving to babysit. When threatened, the females will risk themselves and form defensive positions to protect young whales or other members of the group. Young males leave their family groups between the ages of 4 and 21 to form fluid bachelor groups that venture into higher latitudes. Only adult male sperm whales are found near the North and South Poles. Adult males migrate to the tropics and subtropics where they individually visit female-based groups for breeding. Females stay within a much smaller home range than the adult males that may transverse an ocean basin, but both sexes are more likely to be found in highly productive waters.

Food and foraging

Sperm whales are known for the large volume of prey they ingest. To support such a large body weight they eat approximately three percent of their body weight per day. The total biomass removed by sperm whales as prey from the world's oceans is similar to the total annual catch of all of the world's fisheries, but there is little overlap between the prey of sperm whales and human fisheries. Sperm whale foraging is focused in the deep ocean where they eat primarily cephalopods, including larger squid such as the giant

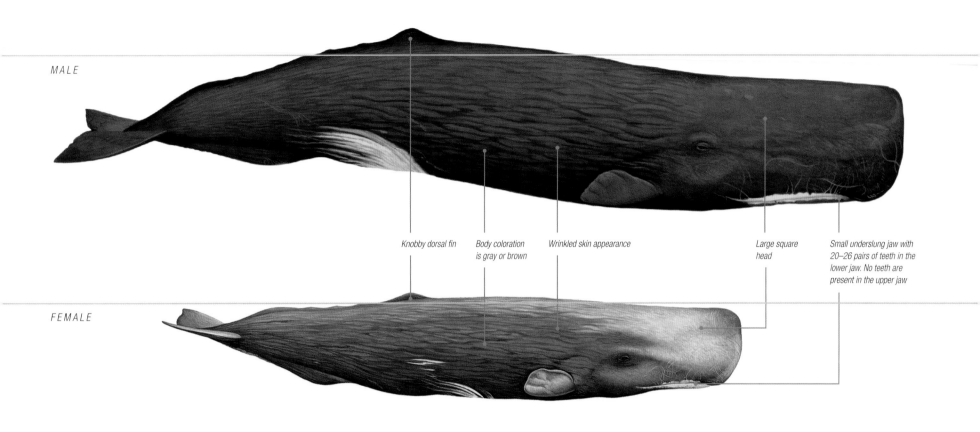

MALE

FEMALE

Knobby dorsal fin

Body coloration
is gray or brown

Wrinkled skin appearance

Large square
head

Small underslung jaw with
20–26 pairs of teeth in the
lower jaw. No teeth are
present in the upper jaw

Spermaceti organ

The sperm whale name comes from the
large spermaceti organ that, along with
other structures comprising the nasal
complex, accounts for approximately a
third of the body length of the whale.
The spermaceti organ is filled with oils
used in sound production.

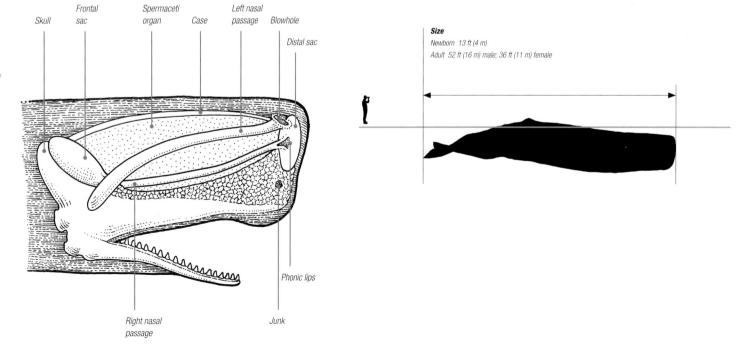

Skull

Frontal
sac

Spermaceti
organ

Case

Left nasal
passage

Blowhole

Distal sac

Phonic lips

Right nasal
passage

Junk

Size
Newborn 13 ft (4 m)
Adult 52 ft (16 m) male; 36 ft (11 m) female

SPERM WHALE

squid and jumbo squid. Sperm whales also forage on fish, especially bottom-dwelling fish. A wide diversity of food and non-food items have been described from the stomach contents of sperm whales examined during whaling or from strandings.

Life history

Sperm whales live for at least 60–70 years and perhaps longer. Females reach sexual maturity around 9 years of age and continue to grow until over 30 years of age. Males undergo a gradual process of maturation, reaching puberty near 10 years of age but not attaining sexual maturity until near 20 years of age. Males continue to grow until they reach a size that is three times the weight of adult females at 45–50 years of age. Females typically give birth to one calf every four to six years, with longer time intervals between births as females age. Gestation lasts for 14–16 months. Although calves may supplement milk with solid food prior to turning one year of age, they continue to suckle for at least two years. Sperm whales were especially vulnerable to the impacts of exploitation due to their low reproductive rate.

Conservation and management

The sperm whale was hunted throughout the world's oceans until a 1988 moratorium. The worldwide abundance estimate of sperm whales pre-whaling is believed to number more than 1.1 million. Approximately 360,000 sperm whales remain in the world's oceans today. The species is undergoing a slow recovery from whaling exploitation and today sperm whales face a wide diversity of threats. These include very small-scale whaling using primitive methods in a few isolated locations, ship strikes, ingestion of oceanic debris including plastic bags that leads to death, interactions with fisheries, underwater noise, pollutants in the ocean, and disease.

Social behavior
Sperm whales are highly social and live in complex societies. A group of sperm whales including adults and a newborn calf are seen socializing below at the Azores in the Atlantic.

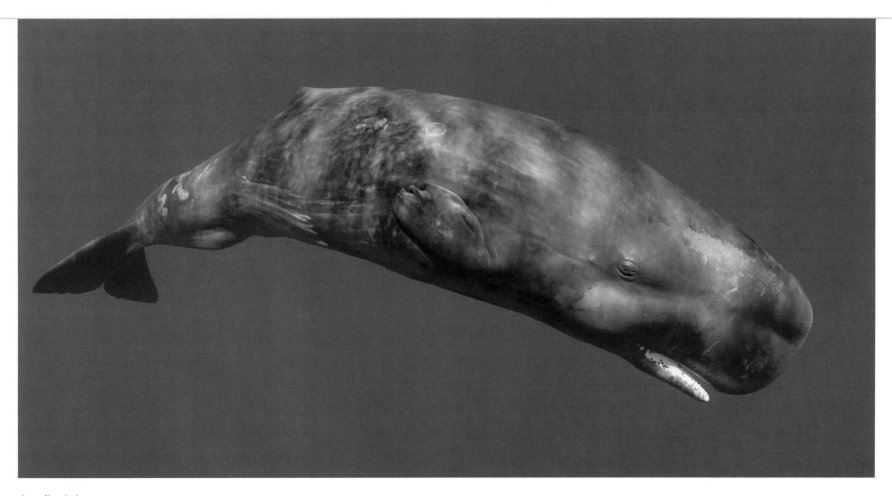

Juvenile whale

A juvenile female sperm whale aged 6–11 years investigates onlookers in the Commonwealth of Dominica in the Caribbean. Juvenile females stay within adult female-based groups as they grow and mature into adults themselves.

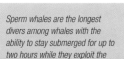

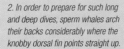

Sperm whales are the longest divers among whales with the ability to stay submerged for up to two hours while they exploit the deep ocean depths for prey.

Dive sequence

1. The blowhole is far forward on the left side of the head and produces a low and angled blow.

2. In order to prepare for such long and deep dives, sperm whales arch their backs considerably where the knobby dorsal fin points straight up.

3. The tail fluke becomes horizontal in the air as the ascent of the dive begins.

Breaching and lobtailing

As part of their highly social society, sperm whales may engage in breaching and lobtailing behaviors, making a loud sound when either the body or powerful tail and fluke hit the surface of the water.

PYGMY SPERM WHALE

Family Kogiidae

Species *Kogia breviceps*

Other common names Pygmy sperm whale

Taxonomy Most closely related to the dwarf sperm whale and more distantly related to the sperm whale

Similar species Dwarf sperm whale

Birth weight 117 lb (53 kg)

Adult weight 515–825 lb (234–374 kg) males; 660–1,060 lb (301–480 kg) females

Diet Cephalopods, fish, and deep water shrimp

Group size 1–3

Major threats Underwater noise, oceanic debris, harpooning, dynamite fishing, occasionally bycatch in driftnet fisheries

IUCN status Data Deficient

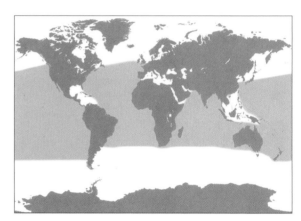

Range and habitat This species inhabits tropical and temperate waters of all major ocean basins. Based on prey remains in South Africa and Taiwan, the pygmy sperm whale likely inhabits oceanic waters beyond the edge of the continental shelf compared to the dwarf sperm whale that may inhabit more coastal waters, including the shelf slope.

Identification checklist

- No teeth in the upper jaw
- Dorsal fin height to body length ratio less than five percent
- Fin position is two-thirds along the back
- Underslung lower jaw
- Head shape is boxy (squarish)

Anatomy

The pygmy sperm whale is larger than the dwarf sperm whale but both have a squarish head shape and underslung lower jaw. The pygmy's dorsal fin is smaller in proportion to body size and is two-thirds of the way back toward the tail. This species has a small number of teeth present only in the lower jaw. The color pattern resembles a false gill slit behind the eye. Both the dwarf and pygmy have a small spermaceti organ similar to that of the sperm whale.

Behavior

Pygmy sperm whales are elusive at sea. Sightings rarely occur except under ideal sea and weather conditions. The species is known for logging at the surface where they lie motionless. They may dive for long periods when approached by boat. When startled, pygmy sperm whales may secrete a dark brown liquid from the anus, forming a cloud of ink that is thought to aid in predator avoidance. Group size ranges between one and three individuals.

Food and foraging

The diet is comprised primarily of cephalopods but pygmy sperm whales also eat deep-water fish and shrimp. The anatomy combined with pairs of tiny teeth on the lower jaw suggests that suction-feeding is used to ingest prey. Based on the depth of prey identified from the stomach contents of stranded individuals, foraging is believed to take place seaward of the continental slope. Pygmy sperm whale foraging overlaps with that of the dwarf sperm whale. In Hawaiian waters, pygmy sperm whales dive to at least 2,600–3,900 ft (800–1,200 m) depths to forage.

Life history

Pygmy sperm whales live only to about 20 years of age. They reach sexual maturity by three to five years of age and may then give birth to a single calf as frequently as every year. Breeding appears to be seasonal in the few locations where the species has been studied. Gestation length is believed to be 11–12 months.

Conservation and management

Pygmy sperm whales may be vulnerable to underwater noise. They are also known to ingest oceanic debris that can lead to death. The species may also be threatened by pelagic driftnets, ship strikes, and harpoon or dynamite fisheries. Overall, the species is very poorly known.

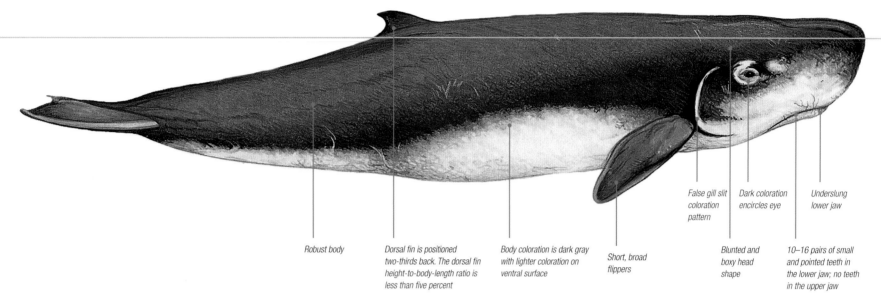

MALE / FEMALE

False gill slit coloration pattern

Dark coloration encircles eye

Underslung lower jaw

Robust body

Dorsal fin is positioned two-thirds back. The dorsal fin height-to-body-length ratio is less than five percent

Body coloration is dark gray with lighter coloration on ventral surface

Short, broad flippers

Blunted and boxy head shape

10–16 pairs of small and pointed teeth in the lower jaw; no teeth in the upper jaw

Skull

Pygmy sperm whales have a characteristic skull that reflects the squarish head shape. The skull is very wide with an extremely short beak. It is asymmetrical and the blowhole is on the left side of the head. No teeth are found in the upper jaw.

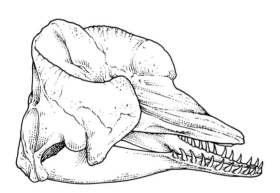

Size
Newborn 4 ft (1.2 m)
Adult 8½–10½ ft (2.6–3.2 m) female; 8–11 ft (2.4–3.3 m) male

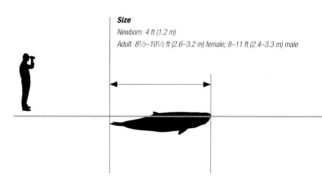

Dive sequence
1. Pygmy sperm whales are known for laying motionless at the surface. This behavior has been described as "logging."

2. They occasionally roll into a dive but more often the species appears to sink vertically and then disappears beneath the surface before beginning to dive.

DWARF SPERM WHALE

Family Kogiidae

Species *Kogia sima*

Other common names Dwarf sperm whale

Taxonomy Most closely related to the pygmy sperm whale and more distantly related to the sperm whale

Similar species Pygmy sperm whale

Birth weight 31 lb (14 kg)

Adult weight 373–582 lb (169–264 kg) female; 245–670 lb (111–303 kg) male

Diet Cephalopods, fish, and deep-water shrimp

Group size 1–8

Major threats Underwater noise, oceanic debris, harpooning, dynamite fishing, occasionally bycatch in driftnet fisheries

IUCN Status Data Deficient

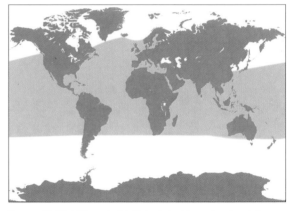

Range and habitat This species inhabits tropical and temperate waters of all major ocean basins. In Hawaii, the dwarf sperm whale is most commonly sighted in 3,330–5,000 ft (1,000–1,500 m) depths. Based on prey remains in South Africa and Taiwan, the dwarf sperm whale is thought to inhabit the continental shelf edge and slope and may be more common in coastal areas of the continental shelf than the pygmy sperm whale.

Identification checklist

- Teeth may be present in the upper jaw
- Dorsal fin height-to-body-length ratio more than five percent
- Fin position is mid-back
- Underslung lower jaw
- Head shape is boxy (squarish)

Anatomy

The dwarf sperm whale is smaller than the pygmy sperm whale but both have a squarish head shape and underslung lower jaw. The dwarf's dorsal fin is larger compared to its body size than the pygmy's and is positioned in the mid-back. Both species have a small number of teeth in the lower jaw but only the dwarf sperm whale may also have teeth in the upper jaw. The coloration pattern resembles a false gill slit behind the eye. Both the dwarf and pygmy have a small spermaceti organ similar to that of the sperm whale.

Behavior

Dwarf sperm whales are elusive at sea. Sightings rarely occur except under ideal sea and weather conditions. The species is known for logging at the surface where they lie motionless. They may dive for long periods when approached by boat. Dwarf sperm whales may release ink from the anus when exhibiting predator avoidance. Group size ranges between one and eight individuals.

Food and foraging

The dwarf sperm whale diet is comprised primarily of cephalopods but they also eat deep-water fish and shrimp. Their anatomy combined with a small number of tiny teeth suggests that suction-feeding is used to ingest prey. Based on the depth of prey identified from the stomach contents of stranded individuals, foraging is believed to take place over the continental shelf and slope. Dwarf sperm whale foraging overlaps with that of the pygmy sperm whale but the dwarf sperm whale may forage in more coastal waters and eat slightly smaller prey.

Life history

Dwarf sperm whales live only to about 20 years of age. They reach sexual maturity between three and five years of age and may then give birth to a single calf as frequently as every year. Breeding appears to be seasonal in the few locations where the species has been studied. Gestation length is believed to be 11–12 months.

Conservation and management

Dwarf sperm whales may be vulnerable to underwater noise. They are also known to ingest oceanic debris that can lead to death. The species may also be threatened by pelagic driftnets, ship strikes, and harpoon or dynamite fisheries. Overall, the species is very poorly known.

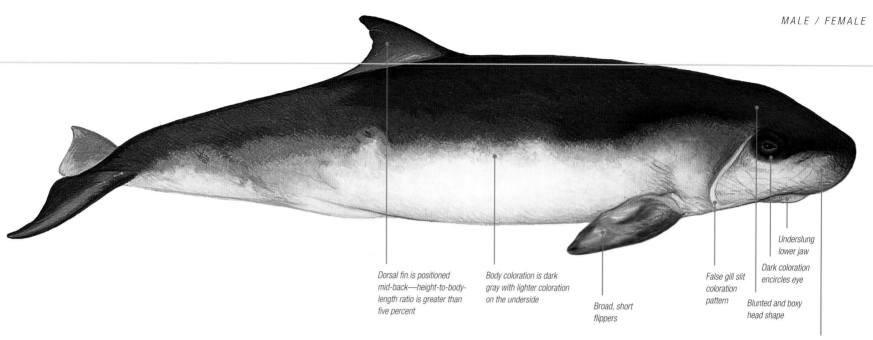

MALE / FEMALE

Dorsal fin is positioned mid-back—height-to-body-length ratio is greater than five percent

Body coloration is dark gray with lighter coloration on the underside

Broad, short flippers

False gill slit coloration pattern

Underslung lower jaw

Dark coloration encircles eye

Blunted and boxy head shape

These are robust whales with a blunted head, small underslung jaw and 7–12 pairs of small, pointed teeth in the lower jaw. A few pairs of teeth may also be present in the upper jaw

Skull shape

Dwarf sperm whales have a characteristic skull that reflects the squarish head shape. The skull is very wide with an extremely short beak. The skull is asymmetrical and the blowhole is on the left side of the head.

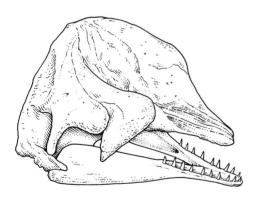

Size
Newborn 3½ ft (1 m)
Adult 7–9 ft (2.1 m–2.7 m) female; 6¼–8½ ft (1.9 m–2.6 m) male

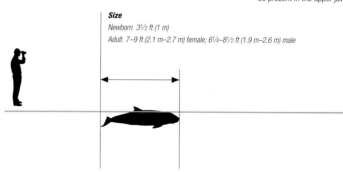

Dive sequence
1. Dwarf sperm whales are known for laying motionless at the surface—behavior known as "logging."

2. Dwarf sperm whales may dive gently below the surface while at other times they appear to sink vertically.

Social animals (right)
Both narwhals and belugas are highly social
animals that seasonally congregate in large groups
during periods of migration, when molting in
estuaries, or when wintering in dense pack ice.
From July to August thousands of belugas are
molting in shallow estuaries in northern Canada.

TOOTHED WHALES

Narwhal & Beluga

Narwhal and beluga are the only two extant members of the family Monodontidae. They are medium-sized toothed whales that occur only in the northern hemisphere. Narwhals are restricted to the Atlantic part of the Arctic and belugas have a wider and almost circumpolar distribution including both Arctic and temperate coastal areas. Both species occur in more or less isolated aggregations. Some of these groups make long-distance migrations in response to sea ice formation and others remain in the same area year round.

The family name Monodontidae, meaning one tooth, refers to the special dentition of narwhals, which have a single canine tooth protruding as a spiraled tusk reaching 8½ ft (2.6 m) from the upper lip on the left side. Belugas have up to 34 teeth—often worn down—that are unsuitable for chewing.

- Both species are born dark brown or gray and change color as they age.
- Both species have dorsal ridges rather than fins that are considered an adaptation for a life in dense sea ice.
- They have exceptionally good hearing and are vocally active using a variety of sounds for communication as well as specialized clicks for echolocation.
- They have thick insulating blubber layers (⅖in; 10 mm) that protect them from the cold waters they inhabit and act as energy stores for periods without feeding activity.
- Monodontids live to considerable age, reaching sexual maturity at a late age, and produce one calf at two- to three-year intervals.
- Both species are gregarious and often travel together in small groups, which may constitute super groups of several hundred animals over larger areas.
- Even though both whales are frequently associated with sea ice they are also known to succumb occasionally in large numbers in ice entrapments.

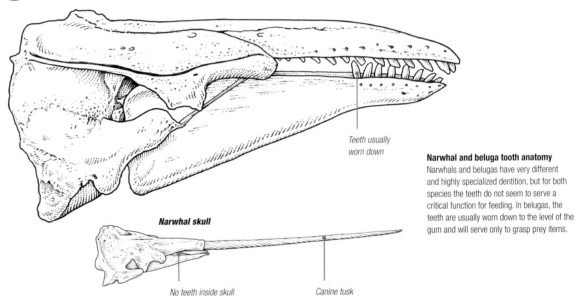

Beluga skull

Teeth usually worn down

Narwhal and beluga tooth anatomy
Narwhals and belugas have very different and highly specialized dentition, but for both species the teeth do not seem to serve a critical function for feeding. In belugas, the teeth are usually worn down to the level of the gum and will serve only to grasp prey items.

Narwhal skull

No teeth inside skull   *Canine tusk*

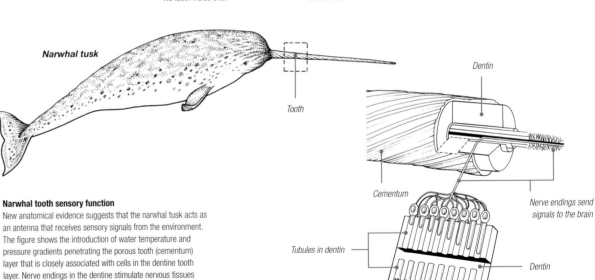

Narwhal tusk

Tooth

Narwhal tooth sensory function
New anatomical evidence suggests that the narwhal tusk acts as an antenna that receives sensory signals from the environment. The figure shows the introduction of water temperature and pressure gradients penetrating the porous tooth (cementum) layer that is closely associated with cells in the dentine tooth layer. Nerve endings in the dentine stimulate nervous tissues connecting the tusk to the brain. This evidence may also play a role in its use in mate selection such as detecting waters where females in estrus may be gathered or foraging.

Dentin

Nerve endings send signals to the brain

Cementum

Tubules in dentin

Dentin

Cementum

Water pressure and temperature gradients penetrate tooth's porous layer

Porous tooth layer (channels through cementum)

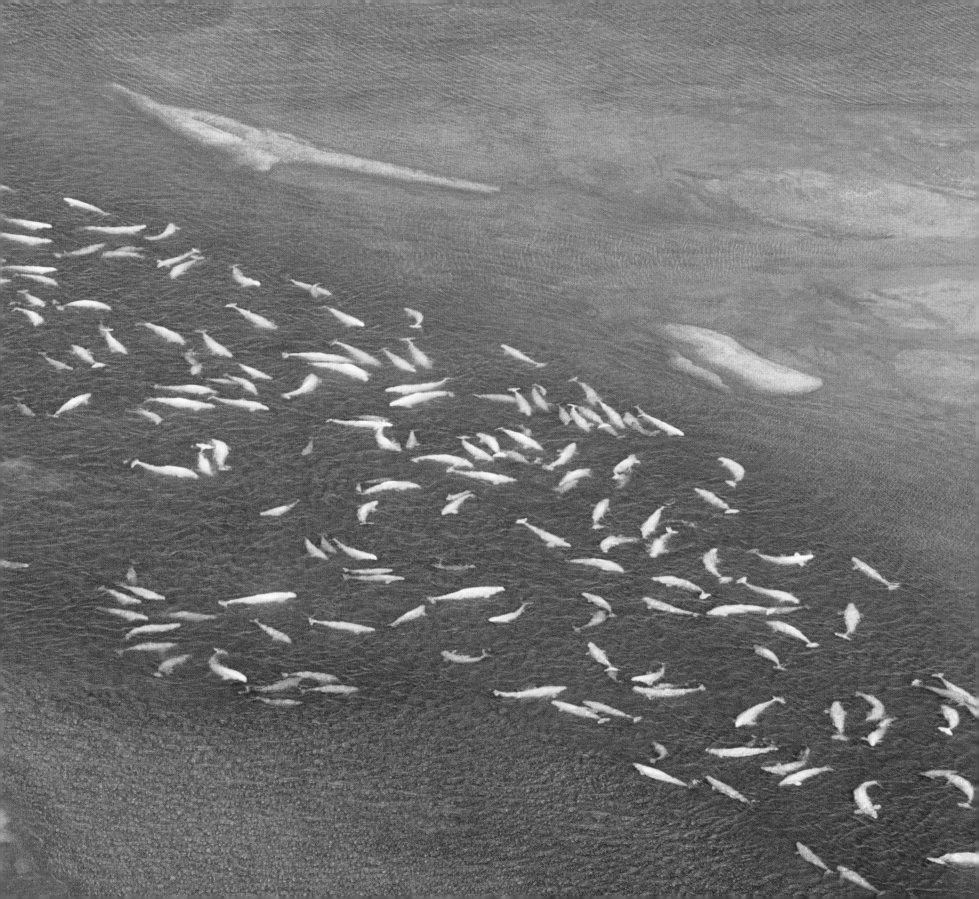

NARWHAL

Family Monodontidae

Species *Monodon monoceros*

Other common names Narwhale

Taxonomy Most closely related to the beluga

Similar species Young animals of less than two years may be confused with beluga calves

Birth weight 330 lb (150 kg)

Adult weight 2,000 lb (900 kg) female; 3,750 lb (1,700 kg) male

Diet Greenland halibut, squid, polar cod, crustaceans

Group size 1–3, up to 10–20 in migrating aggregations

Major threats Hunting, disturbance, fishing, and climate change

IUCN Status Near Threatened

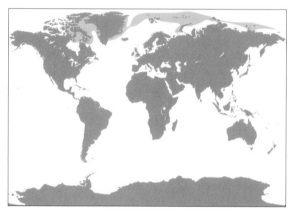

Range and habitat This species occurs in the Atlantic part of the Arctic, north of 60°N. It is found in coastal areas during summer and in deep offshore pack-ice habitats in winter.

Identification checklist

- Gray, brownish, or mottled black body
- White belly
- Males have a spiraled tusk
- No dorsal fin—rather, there is a dorsal ridge
- Short, curved flippers
- Convex margin of each side of the tail fluke

Anatomy

Newborn narwhals are dark brownish-gray and adults are mottled black on the upper side with white fields on the underside. Females have two teeth in the upper jaw that remain inside the skull. In males, the left canine tooth is developed into an elongated tusk that protrudes anteriorly from the rostrum. During growth the tusk spirals to the left. The tusk can attain a size of around 10 ft (3 m) but full-grown males usually carry a tusk of less than 6½ ft (2 m). Some tusks are fairly straight while others are corkscrew-like. Females sometimes develop a tusk, males occasionally have no tusk at all, and males with two tusks—"double tuskers"— are occasionally observed. Recent investigations into the microanatomy of the tusk have revealed open channels from the surface of the tooth to the underlying dentin. Water temperature, pressure, and salinity gradients are thought to illicit a nervous response via the open channels, thereby allowing the tusk to function as an antenna for sensing external environmental stimuli (see page 194).

Behavior

Narwhals are slow compared to other odontocetes (toothed whales) and occasionally rest for long periods at the surface. They dive to some of the greatest depths of any marine mammal at 5,000–6,600 ft (1,500–2,000 m) and up to 25 minutes in duration. Upside-down swimming behaviors have been observed. Narwhals are very skittish and are easily disturbed by boat traffic. Males are occasionally observed gently to cross each other's tusks above the sea surface. The so-called "tusking" behavior may be displays of social dominance, a method for cleaning the tooth, or perhaps serves a sensory role. Narwhals tend to group in pods of 6 to 20 individuals, and groups can be mixed or segregated by gender. On occasion, small groups of narwhals will become entrapped in fjords by ice blocking the whales' escape to the open ocean.

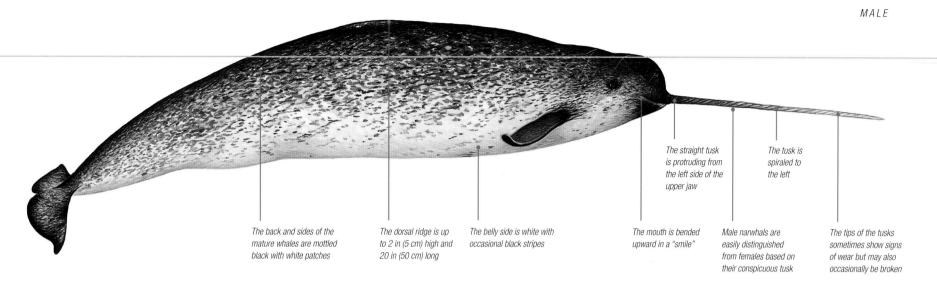

MALE

The back and sides of the
mature whales are mottled
black with white patches

The dorsal ridge is up
to 2 in (5 cm) high and
20 in (50 cm) long

The belly side is white with
occasional black stripes

The straight tusk
is protruding from
the left side of the
upper jaw

The mouth is bended
upward in a "smile"

The tusk is
spiraled to
the left

Male narwhals are
easily distinguished
from females based on
their conspicuous tusk

The tips of the tusks
sometimes show signs
of wear but may also
occasionally be broken

Flukes and flippers
The tail fluke has a notch and the edges of the flukes are
convex. The fluke of males tends to be longer than that
of females, and the median notch is more pronounced.
Upcurled tips of flippers become more prominent with age.

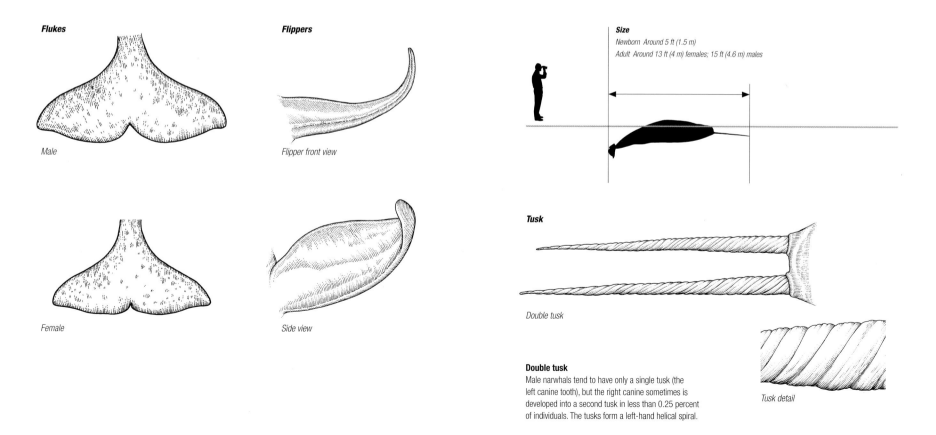

Flukes

Male

Female

Flippers

Flipper front view

Side view

Size
Newborn Around 5 ft (1.5 m)
Adult Around 13 ft (4 m) females; 15 ft (4.6 m) males

Tusk

Double tusk

Tusk detail

Double tusk
Male narwhals tend to have only a single tusk (the
left canine tooth), but the right canine sometimes is
developed into a second tusk in less than 0.25 percent
of individuals. The tusks form a left-hand helical spiral.

NARWHAL

Food and foraging

Most feeding activity takes place in winter when the narwhals are stationary in the pack ice for six to eight months. During this time, the whales may engage in intensive benthic feeding. In contrast, narwhals make few dives below 1,640 ft (500 m) during the summer months. This changes when the whales arrive at the wintering ground where they dive up to 25 times per day to depths exceeding 2,600 ft (800 m) in search for Greenland halibut, squid, crustaceans, and polar cod. Although narwhals participate in deep dives, they tend to forage at all depths.

Life history

Calving likely occurs in May–June with a mating season earlier in spring. Copulation occurs vertically in the water column with the males and females oriented belly-to-belly. Females are generally believed to have one calf every three years, but mothers carrying two fetuses have been reported. Newborn calves are 5–5¾ ft (1.5–1.7 m) long and possess a 10 in (25 cm) layer of blubber. Lactation lasts 1–2 years and the age of sexual maturation is around 8–9 years of age for females and 17 years for males. Maximum longevity is around 100 years.

Conservation and management

Narwhals are considered Near Threatened based on global population estimates, although they have been classified as Data Deficient in the recent past. Humans are the principal predators of narwhals and Inuit hunters in Greenland and Canada hunt the whales for their tusks and skin. However, narwhals have never been targets for commercial whaling. The hunting of narwhals is regulated through bilateral agreements and trade in tusks is limited through international agreements. Narwhals are considered vulnerable to climate change and industrial activities in their habitats.

Exposed tusks

When male narwhals surface during swimming, their canine tusks are exposed above the water's surface. The dark mottled coloration pattern can be observed along the head and back of this individual.

Overhead view
From above, male narwhals are easily identifiable by the pronounced canine tusk and their mottled appearance. The tusk of this individual is nearly half the length as the remainder of the body.

 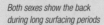

Blow
Blow is straight and about 1¾ ft (50 cm) tall but is hard to see.

Resting at the surface
Male narwhals show their tusk when moving at the surface. Both sexes sometimes lift their head slightly out of the water when surfacing

Both sexes show the back during long surfacing periods

Dive sequence
The back arches slightly when diving
The tail fluke is displayed during deep dives

Tusk display
Males sometimes gently cross or display their tusks above the water surface to communicate social dominance

BELUGA

Family Monodontidae

Species *Delphinapterus leucas*

Other common names White whale

Taxonomy Closely related to the narwhal, there is regional variation in body size but no universally accepted subspecies; may comprise distinct ecotypes

Similar species Morphologically similar to narwhals although lacking the tusks of males; superficially similar to other medium- to small-sized odontocetes that lack a dorsal fin

Birth weight 176–220 lb (80–100 kg)

Adult weight 1,650 lb (750 kg) females; 3,086 lb (1,400 kg) males

Diet Diverse fish (salmonids, arctic cod), cephalopods (squid, octopus), and invertebrates (shrimp and crab)

Group size Variable, from solitary individuals and social groups of up to 20 to herds of 1,000+

Major threats Disturbance, pollution, and habitat loss due to increasing human activities, including oil and gas development, industrial and urban pollution; climate change; small population size

IUCN status Near Threatened: Cook Inlet population Critically Endangered, eastern Hudson Bay and Ungava Bay populations Endangered

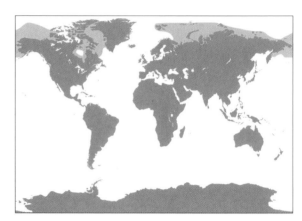

Range and habitat This species inhabits shelf, slope, and abyssal waters of the Arctic Ocean and marginal seas, including the Bering, Chukchi, and Beaufort Seas, Hudson Bay, Davis Strait and the White, Laptev, and Kara Seas. Isolated populations occur in the Okhotsk Sea, Gulf of Alaska, and St. Lawrence Estuary.

Identification checklist

- No dorsal fin
- Calves gray, adults white
- Medium-sized odontocete, up to 17 ft (5.2 m)
- Small head with a prominent melon
- Low and inconspicuous blow

Anatomy

Belugas have no dorsal fin but possess a narrow dorsal ridge. They have a flexible neck and small head with a prominent melon and short beak. The broad fins are spatulate and flukes are wide with a convex, trailing edge. In lieu of a dorsal fin it has been suggested that abdominal fat pads act as vertical stabilizers. Calves are dark gray, becoming white as adults. This species has thick blubber and skin. They grow up to 17 ft (5.2 m) long and are sexually dimorphic with adult males up to 25 percent longer than females.

Behavior

Belugas are equally adept at navigating nearshore open waters as shallow as 5 ft (1.5 m) or plying the dense pack ice of the Arctic Ocean, and diving more than 3,300 ft (1,000 m) below sea ice. Highly gregarious, they often congregate at traditional summering grounds to molt, nurse their young, and feed. In the most northerly populations, belugas migrate up to 2,000 miles (3,000 km) between summering areas and wintering grounds along the southern edge of the pack ice or in areas of predictable open water, termed polynyas, north of the ice edge. Genetic research has revealed that belugas tend to return to the summering ground

of their birth and that this philopatric behavior may extend across generations, such that many discrete summering groups represent demographically distinct sub-populations. Belugas possess one of the most diverse vocal repertoires among cetaceans, and have long been called the "canaries of the sea" by mariners enchanted by their calls. They exhibit a wide variety of behaviors including spyhopping and tail-slapping, and are capable of complex social interactions. Adult males often segregate from females and younger animals forming all-male pods at coastal sites and making more extensive at-sea movements through deep, ice-covered waters.

Food and foraging

Belugas feed on a wide variety of bottom-dwelling and pelagic fish, including Arctic cod, salmon, and capelin. They also consume a diverse array of invertebrates including squid, clams, shrimp, and crab. The foraging behavior of belugas can comprise small-scale daily movements, often in synchrony with the tides, in search of prey fish migrating up rivers in spring and summer; focused feeding bouts at the face of tidewater glaciers; long seasonal migrations; extensive movements across estuarine, shelf, slope, and abyssal waters; and diverse diving behaviors.

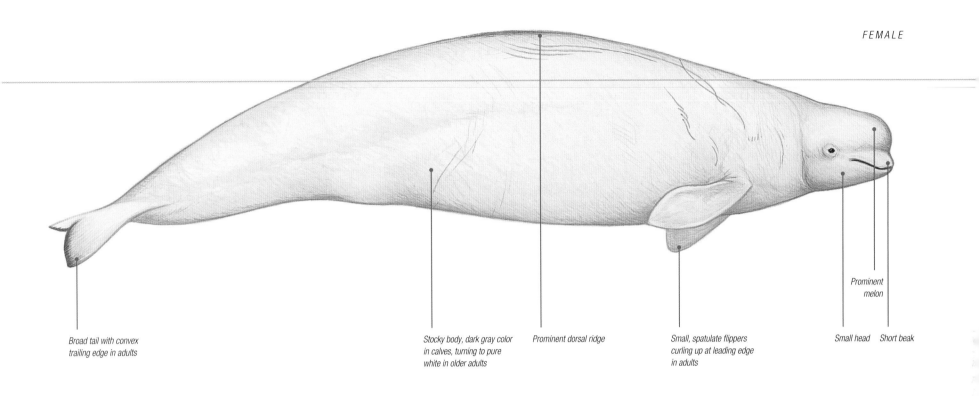

FEMALE

Broad tail with convex
trailing edge in adults

Stocky body, dark gray color
in calves, turning to pure
white in older adults

Prominent dorsal ridge

Small, spatulate flippers
curling up at leading edge
in adults

Small head   Short beak

Prominent
melon

Trailing edge of fluke

The change in the shape (as well as size) of the trailing
edge of the fluke with age is quite distinctive for this
species. The trailing edges of the flukes are straighter
than in older animals. A similar phenomenon occurs
with the pectoral flippers, whereby the tips can curl
progressively upward in older individuals.

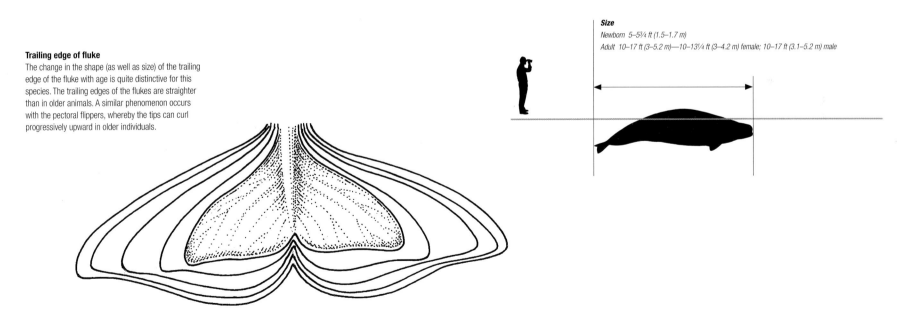

Size
Newborn 5–5¾ ft (1.5–1.7 m)
Adult 10–17 ft (3–5.2 m)—10–13¼ ft (3–4.2 m) female; 10–17 ft (3.1–5.2 m) male

BELUGA

Life history

Belugas reach sexual maturity at between 8 and 12 years of age and can live to more than 80 years. Adult females give birth to a single calf in late spring–early summer and nurse for up to two years. In migratory populations, calves are born during spring migration or on arrival at summering grounds and there is evidence in some populations of reproductive aging in older females. The larger size and higher incidence of interspecific scarring of adult males suggest that belugas have a polygamous mating system where males compete for females. Belugas have few known predators. These include polar bears, killer whales, and man. Despite their adaptations to life in the Arctic and subarctic belugas are prone to getting trapped in sea ice. Often termed *sassats*, a Greenlandic word, several thousand whales have been known to perish entrapped in rapidly forming sea ice or when leads in the ice suddenly close.

Conservation and management

Belugas have been an important subsistence and cultural resource for many northern peoples for thousands of years and in many regions are effectively co-managed by Native communities and government counterparts. A number of small, geographically isolated populations face an uncertain future exacerbated by their close proximity to industrial activities and urban areas. Some populations have been listed as Endangered or Critically Endangered where a failure to recover despite increased protection in recent decades heightens concerns about minimum viable population sizes and risk factors as yet undetermined. All populations face an increasingly uncertain future due to climate-related ecosystem change and increases in human activities.

Dive shapes
Beneath the water, deeper dives have a number of distinct profiles including "square-," "V-," and parabolic-shaped dives.

Parabolic

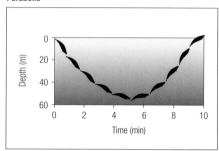

Square

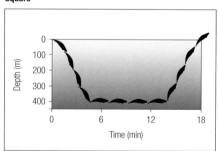

"V"-shaped

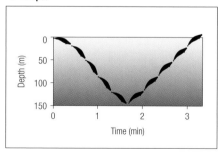

Beluga features
An aerial view of a pod of adult beluga in the Canadian Arctic. Note the pure white color of the adults, the absence of a dorsal fin, and the flexible neck allowing an almost sinuous movement of the body.

Beluga herd

A large herd of beluga whales swim in the Canadian Arctic. In summer, migrating herds of belugas— sometimes numbering several thousand—converge on summering grounds. They congregate in shallow coastal waters to molt, feed, and raise their young.

Shallow dive sequence
1. The head barely breaks the surface to blow.

2. The back of the animal is then exposed.

3. This is followed by the arching of the back.

4. The dive ends with the tail remaining underwater. Belugas typically swim in a slow rolling pattern.

Surface behavior
Deeper dives are often preceded by a series of surfacings and end with the flukes breaking the surface, often rising vertical and clear of the water.

TOOTHED WHALES
Beaked Whales

This family of toothed whales is among the world's least known groups of large mammals. At time of writing, 22 species in six genera are recognized but a few additional species are likely to be discovered and described eventually. Although the name "beaked whale" is properly applied to all of the species in the family, the largest members—in the genera *Berardius* and *Hyperoodon*—are more often referred to as "bottlenose whales." Until recently, everything known about some species came from examinations of stranded animals.

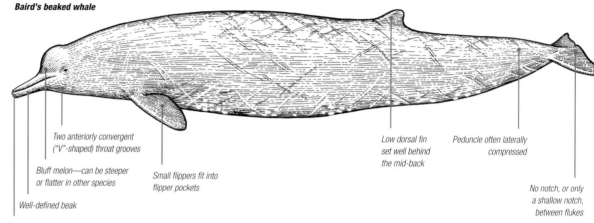

Cuvier's beaked whale (right)

A Cuvier's beaked whale—probably an adult male judging by the white face—breaching explosively high out of the water in the Mediterranean Sea, clearly shows the small dorsal fin and short beak, reminiscent of a goose's (an alternative name of the species is goose-beaked whale). Mass strandings of these whales in recent years have been linked to military sonar exercises—an emergent threat to beaked whales generally.

Baird's beaked whale

Two anteriorly convergent ("V"-shaped) throat grooves

Bluff melon—can be steeper or flatter in other species

Well-defined beak

Erupted teeth at tip of lower jaw—position varies among species

Small flippers fit into flipper pockets

Low dorsal fin set well behind the mid-back

Peduncle often laterally compressed

No notch, or only a shallow notch, between flukes

- Beaked whales, ranging in size from less than 13 ft (4 m) to nearly 43 ft (13 m) long—females tend to be larger than males—occur in all oceans of the world, some as anti-tropical species pairs (northern and southern bottlenose whales, Baird's and Arnoux's beaked whales), some as regional endemics limited to portions of a single basin (for example, Sowerby's beaked whale in the North Atlantic and Stejneger's beaked whale in the northern North Pacific), others as cosmopolitan or pantropical species with relatively little known regional variation (such as Cuvier's beaked whale and Blainville's beaked whale).

- All species have a beak but its length and degree of prominence vary—some species have a steeply rising melon and others a quite flat one. The dorsal fin is always set well behind the middle of the back. All beaked whales have a pair of throat grooves converging toward the front and, at most, a shallow notch in the rear margins of the flukes. Their relatively short flippers fit into slight depressions in the body wall—so-called flipper pockets.

- All but one of the beaked whales (Shepherd's beaked whale) have reduced dentition, with only one and sometimes two pairs of functional teeth in the lower jaw but none in the upper. In most species the teeth erupt only in males as they approach maturity. The teeth of some species become tusk-like—meaning they protrude outside the closed mouth—and may be used as weapons in aggressive male–male interactions. The two teeth of male members of the diverse genus *Mesoplodon* erupt along the lower jaw either just behind the midpoint, at the tip, or somewhere in-between, often causing a strong arch in the mouthline.

- All beaked whales inhabit fairly deep water—usually more than 1,000 ft (300 m)—so their distribution tends to be well away from continental coastlines.

- As adept divers, beaked whales feed primarily on deep-water squid and fish.

- Beaked whales are unusual among the cetaceans in that most species and populations have never been subject to deliberate, large-scale hunting. The only two exceptions are northern bottlenose whales in the northern North Atlantic and Baird's beaked whales off the coast of Japan.

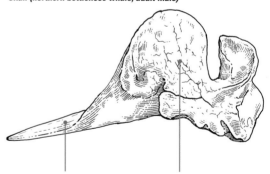

Skull (northern bottlenose whale, adult male)

Teeth are completely absent in the upper jaws of most ziphiids, but one or two pairs are usually present in the mandibles (not shown here)

Prominent maxillary crests help distinguish the northern bottlenose whale from other ziphiids

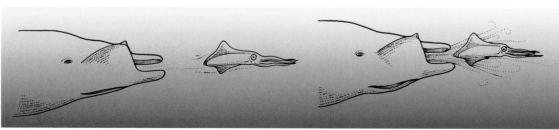

Suction-feeding
Like other toothed cetaceans, they use echolocation for locating and homing in on prey. To secure and ingest prey, however, beaked whales rely on suction. The gular (throat) grooves allow the throat to expand as the tongue is retracted in a piston-like manner, creating an abrupt loss of pressure, or vacuum, inside the mouth and drawing in prey. This would explain why many squid found in beaked whale stomachs are fully intact with no sign of being bitten.

ARNOUX'S BEAKED WHALE

Family Ziphiidae

Species *Berardius arnuxii*

Other common names Giant bottlenose(d) whale, giant beaked whale

Taxonomy No subspecies recognized. Closest relative is Baird's beaked whale, which is similar in appearance. Could be confused with southern bottlenose whales, which share most of this species' range.

Birth weight Unknown

Adult weight Unknown but probably similar to Baird's beaked whale

Diet Probably mainly deep-water fish and squid

Group size Generally in small tight groups of up to about 15, occasionally several dozens

Major threats No major threats known

IUCN status Data Deficient

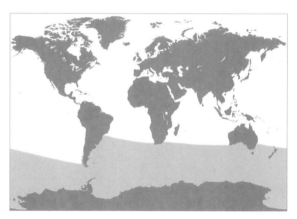

Range and habitat This species is circumpolar in the Southern Ocean all the way south to the Antarctic continent, and is often seen close to the ice edge or in broken sea ice. It is mostly seen south of 40°S, occasionally to as far north as 34°S.

Identification checklist

- Long, relatively slender body
- Small head and long rostrum; front surface of the melon is almost vertical
- Two large triangular teeth exposed at tip of lower jaw (at sexual maturity), which extends slightly farther forward than upper jaw; teeth heavily abraded in old animals and may be colonized by stalked barnacles
- Coloration is slate gray to dark brown, head somewhat lighter; numerous linear scars (tooth rakes) on the melon, back, and sides of older individuals.
- Small, triangular or slightly falcate dorsal fin (rounded at tip) set far behind the mid-back

Anatomy

Virtually identical to Baird's beaked whale although possibly smaller. Females grow slightly larger than males. Most features are typical of beaked whales: "V"-shaped throat grooves, small flippers that fit into depressions on the body, and no notch in the rear edge of the flukes. The long beak, bulbous melon, and small triangular dorsal fin set well behind the mid-back are characteristic. The front pair of large, flat, triangular teeth at the tip of the lower jaw erupt in both adult males and females and at close range can be visible when the mouth is closed. The long, slender body is extensively scarred by tooth rakes.

Behavior

This species is usually seen in schools of up to about 15 individuals that surface together and produce low, bushy blows. Larger schools of 40–80 are sometimes seen but often split into subgroups, possibly for foraging. They can be active at the surface, slapping the water with their tails and occasionally breaching. Acoustic recordings have shown these whales to be extremely vocal. Their sounds include clicks, click trains (a series of clicks), and whistles.

Food and foraging

This species' diet may be similar to that of Baird's beaked whales in the North Pacific, consisting mainly of deep-sea bottom-dwelling or pelagic fishes and squids. They are capable of prolonged dives lasting well over an hour. Although often described as oceanic, Arnoux's beaked whales have been seen in coastal Antarctic waters less than 3,300 ft (1,000 m) deep.

Life history

The life history of this species is probably similar to Baird's beaked whale, which has been comparatively well studied. Therefore, gestation may last for nearly a year and a half, and females may give birth once every two years.

Conservation and management

There are no good estimates of abundance for this species. It has never been hunted to a significant extent, nor is it threatened by human activities as far as is known. Although considered less common than southern bottlenose whales, Arnoux's beaked whales are observed fairly frequently in Cook Strait during the summer as well as south of New Zealand and South America.

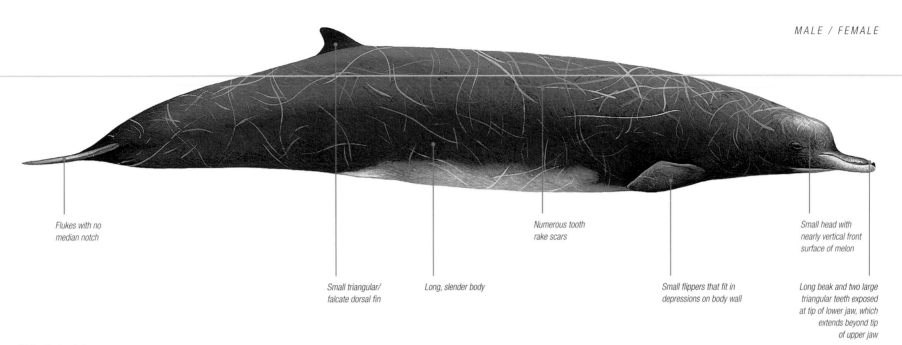

MALE / FEMALE

Flukes with no
median notch

Small triangular/
falcate dorsal fin

Numerous tooth
rake scars

Long, slender body

Small flippers that fit in
depressions on body wall

Small head with
nearly vertical front
surface of melon

Long beak and two large
triangular teeth exposed
at tip of lower jaw, which
extends beyond tip
of upper jaw

Distinctive head shape
The small head of Arnoux's beaked whale is fairly
distinctive with the steep front surface of the melon, the
long beak, and the two exposed teeth at the tip of the
lower jaw, which protrudes beyond that of the upper jaw.

Size
Newborn Unknown but possibly 13 ft (4 m) or up to 14¾ ft (4.5 m) as for Baird's beaked whale
Adult Possibly around 33ft (10 m)

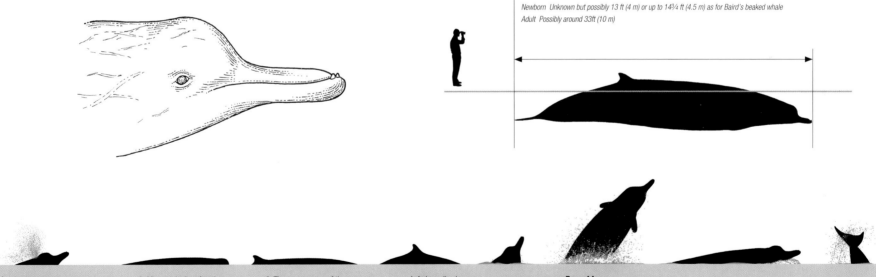

Dive sequence
1. Not always, but occasionally, one
animal in a group of Arnoux's beaked
whales tilts its head at the surface
after the blow, revealing the long
beak and bluff melon, then simply
sinks and stays near the surface.

2. More typically, after the
blow the long dark back and
perhaps the bulging melon
becomes visible.

3. The appearance of the
small, posteriorly positioned
dorsal fin follows quickly.

4. A deep dive is
signaled by a strong
arching of the back.

Breaching
Arnoux's beaked whales do not
often breach, but when they do
it provides a dramatic view of the
long, slender body, small flippers,
and long beak.

BAIRD'S BEAKED WHALE

Family Ziphiidae

Species *Berardius bairdii*

Other common names Giant bottlenose(d) whale, giant beaked whale

Taxonomy No subspecies are recognized—populations in the eastern and western North Pacific are generally assumed to be separate, and there may be multiple populations in the west. The closest relative is the southern hemisphere Arnoux's beaked whale.

Similar species Arnoux's beaked whale is very similar in appearance. Other beaked whales are unlikely to be confused with Baird's beaked whales, given their large size, prominent rounded melon, and long beak with exposed teeth at the tip. At a distance, confusion with sperm whales is a possibility.

Birth weight Unknown

Adult weight 17,650–24,250 lb (8,000–11,000 kg)

Diet Mainly deep-water fish and squid

Group size Generally from 3 to about 10 but can be up to 50

Major threats No major threats but a regular hunt takes place in Japan with an annual catch limit of around 60

IUCN status Data Deficient

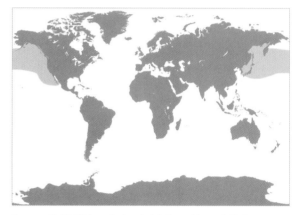

Range and habitat This species is endemic to the cool-temperate North Pacific mainly north of 30°N, particularly along continental slopes and over submarine escarpments and seamounts, and at preferred depths of 3,300–10,000 ft (1,000–3,000 m). They are migratory but are seen occasionally in sea ice in the Sea of Okhotsk in winter and spring.

Identification checklist

- Body long and slender
- Small head and long beak; the melon has a front surface that is almost vertical
- Two teeth exposed at tip of lower jaw (at sexual maturity), which extends slightly farther forward than the upper jaw; teeth heavily abraded in old animals and often colonized by stalked barnacles
- Coloration evenly dark brown to black, paler at the sides with irregular white areas on the belly; numerous linear scars (tooth rakes) and cookie-cutter shark bite scars
- Small dorsal fin, either falcate or triangular, rounded at the tip and set far behind the mid-back

Anatomy

The largest of the beaked whales. Females are slightly larger than males. Most features are typical of beaked whales—"V"-shaped throat grooves, small flippers that fit into depressions on the body, no notch in the rear margin of the flukes, and a small triangular dorsal fin far behind mid-back. The front pair of large, flat, teeth at the tip of the lower jaw erupt in both adult males and females and are visible when the mouth is closed. The long, slender body is scarred by tooth rakes and cookie-cutter shark bites.

Behavior

At sea, Baird's beaked whales are usually seen in tight schools of up to about 10 individuals that surface at the same time and produce low, bushy blows. They occasionally breach, slap the surface with the tail, or spyhop.

Food and foraging

The diet consists mainly of benthic fish. These whales make deep, prolonged dives—some to depths of greater than 4,900 ft (1,500 m) and lasting 45 minutes to longer than an hour.

Life history

It has been inferred that gestation lasts for about 17 months. Females probably begin ovulating at 10–15 years of age and thereafter ovulate once every 2 years. Males apparently mature a little earlier—at 6–11 years. The whales become physically mature by 15 years of age. Strangely, males seem to experience a lower rate of natural mortality than females and they live longer—to about 84 years of age compared with about 54 years for females.

Conservation and management

Around 300 Baird's beaked whales were killed in Japan annually after World War II but a catch limit of around 60 per year is now in force. Small numbers (less than 100 total) were killed by shore-based whalers off North America—from California to Alaska—between 1915 and 1966 but the species has been fully protected in the eastern Pacific for the last half-century. Abundance estimates in parts of the range are about 7,000 whales in Japan and 1,000 off western North America. These whales are occasionally caught accidentally in fishing gear (especially offshore drift gill nets) and struck by ships, and like other beaked whales they likely are affected by anthropogenic noise.

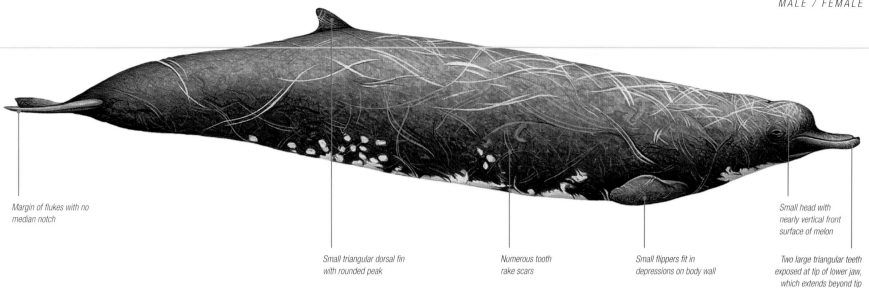

MALE / FEMALE

Margin of flukes with no
median notch

Small triangular dorsal fin
with rounded peak

Numerous tooth
rake scars

Small flippers fit in
depressions on body wall

Small head with
nearly vertical front
surface of melon

Two large triangular teeth
exposed at tip of lower jaw,
which extends beyond tip
of upper jaw

Adult head appearance

The small head of Baird's beaked whale is fairly
distinctive with the steep front surface of the
melon, the long beak, and the two conspicuous
exposed teeth at the tip of the lower jaw, which
protrudes beyond that of the upper jaw.

Size

Newborn 14¾–15 ft (4.5-4.6 m)

Adult 42 ft (12.8 m) female; 39½ ft (12 m) male

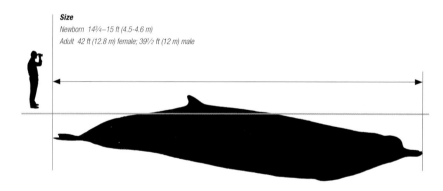

Dive sequence

1. The initial surfacing after a long
dive may be fairly steep with the
beak and head appearing first at the
same time as an explosive blow.

2. However, these large whales usually
convey a long-bodied shape, with the small
domed head leading the way, followed by the
very long dark back (usually covered with
linear scars and cookie-cutter shark bites).

3. Finally, well back on the body,
can be seen the small triangular
dorsal fin that is more often round
than pointed at the tip.

Fluking

*They sometimes raise their flukes,
with the characteristically
unnotched rear margin, and slap
the surface with their tail (lobtail).*

Spyhopping

*And they also occasionally
spyhop—raising the head
vertically such that the long
beak can be seen clearly.*

NORTHERN BOTTLENOSE WHALE

Family Ziphiidae

Species *Hyperoodon ampullatus*

Other common names North Atlantic bottlenosed whale, flathead, bottlehead, steephead

Taxonomy One well-characterized distinct population is recognized in "The Gully," a deep submarine canyon off Nova Scotia, Canada. Other similarly distinct populations may exist in other parts of the North Atlantic.

Similar species Cuvier's and Sowerby's beaked whales share similar temperate portions of range but are much smaller with less bulbous foreheads

Birth weight Unknown

Adult weight 12,800–16,500 lb (5,800–7,500 kg)

Diet Dominated by one species of squid, *Gonatus fabricii*; also fish such as herring and redfish as well as deep-sea prawns and echinoderms

Group size Averages about 4, seldom larger than about 10

Major threats Intensively hunted in the past, now protected; incidental mortality in longline fisheries; disturbance from underwater noise, especially from seismic surveys and naval sonar

IUCN status Data Deficient

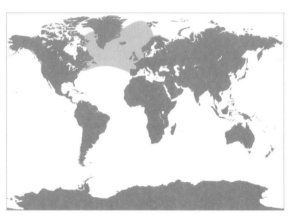

Range and habitat This species is endemic to cool temperate and sub-Arctic North Atlantic mainly north of 40°N, particularly in canyons, along continental slopes, and in waters more than 3,330 ft (1,000 m) deep. It occurs near ice edge and in broken pack ice. It is apparently resident year round in some areas but there is also evidence of long-distance migratory movements.

Identification checklist

- Body long but robust
- Thick, short, well-defined beak
- Prominent bluff melon, especially steep and flattened (squared off) in front in adult males
- Two teeth erupt at tip of lower jaw of adult males, oriented slightly forward, occasionally with stalked barnacles attached— rarely visible on live individuals
- Coloration gray or chocolate-brown on top and lighter underneath (countershaded)
- Entire beak and front part of the head of adult males are white
- Falcate dorsal fin positioned well behind the mid-back

Anatomy

The most striking feature of this species' anatomy is the adult male's head, which is characterized by a bulbous forehead that becomes flat and squared-off in front. This is a result of the formation of large maxillary crests on the upper surface of the skull. These may have an acoustic function and the dense bone underlying the forehead could be useful in head-butting contests between males. Unlike most other ziphiid species, males grow larger than females.

Behavior

Groups of 4–10 are typical. These can be all-male groups, groups consisting of adult females and juveniles, or mixed groups. Surfacing behavior is variable, with the animals either remaining still at the surface or swimming rapidly in different directions. Whalers remarked on the tendency of bottlenose whales to approach stationary or slow-moving vessels, possibly out of curiosity, making them vulnerable to being shot. These whales were also known for their reluctance to leave a wounded companion until it had died—another characteristic used by whalers to their advantage.

Food and foraging

Their diet is dominated by a single squid species, *Gonatus fabricii*, which occurs across the northern bottlenose whale's range. These whales are deep divers, routinely diving to more than 4,600 ft (1,400 m). Dives lasting longer than an hour are not unusual. It appears that some whales remain within a limited region year round whereas others migrate seasonally.

Life history

Females reach sexual maturity at 11 years of age, males somewhat earlier. The gestation period is at least a year and the interval between births is two or more years. The lifespan is at least 37 years.

Conservation and management

More than 65,000 northern bottlenose whales were killed by whalers between the 1850s and 1970s. The species has been protected for the last 40 years. Fairly high densities of these whales have been observed recently around Iceland and the Faeroes, so some recovery is likely. The greatest threats are now incidental mortality in fisheries and disturbance from noise associated with offshore oil and gas development and military activities.

MALE

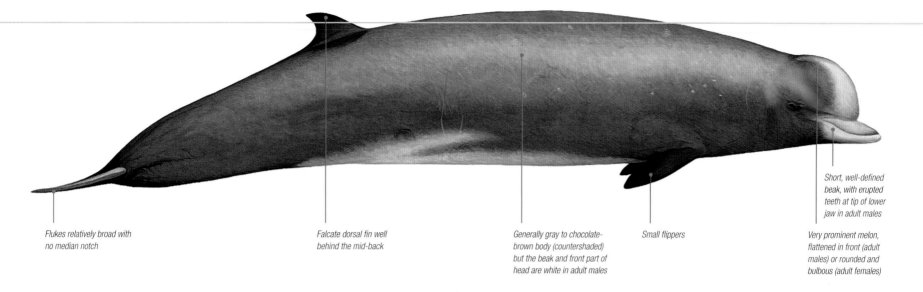

Flukes relatively broad with
no median notch

Falcate dorsal fin well
behind the mid-back

Generally gray to chocolate-
brown body (countershaded)
but the beak and front part of
head are white in adult males

Small flippers

Short, well-defined
beak, with erupted
teeth at tip of lower
jaw in adult males

Very prominent melon,
flattened in front (adult
males) or rounded and
bulbous (adult females)

Adult male head features
The head of the adult male is a striking feature, with
its white and bluff melon. The two small erupted teeth
at the tip of the lower jaw are sometimes colonized by
stalked barnacles. These teeth are rarely visible on live
animals observed at sea.

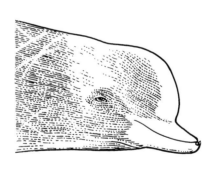

Size
Newborn 10–11½ ft (3–3.5 m)
Adult 28½ ft (8.7 m) female; 32 ft (9.8 m) male

Dive sequence
*1. The bushy blow can often be seen
briefly along with the prominent, rounded,
lightly pigmented melon as a bottlenose
whale breaks the sea surface. In good
sea conditions and with luck, the beak
may also be visible.*

*2. Typically, as the
blow dissipates, the
long back and falcate
dorsal fin appear.*

*3. Only the dorsal fin
and peduncle are visible
as the animal dives.*

Fluking and lobtailing
*Bottlenose whales show their
flukes only occasionally at the
beginning of a deep dive but they
lobtail fairly often—slapping the
sea surface with their flukes.*

Breaching
*Bottlenose whales breach occasionally.
The social inclinations of bottlenose
whales mean that small groups are often
seen in close company, interacting at the
surface for substantial periods as they
recover from a long dive.*

SOUTHERN BOTTLENOSE WHALE

Family Ziphiidae

Species *Hyperoodon planifrons*

Other common names Antarctic bottlenosed whale, flathead

Taxonomy There are no subspecies but studies suggest there is population structure, possibly at the subspecies or even species level

Similar species May be confused with Arnoux's beaked whale, which has a similar range but Arnoux's beaked whale is larger with a longer beak, less bulbous melon, and smaller dorsal fin

Birth weight Unknown

Adult weight Maximum 8,800 lb (4,000 kg)

Diet Squid

Group size Small, usually no more than about 10 individuals

Major threats Never subject to significant exploitation; incidental mortality in longline fisheries; disturbance from underwater noise, especially from seismic surveys and naval sonar

IUCN status Least Concern

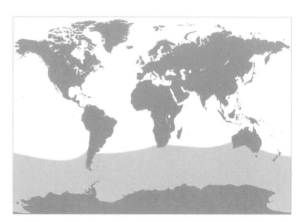

Range and habitat This species is widely distributed in the southern hemisphere between the Antarctic Convergence and the pack-ice edge but also to as far north as approximately 30°S along the coasts of Australia, Africa, and South America.

Identification checklist

- Body long and robust
- Thick, well-defined beak
- Prominent bluff melon, more so in adult males
- Two teeth erupt at tip of lower jaw of adult males, oriented slightly forward
- Coloration is dark dorsally and lighter ventrally (countershaded)
- Beak and head lighter than rest of body especially in adult males in which it may appear almost white at sea
- Falcate dorsal fin positioned well behind the mid-back

Anatomy

Although the southern bottlenose whale has maxillary crests much like those of the northern species, they are generally flatter and less well-developed. Therefore the adult male's head is less dramatically squared-off in front than is true of the northern species. Southern bottlenose whales also seem to grow to a smaller size than their northern relatives. Otherwise the two species are anatomically very similar.

Behavior

Much of what is believed about the southern bottlenose whale's behavior and life history is inferred by analogy with its northern counterpart, which has been studied in detail. Groups tend to be small, usually fewer than 10 individuals. The bushy blow is usually visible and these whales breach occasionally.

Food and foraging

Southern bottlenose whales are presumably squid specialists although there is little direct information on diet or foraging behavior. Sea quirts and skeletal material from Patagonian toothfish have been found in the stomachs of bottlenose whales stranded in

Australia. Southern bottlenose whales are likely deep divers although no actual data on dive times or other aspects of their diving are available from direct observations.

Life history

By analogy with the much better-studied northern bottlenose whale, females probably reach sexual maturity at around 11 years of age, males possibly somewhat earlier. The gestation period is at least a year and the interval between births is two or more years. Judging by known longevity of their northern counterparts, these whales can live for more than 35 years.

Conservation and management

These whales have never been exploited on a significant scale anywhere in their range. A crude estimate of numbers based on ship surveys of the Southern Ocean in the 1970s–80s was around half a million. No specific information on threats is available but mortality in pelagic driftnets has been documented and there is general concern that, like other beaked whales, southern bottlenose whales could be affected by disturbance from the noise associated with offshore oil and gas development and military activities.

FEMALE

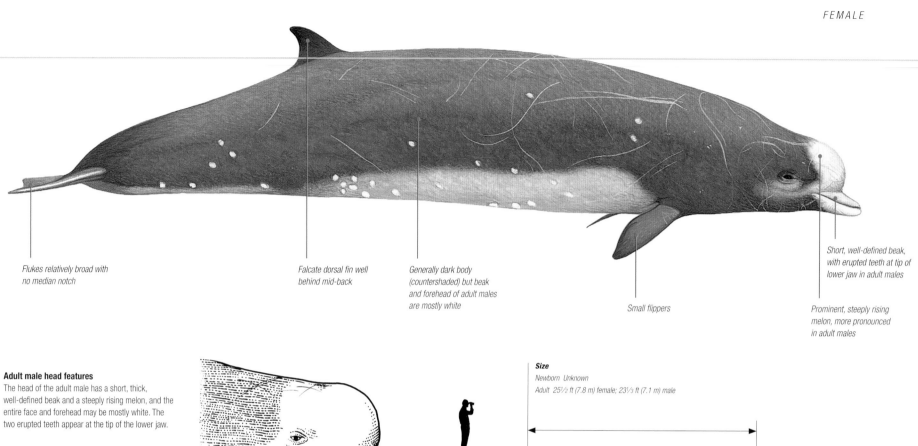

Flukes relatively broad with
no median notch

Falcate dorsal fin well
behind mid-back

Generally dark body
(countershaded) but beak
and forehead of adult males
are mostly white

Small flippers

Short, well-defined beak,
with erupted teeth at tip of
lower jaw in adult males

Prominent, steeply rising
melon, more pronounced
in adult males

Adult male head features
The head of the adult male has a short, thick,
well-defined beak and a steeply rising melon, and the
entire face and forehead may be mostly white. The
two erupted teeth appear at the tip of the lower jaw.

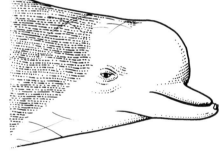

Size
Newborn Unknown
Adult 25½ ft (7.8 m) female; 23⅓ ft (7.1 m) male

Dive sequence
1. The bushy blow can often be seen briefly
along with the prominent, rounded, lightly
pigmented melon as a bottlenose whale
breaks the sea surface. In good sea
conditions, and with luck, the beak may
also be visible.

2. Typically, after the blow and most
of the head disappear, the long back
and falcate dorsal fin appear.

3. Only the dorsal fin and
peduncle are visible and the
animal disappears into the depths.

Fluking
Bottlenose whales show their
flukes only occasionally at the
beginning of a deep dive.

Breaching
They breach occasionally, sometimes almost vertically
and at other times simply by porpoising clear of the
sea surface. The social organization of bottlenose
whales means that small groups are often seen in
close company, interacting at the surface for
substantial periods as they recover from a long dive.

LONGMAN'S BEAKED WHALE

Family Ziphiidae

Species *Indopacetus pacificus*

Other common names Indo-Pacific beaked whale

Taxonomy No subspecies or separate populations are recognized

Similar species Confused for a time in the past with the southern bottlenose whale—referred to as the "tropical bottlenose whale"—now recognized as belonging to a separate genus

Birth weight Unknown

Adult weight Unknown

Diet Squid

Group size Larger than most other beaked whales in the Indo-Pacific, average of 7–30 but can be considerably more (up to 100)

Major threats No major threats are known although driftnet entanglement, noise (naval sonar), and ingestion of plastic debris are concerns

IUCN status Data Deficient

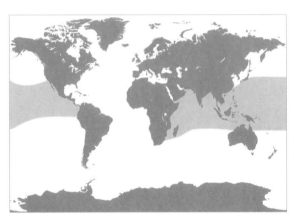

Range and habitat This species is endemic to tropical Indo-Pacific from the west coast of Mexico to the east coast of Africa and the Gulf of Aden. It is primarily found in oceanic waters deeper than 6,600 ft (2,000 m) with sea-surface temperatures of at least 79°F (26°C).

Identification checklist

- Long beak and prominent rounded melon
- Low, bushy blow, generally visible and angled slightly forward
- Large, triangular, and somewhat falcate dorsal fin, set two-thirds of the way back on the body
- Two teeth at tip of lower jaw, which erupt only in adult males
- Coloration evenly brown or gray but the head region is often pale tan
- Body often bears numerous white, oval cookie-cutter shark bite scars and adult males also have linear scars from tooth raking

Anatomy

The single pair of teeth at the tip of the lower jaw, which erupt only in adult males, are unlikely to be visible except when the animals are observed at close range in ideal conditions. The bluff, light-colored melon, fairly long beak, relatively large dorsal fin and intrusions of white or light gray from the ventral region onto the sides of the front part of the body can be useful field marks.

Behavior

These whales are gregarious and usually seen in close-knit groups averaging 7–30 individuals, including adults of both sexes as well as calves. Groups of up to 100 have been reported. When traveling fast at the surface, these whales often bring the head high out of the water and even sometimes porpoise slightly above the surface, much like large dolphins. They also breach occasionally. Sometimes Longman's beaked whales have been seen in association with either pilot whales or bottlenose dolphins, or both.

Food and Foraging

Almost nothing is known about diet or foraging behavior, or about movements of individuals. The two stomachs that have been examined contained the remains of squid. These whales are thought to be deep divers, with measured dive times of longer than a half hour.

Life history

Virtually nothing is known about the life history of this species.

Conservation and management

Judging by the infrequency of reported observations, Longman's beaked whales are fairly rare. Rough estimates of numbers are available for waters around Hawaii—about a thousand—and the eastern tropical Pacific, about 300. No specific threats are known. However, driftnet and longline fisheries exist in large parts of their range and entanglements undoubtedly occur. Like other beaked whales, Longman's beaked whales are probably vulnerable to disturbance from underwater noise, such as naval sonar and seismic surveys.

FEMALE

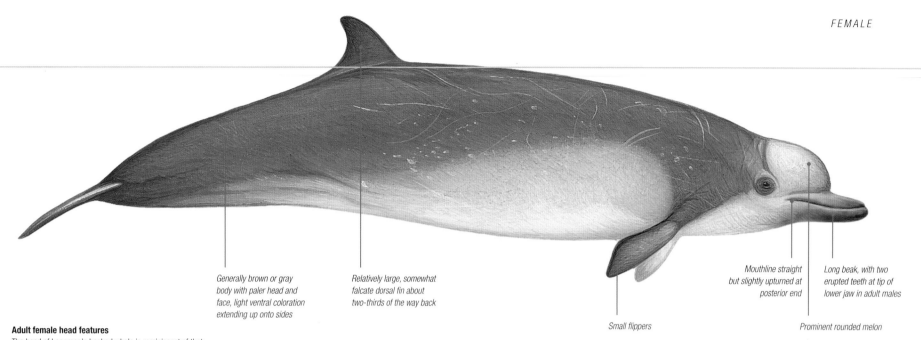

Generally brown or gray body with paler head and face, light ventral coloration extending up onto sides

Relatively large, somewhat falcate dorsal fin about two-thirds of the way back

Mouthline straight but slightly upturned at posterior end

Long beak, with two erupted teeth at tip of lower jaw in adult males

Small flippers

Prominent rounded melon

Adult female head features

The head of Longman's beaked whale is reminiscent of that of bottlenose whales, with a prominent melon and a fairly long beak—this image shows an adult female (no good photograph of an adult male was available, and the features of adult females may vary as indicated by the differences in the two images provided here). The single pair of teeth, at the tip of the lower jaw, erupt only in adult males and project somewhat forward. They are not particularly conspicuous.

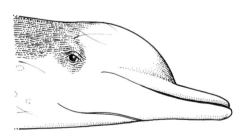

Size
Newborn 9½ ft (2.9 m)
Adult 18½–21½ ft (5.6–6.5 m)

Dive sequence
1. Longman's beaked whales are large-bodied and tend to occur in relatively large groups. They often approach the surface energetically, with the beak and melon becoming visible along with the blow.

2. The contrast between the light top of the head and the rest of the gray or brown dorsal surface can be evident as the blow dissipates.

3. Soon only the long back and prominent dorsal fin are visible.

4. If the animal is beginning a deep dive, the back may then arch.

5. Lastly, only the dorsal fin and peduncle are visible and the animal disappears into the depths for 45 minutes or longer.

SOWERBY'S BEAKED WHALE

Family Ziphiidae

Species *Mesoplodon bidens*

Other common names North Atlantic beaked whale, North Sea beaked whale

Taxonomy Genetic analyses indicate it is most closely related to True's beaked whale

Similar species Difficult to distinguish at sea from other North Atlantic beaked whales, though the position of the exposed teeth in adult male Sowerby's differs from that of male True's beaked whales (at the tip of the lower jaw), male Blainville's beaked whales (in an elevated portion of lower jaw), and Gervais' beaked whales (only slightly visible)

Birth weight 375 lb (170 kg)

Adult weight 2,200–2,900 lb (1,000–1,300 kg)

Diet Squid and small deepwater fish

Group size A few up to about 10 individuals

Major threats Entanglement in pelagic driftnets and longlines, probably also noise and debris ingestion

IUCN status Data Deficient

Range and habitat This species is endemic to the North Atlantic, mainly in cold-temperate to sub-Arctic waters. It is seen as far north as 71°N in the Norwegian Sea. It inhabits primarily deep shelf and slope waters including submarine canyons. Nearly all records are from waters with surface temperatures of 43–72.5°F (6–22.5°C).

Identification checklist

- Relatively long beak without a pronounced melon
- Forward-tilted tips of erupted teeth (tusks) visible in males on elevated portion of lower jaw, behind the mid-beak
- Prominent, falcate dorsal fin behind the mid-back
- Coloration fairly nondescript— dark gray on the upper side and lighter underside
- Back and sides, especially of adult males, marked by long linear tooth rakes

Anatomy

Sowerby's beaked whales conform to the typical pattern of a long body tapered at both ends. The beak is relatively long and the melon (forehead) slightly convex but not bluff, sloping smoothly onto the rostrum. The coloration is dark gray on top and lighter gray to white underneath, with darkened areas around the eyes. The mouthline is fairly straight in females but there is a noticeable rise toward the rear in adult males, with the tips of the two erupted teeth emerging as small tusks from the lower jaws.

Behavior

This species is only observed in fairly small groups from several up to about 10 individuals. Such groups can include adult males, adult females, and calves. Generally described as shy and elusive, these whales are rarely observed at close quarters. The blow is inconspicuous under most sea conditions. Long linear scars on the back and sides, especially of adult males, suggest that they engage in combat, using their small pointed tusks as weapons.

Food and foraging

Most of what is known about the diet of this species comes from analyses of stomach contents from stranded or bycaught individuals. These whales appear to rely primarily on squid and deep-water fish but they also consume swimming crabs and cuttlefish. Dives can last at least a half hour and they presumably reach depths of at least 3,300 ft (1,000 m).

Life history

Virtually nothing is known about the life history of this species.

Conservation and management

Abundance estimates indicate that there are at least a few thousand of these whales in the central North Atlantic around the Faroe Islands as well as in the western North Atlantic. Pelagic driftnets are deadly for them as was shown during a small driftnet fishery along the continental shelf edge of the eastern United States that killed at least 24 Sowerby's from 1989–98. Incidental mortality also occurs in longline fisheries. There is concern that, like other beaked whales, Sowerby's are highly sensitive to underwater noise, especially that from naval sonar and seismic surveys.

MALE

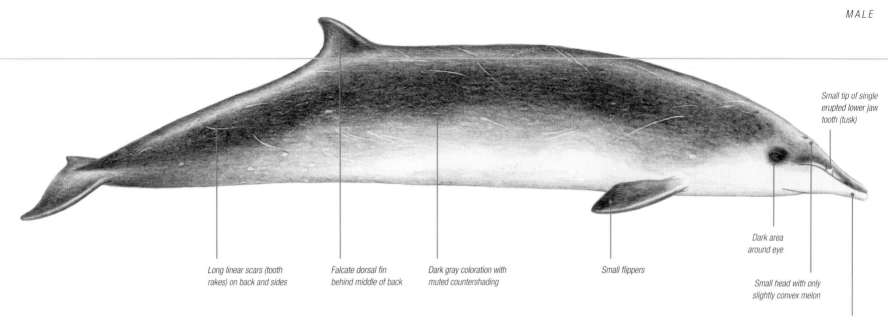

Small tip of single
erupted lower jaw
tooth (tusk)

Dark area
around eye

Small head with only
slightly convex melon

Relatively long beak;

Long linear scars (tooth
rakes) on back and sides

Falcate dorsal fin
behind middle of back

Dark gray coloration with
muted countershading

Small flippers

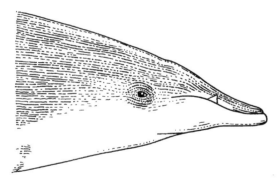

Adult male head features

The small head includes a fairly smooth,
sloping transition from the slightly rounded
melon onto a long beak. The two teeth in the
lower jaws erupt only in adult males, protruding
as small, triangular tusks from raised areas
toward the back of the mouthline.

Size

Newborn 8 ft (2.4 m)
Adult 16½–18 ft (5–5.5 m)

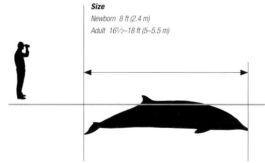

Dive sequence

1. The long narrow beak of Sowerby's
may be the first thing to appear above
the surface, rising at an angle of 30–45
degrees. The head is proportionally small
and the forehead not at all bulbous.

2. The beak may at least start to
dip back into the water before the
blow, which is often visible but not
particularly dense or conspicuous.

3. As the head disappears, it is
followed by the back with the
relatively prominent, pointed, and
somewhat falcate dorsal fin.

Surface behavior

Sowerby's occur in small, fairly
close-knit groups and are
undemonstrative at the surface
although tail-slapping of the sea
surface happens occasionally.

ANDREWS' BEAKED WHALE

Family Ziphiidae

Species *Mesoplodon bowdoini*

Other common names Deep-crest beaked whale

Taxonomy No subspecies recognized, and nothing is known about population structure

Similar species Probably confused with other "white-beaked" whales in its southern hemisphere range, especially strap-toothed, Gray's beaked, and Hector's beaked whales

Birth weight Unknown

Adult weight Unknown

Diet Unknown but probably squid and possibly fish

Group size Unknown

Major threats Possibly noise, especially from seismic surveys and naval sonar, entanglement in offshore drift gill nets and longlines, debris ingestion

IUCN status Data Deficient

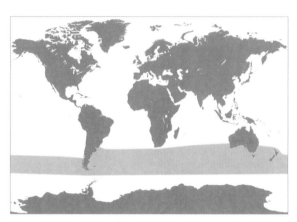

Range and habitat This species is limited to cool-temperate waters of the southern hemisphere, probably circumpolar between 32°S and 54°30'S. The vast majority of strandings have been in Australia and New Zealand but some have also occurred in Uruguay, Argentina (Tierra del Fuego), the Falkland Islands, and Tristan da Cunha. Nothing is known about habitat but presumably it is offshore and in deep water.

Identification checklist

- Short head with melon that is slightly bulging
- Short, thick beak, with strong arch in rear half of lower jaw
- Single tooth on each side of lower jaw, which forms a tusk in adult males
- Small, somewhat falcate dorsal fin behind the mid-back
- Adult males have a mostly white beak but are otherwise dark with no bold patterning
- Adult females said to have less white on the beak and somewhat lighter tones than males elsewhere on the body
- Body often has long linear tooth-rake scars, especially on adult males

Anatomy

Andrews' beaked whale is one of the smaller of the beaked whales but also one of the most poorly known. The melon has a slight bulge but slopes smoothly onto the rostrum. The beak is short and thick, with an arch in the lower jaw where—in adult males—the erupted tooth (tusk) is situated, extending above the plane of the upper jaw. These tusks, one on each side, probably account for the long white linear scars present on the bodies of adults. Coloration features are described on the basis of observations of stranded animals and are accordingly suspect because of post-mortem changes. However, the white front portion of the beak of adult males is certain; apparently there is less white on the beak of females. Males are said to be darker overall than females.

Behavior

In the absence of any confirmed sightings at sea, nothing is known about the behavior or social organization.

Food and foraging

No information available, although they are assumed to feed primarily on squid like other members of the genus.

Life history

Almost nothing is known. The calving season, at least in New Zealand waters, is believed to be summer and fall.

Conservation and management

These whales could be naturally rare although the lack of at-sea sightings could also reflect the difficulty of detecting and identifying them. The fact that most strandings have occurred in New Zealand and Australia could be misleading, since relatively more effort has been made in those countries to detect, report, and investigate cetacean strandings. In any event, Andrews' beaked whales have not been hunted anywhere and there is no evidence to indicate that they have been significantly affected by fisheries (as bycatch, for example), noise (no mass strandings reported), or marine debris.

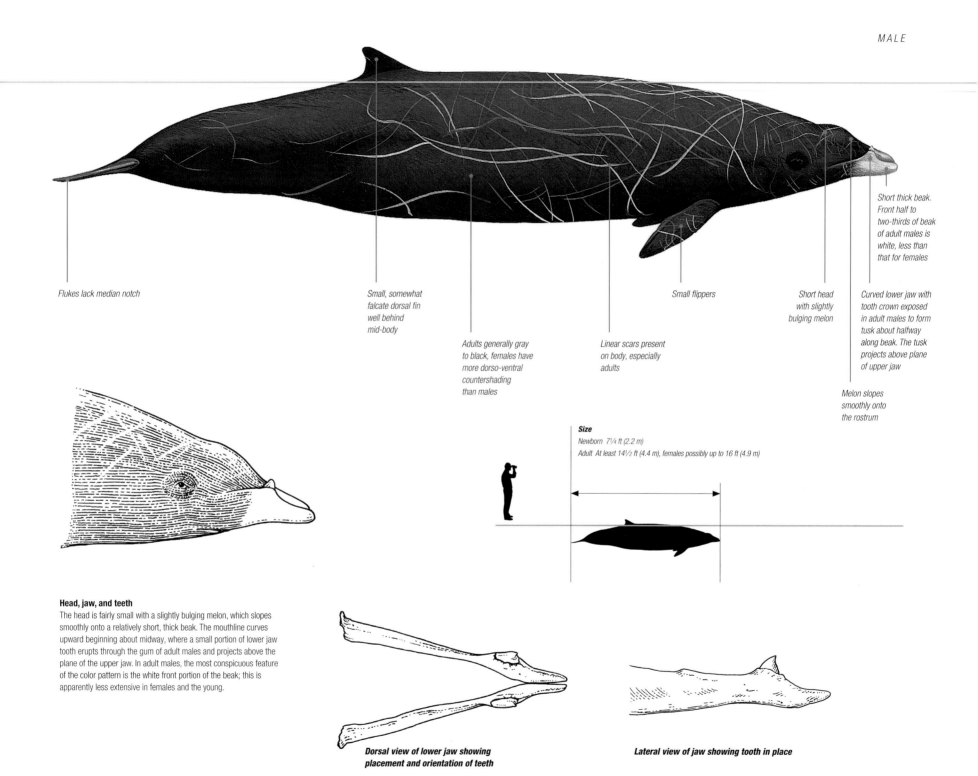

MALE

Flukes lack median notch

Small, somewhat
falcate dorsal fin
well behind
mid-body

Adults generally gray
to black, females have
more dorso-ventral
countershading
than males

Linear scars present
on body, especially
adults

Small flippers

Short head
with slightly
bulging melon

Short thick beak.
Front half to
two-thirds of beak
of adult males is
white, less than
that for females

Curved lower jaw with
tooth crown exposed
in adult males to form
tusk about halfway
along beak. The tusk
projects above plane
of upper jaw

Melon slopes
smoothly onto
the rostrum

Size
Newborn 7¼ ft (2.2 m)
Adult At least 14½ ft (4.4 m), females possibly up to 16 ft (4.9 m)

Head, jaw, and teeth

The head is fairly small with a slightly bulging melon, which slopes
smoothly onto a relatively short, thick beak. The mouthline curves
upward beginning about midway, where a small portion of lower jaw
tooth erupts through the gum of adult males and projects above the
plane of the upper jaw. In adult males, the most conspicuous feature
of the color pattern is the white front portion of the beak; this is
apparently less extensive in females and the young.

**Dorsal view of lower jaw showing
placement and orientation of teeth**

Lateral view of jaw showing tooth in place

HUBBS' BEAKED WHALE

Family Ziphiidae

Species *Mesoplodon carlhubbsi*

Other common names Arch-beaked whale

Taxonomy There are no subspecies or separate populations—its closest modern relative may be Andrews' beaked whale

Similar species Stejneger's, Cuvier's, and ginkgo-toothed beaked whales all occur in the same range—though the characteristic white "cap" on the head of the adult male Hubbs' beaked whale makes easy to distinguish

Birth weight Not known

Adult weight About 3,300 lb (1,500 kg)

Diet Squid and fish at depths of around 660–3,300 ft (200–1,000 m)

Group size Probably small, although too few at-sea observations to know with certainty

Major threats Occasional reports of deliberate kills in Japan and entanglements in driftnets; concerns about impacts of underwater noise and ingestion of marine debris

IUCN status Data Deficient

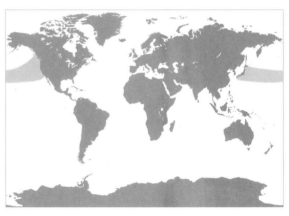

Range and habitat This species is endemic to the cool-temperate North Pacific Ocean, mainly between 33°N and 54°N. Most of what is known is based on the locations of strandings, which have occurred in the eastern Pacific between San Diego, California, and Prince Rupert, British Columbia, and in the western Pacific only in Honshu, Japan.

Identification checklist

- Rotund body with typical *Mesoplodon* features, including small flippers and moderate-sized dorsal fin, which is slightly falcate and positioned behind the mid-back
- Moderately long beak, white in adult males and light-colored in females
- Strongly arched mouthline of adult males with a prominent tusk at the apex of the arch on each side
- Very conspicuous white "cap" on the somewhat bulbous melon of adult males
- Body of adult males dark gray to black with very heavy scarring, females and juveniles dark upper side and light underside

Anatomy

The most striking features of this beaked whale are found on the adult male. The well-defined beak is almost completely white as is a sort of skull cap that sets off the melon from the rest of the dark body. The large, laterally compressed, pointed teeth erupt from a prominent arch midway along either side of the mouthline. These teeth protrude above the rostrum when the mouth is closed, and this helps explain the extensive linear (tooth-rake) scarring that covers the bodies of adult males. That is, males, mouth closed, probably press their rostrum hard against an opponent in hopes of inflicting pain, and this results in the long linear wounds.

Behavior

Since this species has been seen and identified so rarely, virtually nothing is known about its behavior. However, the scarring on adult males has been interpreted as evidence of fighting, perhaps to gain social dominance. The linear scars, often in parallel pairs, can be up to 6½ ft (2 m) long.

Food and foraging

The diet has only been studied from the stomach contents of stranded individuals. On that basis, these whales are believed to rely principally on deep-water squid and fish. It was, in part, from observations of two live-stranded Hubbs' beaked whale calves that anatomists determined the mechanism for suction-feeding in beaked whales.

Life history

Nothing is known about this species' life history.

Conservation and management

Of eight *Mesoplodon* whales documented to have died in California's large-mesh driftnet fishery in the first half of the 1990s, five of them were Hubbs' beaked whales. Since this was only a sample of the bycatch in that fishery, many more likely died. Driftnetting in deep waters seaward of the continental shelf is clearly a threat to this species as well as other beaked whales. Also, disturbance from underwater noise, especially that from naval sonar and seismic surveys, and ingestion of marine debris are potential problems for Hubbs' beaked whales.

MALE

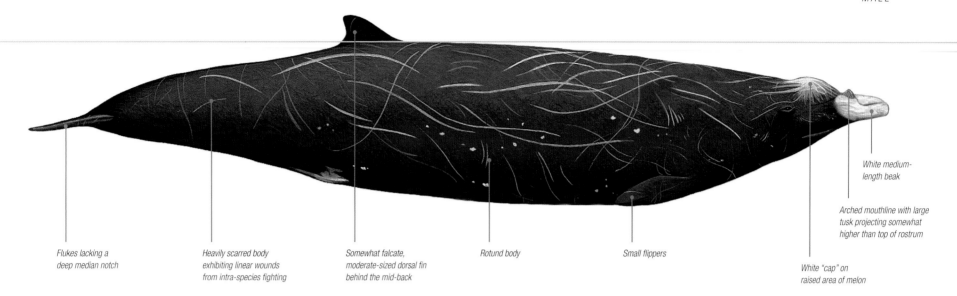

Flukes lacking a
deep median notch

Heavily scarred body
exhibiting linear wounds
from intra-species fighting

Somewhat falcate,
moderate-sized dorsal fin
behind the mid-back

Rotund body

Small flippers

White "cap" on
raised area of melon

White medium-
length beak

Arched mouthline with large
tusk projecting somewhat
higher than top of rostrum

Adult head features

Although both males and females have a white or partially white to
light gray beak, certain features of the head differ markedly between
the sexes. The mouthline of adult males has a prominent rise midway
along, where the tusk protrudes from the lower jaw. In contrast, the
female mouthline is long and smoothly contoured with no such rise
and no visible teeth. Another conspicuous difference is the white cap
atop the male's rounded head.

Male

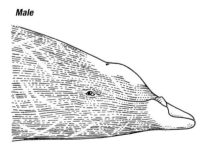

Female

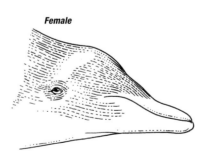

Size
Newborn Possibly up to 8¼ ft (2.5 m)
Adult Up to 18 ft (5.4 m)

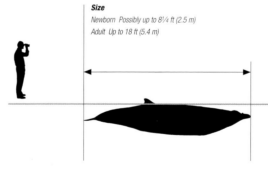

Surface behavior

The rarity of sightings at sea
suggests that these whales are not
easily detected and identified and
therefore their profile at the surface
may be relatively inconspicuous—
small groups, blow hard to discern,
little time spent at surface etc.

Dive sequence

*1. The white medium-length
beak, with strongly arched
lower jaw, as well as the
prominent white "beanie"
on the head of an adult male
may be seen first as the
animal breaks the surface.*

*2. Immediately afterward,
the blow may or may not be
visible as the top of the head
and back appear. Density and
visibility of the blow depend
on atmospheric conditions.*

*3. As the head disappears, the
back and medium-sized dorsal
fin appear. It is unlikely that these
whales regularly show flukes
when beginning a dive.*

BLAINVILLE'S BEAKED WHALE

Family Ziphiidae

Species *Mesoplodon densirostris*

Other common names Dense-beaked whale

Taxonomy No subspecies recognized; distinct populations almost certainly exist because the species' distribution precludes animals in the Atlantic and Pacific from mixing

Similar species Female and juvenile *Mesoplodon* species and Cuvier's beaked whales. Adult male Blainville's beaked whales are distinctive owing to an elevated portion of the lower jaw and single erupted tooth on either side that is tilted forward.

Birth weight 133 lb (60 kg)

Adult weight 1,760–2,200 lb (800–1,000 kg)

Diet Squid and small fish

Group size 2 or 3, up to about 10 individuals

Major threats Noise especially from naval sonar, gill-net entanglement, debris entanglement and ingestion

IUCN status Data Deficient

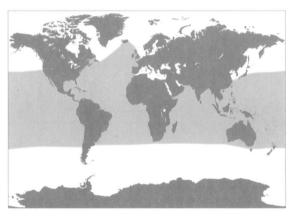

Range and habitat Blainville's beaked whales is nearly cosmopolitan in tropical and warm-temperate waters. They are deep-water animals generally associated with slope waters from 1,640 ft (500 m) to many thousands of feet deep, with island-associated populations that appear to be long-term local residents.

Identification checklist

- Melon is relatively flat and the beak long, with an abrupt rise of the lower jaw just behind the tip, a feature that is extremely prominent on adult males
- Small pointed crown of the large mandibular tooth on each side on elevated lower jaw
- Small- to medium-sized, somewhat falcate dorsal fin behind the mid-back
- Countershaded color pattern, dark dorsally (including around the eyes) and light ventrally including the lower jaw
- Body often covered with cookie-cutter shark bites and scars, as well as long linear tooth-rake scars on adult males

Anatomy

The body is robust and the melon is almost flat. By far the most distinctive feature in comparison to other *Mesoplodon* species is the extreme development of the elevated parts of the lower jaws of adults; these elevations dominate the face and are most prominent in adult males. In adult males, the crown of a single large tooth erupts at the front of each elevated portion and is usually colonized by at least one stalked barnacle that can be more conspicuous than the tooth crown itself. Another exceptional feature is the upper jawbone, the density of which is unmatched in animals (it is denser than elephant ivory).

Behavior

Blainville's beaked whales typically occur in small groups of 2–4 up to 10–12 individuals. The small group sizes and lack of aerial activity make them hard to detect in less than ideal sea conditions.

Food and foraging

Like most other beaked whales, Blainville's consume mainly squid and small deep-sea fish. They appear to prefer waters at least 1,640 ft (500 m) deep. Largely resident populations of perhaps a few hundred individuals are associated with certain oceanic islands, such as the Hawaiian Islands, where deep waters occur close to shore. In Hawaii, where the species has been relatively well studied, an island-associated population tends to remain in waters 1,640–3,300 ft (500–1,000 m) deep while a separate oceanic population is centered in waters around 11,500–13,100 ft (3,500–4,000 m) deep. The average duration of deep foraging dives is close to an hour.

Life history

Virtually nothing is known about life history.

Conservation and management

One of the more widespread beaked whales but also quite rare, Blainville's ranks behind Cuvier's beaked whale as the ziphiid most frequently involved in mass strandings caused by naval exercises using intense mid-frequency sonar. Such strandings have been documented in the Bahamas and Canary Islands. Blainville's are also known to die in offshore drift gill nets and on pelagic longlines, and they are known to ingest marine debris. They have not been significantly affected by whaling as far as is known.

MALE

Small head with flat
melon, fairly long beak

Flukes lacking median notch

Numerous cookie-cutter
shark bite scars

Falcate dorsal fin about
two-thirds of the way back
on the body

Robust body

Small flippers

Large, conspicuous
raised area on lower
jaw (in both sexes)

Small tooth crown
exposed above
elevated area of
lower jaw

Adult head features

The adult head shape is somewhat bizarre for both male
and females. In the front view, the female appears almost
comical, with the lower jaws arching strongly toward the
back and pressing against (as though "pinching") the
rostrum (also note the "V"-shaped throat grooves). Side-on,
the male's relatively flat (non-bulbous) melon is evident, as
are the remarkably prominent arches in the lower jaws, and
the erupted teeth, often infested with stalked barnacles.

Female front view

Size
Newborn 6½ ft (2 m)
Adult 15½ ft (4.7 m) maximum

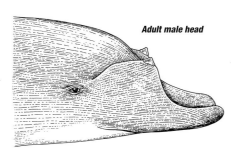

Adult male head

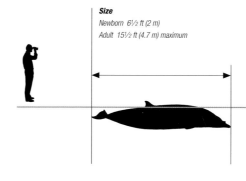

Dive sequence
1. As an adult male Blainville's
comes to the surface, it is likely
that the first thing noticed will be
its bizarre face, dominated by the
elevated area of the lower jaw
that rises well above the rostrum.

2. The melon is flatter on this
species than on many other
beaked whales. The erupted tooth
at the top front edge of the raised
area on each side of the lower jaw
is often obscured by the presence
of one or more stalked barnacles.

3. As the whale rolls following
the blow, it displays a prominent,
falcate and usually pointed
dorsal fin that is situated well
behind the mid-body.

Surface behavior
Blainville's generally do not
show their flukes as they
dive, nor do they regularly
engage in aerial behavior,
such as breaching.

GERVAIS' BEAKED WHALE

Family Ziphiidae

Species *Mesoplodon europaeus*

Other common names Antillean beaked whale, Gulf Stream beaked whale, European beaked whale

Taxonomy No subspecies recognized; nothing known about population structure

Similar species Easily confused with female and juvenile *Mesoplodon* species as well as Cuvier's beaked whales (though Gervais' beaked whales have a longer beak) and Blainville's beaked whales (though they lack the elevated portion of the lower jaw of adult Blainville's)

Birth weight Unknown

Adult weight 2,650 lb (1,200 kg) or more

Diet Squid

Group size Small, probably no more than 10 individuals and usually fewer

Major threats Noise, especially from naval sonar; entanglement in offshore drift gill nets and possibly longlines; debris entanglement and ingestion

IUCN status Data Deficient

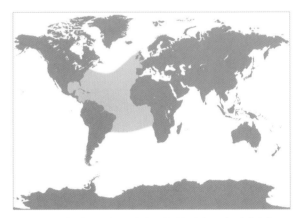

Range and habitat This whale is endemic to the Atlantic Ocean, largely limited to tropical and warm-temperate waters, and more common in the west than the east. It has been recorded from the east coast of the United States southward to southern Brazil, from Ascension Island in the mid-Atlantic and from the North Sea southward to West Africa. This species mostly occurs in deep waters along and seaward of the continental shelf edge.

Identification checklist

- Head small with slightly bulging melon and long slender beak; mouthline almost straight with no rise of the mandibles
- Small pointed crown of the mandibular tooth erupts on each side well behind tip of jaw, sometimes with stalked barnacles attached
- Small, somewhat falcate dorsal fin behind the mid-back
- Countershaded color pattern: dark dorsally (including around the eyes) and light ventrally, including the lower jaw
- Body often has long linear tooth-rake scars, especially on adult males, and females often have white or light gray markings in genital area and around mammary slits

Anatomy

This species has a relatively small head compared to other *Mesoplodon* species and the mouthline has no rise in the lower jaw, even in adult males. The typical erupted tooth crowns, one on each side, are large enough to be evident behind the tip of the beak in adult males but otherwise there is less sexual dimorphism in these whales than in those *Mesoplodon* species with more formidable tusks or more distinctive color patterns.

Behavior

Gervais' beaked whales typically occur in pairs or small groups. They are relatively inconspicuous at sea and therefore are rarely sighted, identified, and reported. The linear scars on adult males are probably made by wounds from the tusks of other adult male of this species.

Food and foraging

Like most other beaked whales, Gervais' whales consume mainly squid but also prey on deep-sea fish and shrimp. They are deep and prolonged divers. There is no evidence to suggest they make long-distance migrations.

Life history

Virtually nothing is known about life history.

Conservation and management

Gervais's may be among the least abundant ziphiids, given their limited range. They are relatively common in at least portions of the Caribbean region and in the Gulf Stream and possibly the North Atlantic Current although it is generally believed (based primarily on stranding records) that they are much more numerous in the western than the eastern Atlantic. They strand fairly frequently in the southeastern United States but have generally not been involved in mass-stranding events associated with naval sonar. However, this does not mean they are not susceptible to the effects of noise. Similarly, although they are not often recorded as fishery bycatch, they are certainly vulnerable to entanglement in gill nets, probably longlines, and even pound nets. A young individual that stranded in Puerto Rico had a large quantity of plastic bags in its stomach. Fortunately, Gervais' whales have not been regular targets of commercial or subsistence whaling.

MALE

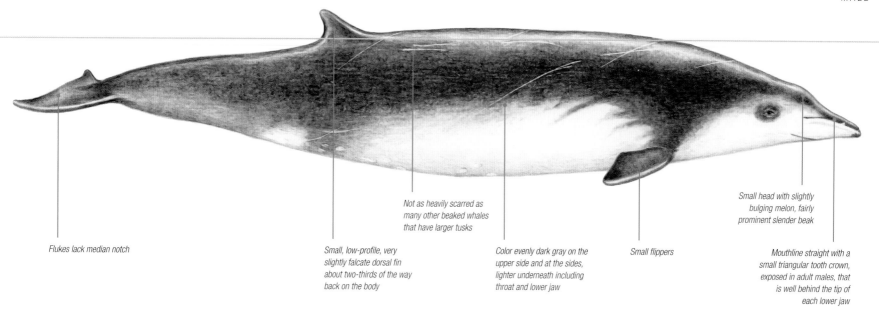

Flukes lack median notch

Not as heavily scarred as many other beaked whales that have larger tusks

Small, low-profile, very slightly falcate dorsal fin about two-thirds of the way back on the body

Color evenly dark gray on the upper side and at the sides, lighter underneath including throat and lower jaw

Small flippers

Small head with slightly bulging melon, fairly prominent slender beak

Mouthline straight with a small triangular tooth crown, exposed in adult males, that is well behind the tip of each lower jaw

Adult male head features
The head of the Gervais' beaked whale is quite small and has a streamlined, somewhat elongated appearance. The beak is long and slender, the mouthline straight and interrupted only on adult males by the small triangular crown of an exposed tooth that is well behind the tip of the lower jaw on each side. The color pattern is simple, mainly dark dorsally and light ventrally, with the typical dark patch around the eye.

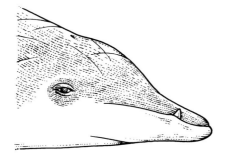

Size
Newborn 7 ft (2.1 m)
Adult 14¾–17 ft (4.5–5.2 m)

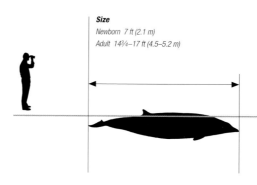

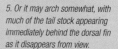

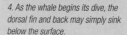

Dive sequence
1. Gervais' beaked whales often lead with the beak as they surface, and in suitably calm conditions the tip of the beak may be the first thing to break the surface.

2. This is quickly followed by the melon and front part of the back. Depending on the conditions, the blow may or may not be visible.

3. As the head disappears, it is followed by the long back and prominent dorsal fin, which is falcate, sometimes with a slightly hooked peak.

4. As the whale begins its dive, the dorsal fin and back may simply sink below the surface.

5. Or it may arch somewhat, with much of the tail stock appearing immediately behind the dorsal fin as it disappears from view.

GINKGO-TOOTHED BEAKED WHALE

Family Ziphiidae

Species *Mesoplodon ginkgodens*

Other common names Japanese beaked whale

Taxonomy No subspecies recognized, and nothing is known about population structure—the species name is inspired by the resemblance of the tooth shape to the leaf of the ginkgo tree, *Gingko biloba*. A member of True's beaked whale clade that also includes *Mesoplodon mirus* and *M. europaeus*.

Similar species The recently described Deraniyagala's beaked whale is virtually indistinguishable at sea

Birth weight Unknown

Adult weight Unknown

Diet Squid and fish

Group size Small, up to five

Major threats Incidental mortality in fisheries including in gill nets, set nets (Japan), and longlines—probably also vulnerable to noise from seismic surveys and naval sonar

IUCN status Data Deficient

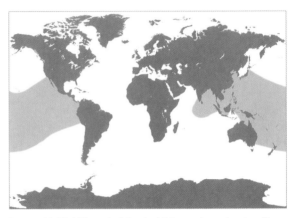

Range and habitat This species is found mainly in warm-temperate waters of the western Pacific Ocean but is also known from the Galapagos and there are a few records from cold-temperate waters off California, New Zealand, and Australia. It is especially common in the Kuroshio Current off southern Japan and Taiwan. It occurs mainly in deep water but nothing is known in detail about its habitat largely because ginkgo-toothed beaked whales are rarely identified at sea.

Identification checklist

- Fairly prominent beak
- No distinct melon as the forehead slopes smoothly and rather steeply to merge with the rostrum
- Lower jaw elevated toward the rear. Small portion of tooth crown erupts from lower jaw of adult males at vertex of the raised area
- Flippers and flukes are relatively small
- Small, falcate dorsal fin about two-thirds of the way back on body
- Body mostly dark with subtle gray shading and often some light areas on throat
- Body generally not heavily scarred other than in the ano-genital area where cookie-cutter shark bites are evident

Anatomy

It has been stated that this is the only mesoplodont in which white scar tissue does not form over wounds, which means the usual linear scars typical of adult males, in particular, are not usually evident on the bodies of ginkgo-toothed beaked whales. The almost complete absence of such scars might also be related to the fact that the lower jaw tusks protrude only slightly through the gum and do not extend above the plane of the rostrum. In most other respects, these whales have the typical *Mesoplodon* features— "V"-shaped throat grooves, small flippers low on the body, at most a small notch in the rear margin of the flukes, and a small falcate dorsal fin well-behind the mid-back.

Behavior

Groups of up to five have been reported (tentative identification).

Food and foraging

There is no definite information on diet but it includes squid and possibly fish.

Life history

Virtually nothing is known.

Conservation and management

There are no estimates of abundance for ginkgo-toothed beaked whales. They were historically killed in small numbers by hunters of small whales in Japan, but in recent years most known deaths have been the result of incidental entanglement or entrapment in fishing gear, including gill nets, longlines, and set nets (in Japan).

MALE

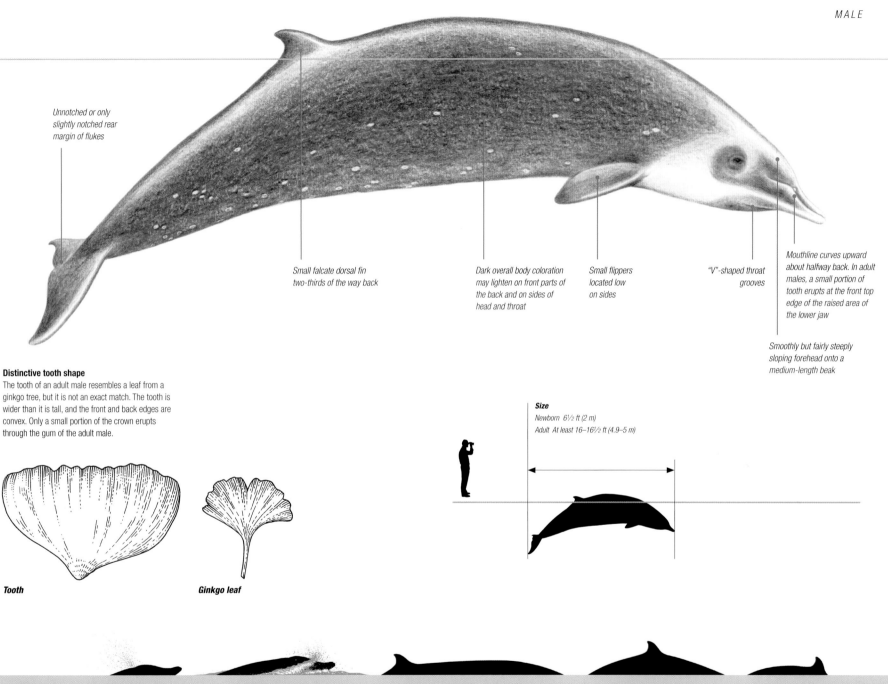

Unnotched or only slightly notched rear margin of flukes

Small falcate dorsal fin two-thirds of the way back

Dark overall body coloration may lighten on front parts of the back and on sides of head and throat

Small flippers located low on sides

"V"-shaped throat grooves

Mouthline curves upward about halfway back. In adult males, a small portion of tooth erupts at the front top edge of the raised area of the lower jaw

Smoothly but fairly steeply sloping forehead onto a medium-length beak

Distinctive tooth shape
The tooth of an adult male resembles a leaf from a ginkgo tree, but it is not an exact match. The tooth is wider than it is tall, and the front and back edges are convex. Only a small portion of the crown erupts through the gum of the adult male.

Tooth

Ginkgo leaf

Size
Newborn 6½ ft (2 m)
Adult At least 16–16½ ft (4.9–5 m)

Because these whales have not been reliably identified at sea, this sequence of images is speculative.

Dive sequence
1. The blow is likely invisible or faint, depending on the conditions.

2. Following the blow, the melon and beak may remain exposed above the surface for a short time before the dorsal fin appears.

3. As with other mesoplodonts, the dorsal fin is well back on the body and may not be seen until the head has submerged.

4. Depending on the depth of a dive, the whale may or may not arch its back high before submerging.

5. As far as is known, these whales do not lift their flukes as they slip beneath the surface.

GRAY'S BEAKED WHALE

Family Ziphiidae

Species *Mesoplodon grayi*

Other common names Scamperdown whale

Taxonomy No subspecies recognized, nothing known about population structure.

Similar species Could be confused with other ziphiids that occur in its range, especially strap-toothed and Andrews' beaked whales—although the exceptionally long all-white beak and small head of Gray's beaked whale help distinguish it

Birth weight Unknown

Adult weight Probably at least 2,650 lb (1,200 kg)

Diet Fish and squid

Group size Usually 5 or fewer but occasionally up to at least 10 individuals

Major threats None known with certainty although possible threats include noise— especially from naval sonar and seismic exploration—entanglement in fishing gear or debris, and ingestion of plastics

IUCN status Data Deficient

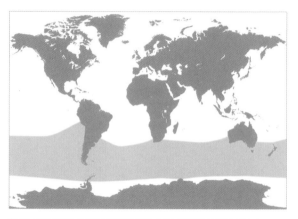

Range and habitat This species is endemic to the southern hemisphere and normal distribution is generally limited to cool-temperate waters south of 30°S, with strandings reported from the coasts of Australia, New Zealand, South America, and southern Africa. There have been a few sightings in sub-Antarctic and Antarctic waters, and strandings have occurred as far north as Namibia and Peru. The species is mainly found in deep offshore waters but these whales are occasionally seen in shallow coastal waters as well.

Identification checklist

- Small head with exceptionally long beak
- Melon is rounded but low; merges smoothly with beak
- Mouthline straight with a small portion of tooth crown erupting from lower jaw (adult males and sometimes females) more than halfway back from tip of beak
- Small, falcate dorsal fin about two-thirds back on body
- White or light gray beak (can be stained yellowish by diatoms) and front part of face but otherwise gray-bodied with dark areas around the eyes and generally dark on top and lighter underneath
- Some tooth-rake scars especially on adult males and cookie-cutter shark bite scars

Anatomy

Gray's beaked whales have an unusual head shape, with a long slender beak—longer in females than males—and low, rounded melon that slopes smoothly onto the rostrum. The mouthline is straight but interrupted in adult males and sometimes females by the slightly protruding tip of a flattened tooth (tusk) on each side at the middle of the lower jaw. The tusks may be colonized by stalked barnacles. Another unusual feature of the Gray's beaked whale is the presence, often in both sexes, of small, dolphin-like teeth along both sides of the upper jaw inside the mouth, numbering 4–19 on each side.

Behavior

Groups usually consist of 5 or fewer individuals but larger groups of at least 10 have been documented. The largest mass stranding consisted of 28 whales, at least 3 of which were confirmed as Gray's beaked whales. Mothers tend to remain segregated from larger groups until their calf is more than 10 ft (3 m) long. They may move into shallower water during the early nursing period.

Food and foraging

What is known about the diet is based entirely on analyses of stomachs from stranded animals. Only fish remains have been found in the animals from southern Africa and South America whereas those from New Zealand appeared to prefer small squid. These whales are presumably deep divers but their diving behavior has not been studied directly.

Life history

Little is known about life history. A female that stranded with a calf in New Zealand was both lactating and pregnant. Also, based on stranding patterns, summer is thought to be the calving season.

Conservation and management

There are no estimates of abundance for Gray's beaked whales but they commonly strand in New Zealand and Australia and therefore may be fairly numerous in those parts of their range. Like other beaked whales, they would be susceptible to bycatch in any part of their range where driftnet fisheries are conducted, and also to the harmful effects of underwater noise, such as that from seismic surveys and naval sonar.

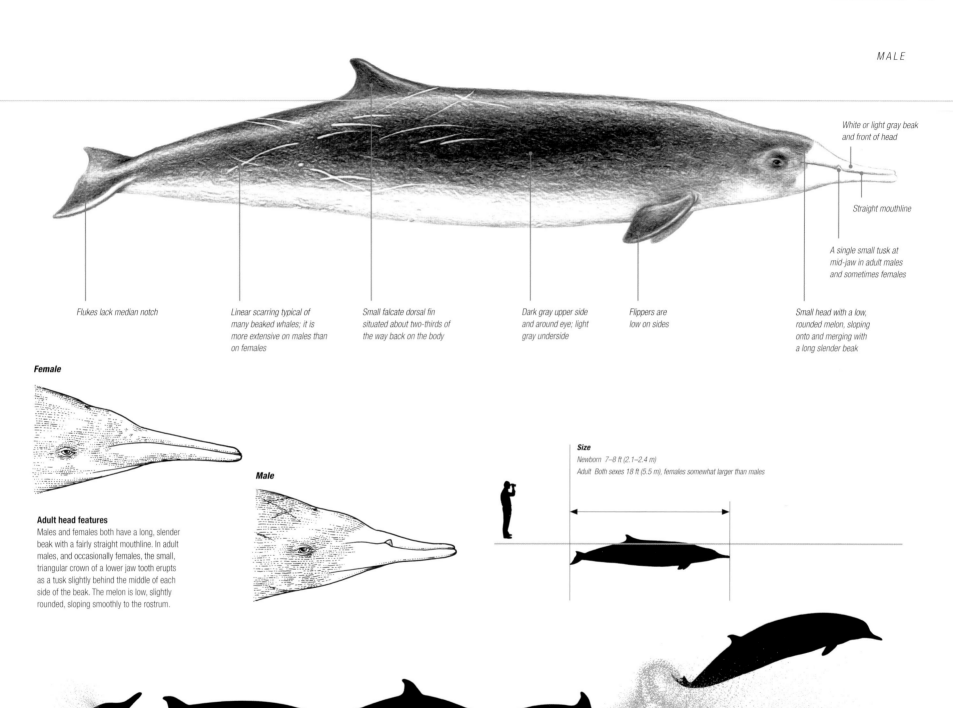

MALE

White or light gray beak and front of head

Straight mouthline

A single small tusk at mid-jaw in adult males and sometimes females

Flukes lack median notch

Linear scarring typical of many beaked whales; it is more extensive on males than on females

Small falcate dorsal fin situated about two-thirds of the way back on the body

Dark gray upper side and around eye; light gray underside

Flippers are low on sides

Small head with a low, rounded melon, sloping onto and merging with a long slender beak

Female

Adult head features

Males and females both have a long, slender beak with a fairly straight mouthline. In adult males, and occasionally females, the small, triangular crown of a lower jaw tooth erupts as a tusk slightly behind the middle of each side of the beak. The melon is low, slightly rounded, sloping smoothly to the rostrum.

Male

Size
Newborn 7–8 ft (2.1–2.4 m)
Adult Both sexes 18 ft (5.5 m), females somewhat larger than males

Dive sequence
1. These whales often thrust their long, white beak clear of the surface as they end a dive, and their blows are low and diffuse.

2. The animals may then arch slightly so that the head is no longer in view but the long back and dorsal fin are visible.

3. When the animals begin a long dive, the back is arched high but the flukes are not normally lifted above the surface.

Breaching
Gray's beaked whales sometimes breach high out of the water and re-enter with a turbulent splash.

HECTOR'S BEAKED WHALE

Family Ziphiidae

Species *Mesoplodon hectori*

Other common names New Zealand beaked whale

Taxonomy First recognized as a species in the 1860s, but until genetic analyses in the 1990s settled the matter, it was confused with Perrin's beaked whale, *Mesoplodon perrini*

Similar species Outwardly indistinguishable from Perrin's beaked whale, but the two species inhabit different hemispheres so they should not be mistaken for one another

Birth weight Not known

Adult weight Not known

Diet Assumed to be mainly squid and possibly some fish

Group size Nothing known from direct observation but probably small

Major threats None known but entanglement in pelagic driftnets and longlines may occur, and noise and debris ingestion are possible threats

IUCN status Data Deficient

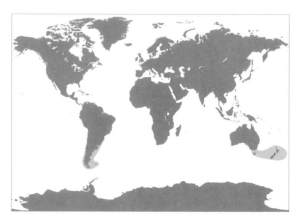

Range and habitat This species is endemic to the southern hemisphere, mainly in cool-temperate waters between 35 and 55° S. It is known primarily from strandings and a very few sightings around New Zealand, Australia, southern Africa, and both the east and west coasts of South America (including the Falklands).

Identification checklist

- Tapered head with smoothly sloping melon and moderately long beak
- An exposed tooth (tusk) is visible on each side just behind tip of lower jaw in adult males
- Dorsal fin is well behind the mid-back and relatively small and only slightly falcate
- Coloration is fairly nondescript—mostly gray with white or very light gray elements in front of the flippers and on the undersides including the lower jaw, with a dark area around the eyes
- Back and sides, especially of adult males, have linear tooth rakes and cookie-cutter shark scars but apparently are not as heavily marked in this manner as some other beaked whales

Anatomy

Hector's beaked whale is one of the smallest ziphiids, probably reaching a maximum length of about 13¼ ft (4.2 m). It conforms to the typical mesoplodont pattern of a long body tapered at both ends. The beak is well-defined but not especially long, with a smooth transition from the non-bulbous melon to the rostrum. The body is generally gray on the upper side and lighter underneath. It has a fairly complex mixture of lighter gray to white areas in front of the flippers, a white lower jaw and tip of upper jaw, and darkened areas around the eyes. The mouthline lacks the rise toward the rear that is observed in adult males of some other beaked whales. The triangular upper portions of the two teeth emerge as small tusks from the lower jaws of adult males.

Behavior

These are likely deep-water, offshore animals that spend most of their time submerged and little is known about their behavior. The few sightings suggest that they are not gregarious.

Food and foraging

Virtually nothing is known about the diet of Hector's beaked whales but they almost certainly feed on mid- and deep-water squid, and probably some deep-water fish. Given their offshore distribution they likely dive deep for prey and are capable to prolonged submergence.

Life history

Virtually nothing is known about the life history of this species.

Conservation and management

It is impossible to identify with any certainty what kinds of threats these whales may face. However, like other beaked whales, they are likely susceptible to entanglement in fishing gear deployed in areas where they occur. Also, disturbance from underwater noise, especially that from naval sonar and seismic surveys, and ingestion of marine debris are potential problems.

MALE

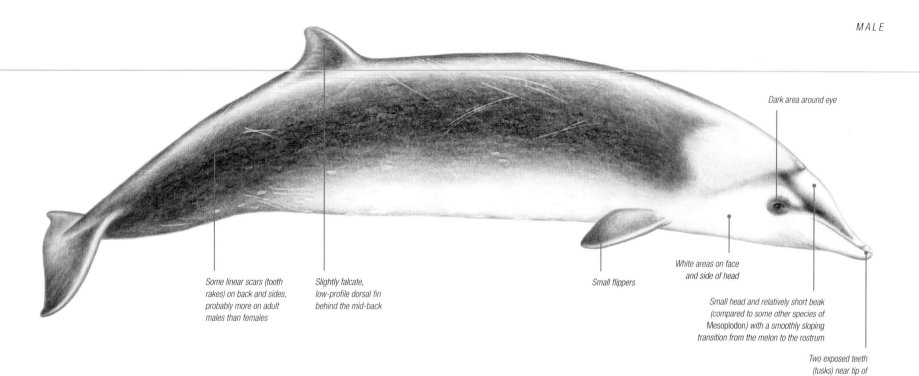

Dark area around eye

Some linear scars (tooth rakes) on back and sides, probably more on adult males than females

Slightly falcate, low-profile dorsal fin behind the mid-back

Small flippers

White areas on face and side of head

Small head and relatively short beak (compared to some other species of Mesoplodon) with a smoothly sloping transition from the melon to the rostrum

Two exposed teeth (tusks) near tip of lower jaw

Adult male head features

The small head includes a smooth, sloping transition from the melon onto a relatively short beak. The white lower jaw and tip of upper jaw and the dark eye patch are part of the complex head coloration. The two teeth erupt only in adult males, protruding as small, triangular tusks slightly behind the tip of the lower jaw.

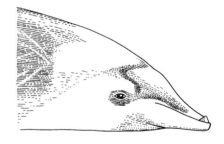

Size

Newborn Unknown but probably around 6 ft (1.8 m)

Adult Up to 13¼ ft (4.2 m)

Dive sequence

1. The relatively short but well-defined beak appears first as the animal ascends from a long dive.

2. Immediately afterward, the blow may or may not be visible as the top of the head and back appear. Density and visibility of the blow depend on atmospheric conditions.

3. As the head disappears, the back and small, fairly low-profile dorsal fin appear. It is unlikely that these whales regularly show flukes when beginning a dive.

Breaching

A low breach by a male Hector's beaked whale was photographed off Western Australia. It is uncertain how commonly these whales exhibit such behavior as this was an exceptional circumstance, with the animal swimming near a boat in very nearshore waters.

DERANIYAGALA'S BEAKED WHALE

Family Ziphiidae

Species *Mesoplodon hotaula*

Other common names None

Taxonomy No subspecies recognized, and nothing is known about population structure

Similar species Virtually indistinguishable at sea from the ginkgo-toothed beaked whale, and the two species have overlapping distributions—DNA evidence is required to tell the two species apart with certainty

Birth weight Unknown

Adult weight Unknown

Diet Probably squid and fish

Group size 2–3 individuals

Major threats Probably vulnerable to entanglement in nets and longlines as well as noise from seismic surveys and naval sonar

IUCN status Data Deficient

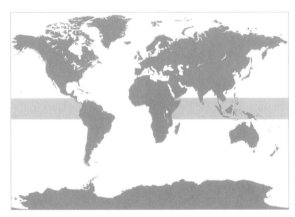

Range and habitat This species is found in the equatorial Indo-Pacific. It is definitely known from Sri Lanka, the Tungaru (Gilbert) Islands in Kiribati, the US Line Islands, the Maldives, and the Seychelles. Little is known about habitat preferences but the animals likely occur mainly in deep water.

Identification checklist

- Fairly prominent beak
- No distinct melon as the forehead slopes smoothly and rather steeply to merge with the rostrum
- Lower jaw is elevated toward the rear (mouthline curves upward about halfway along it)
- Portion of tooth crown erupts from lower jaw of adult males at vertex of the raised area
- Flippers and flukes are relatively small
- Small, falcate dorsal fin set about two-thirds of the way back on the body
- Body mostly dark with subtle gray shading and often some light areas on throat
- Body apparently not heavily scarred but cookie-cutter shark bite scars can be extensive

Anatomy

The external appearance is very similar to that of the ginkgo-toothed beaked whale, with which it has long been confused. At present, it is not possible to distinguish the two species at sea (other than possibly in the case of very experienced expert observers). Even for carcasses and bone material, DNA evidence is necessary to confirm species identification. The lower jaw tusks protrude only slightly through the gum and do not extend above the plane of the rostrum. As is often true of beaked whale tusks, those of Deraniyagala's beaked whales are often colonized by stalked barnacles. In most other respects, these whales have the typical *Mesoplodon* features—"V"-shaped throat grooves, small flippers low on the body, at most a small notch in the rear margin of the flukes, and a small, falcate dorsal fin positioned well behind the mid-back.

Behavior

Groups of up to five have been reported (tentative identification). Also, on one occasion, several individuals were observed to breach clear of the surface.

Food and foraging

There is no definite information on diet but it probably includes squid and possibly deep-water fish.

Life history

Virtually nothing is known.

Conservation and management

There are no estimates of abundance and nothing is known about conservation status. There are concerns, however, about the potential for these whales to be affected by fisheries, such as longline fisheries, as well as naval sonar activities in parts of their range. Also, there is some evidence to indicate that these whales are occasionally hunted by local people in Kiribati.

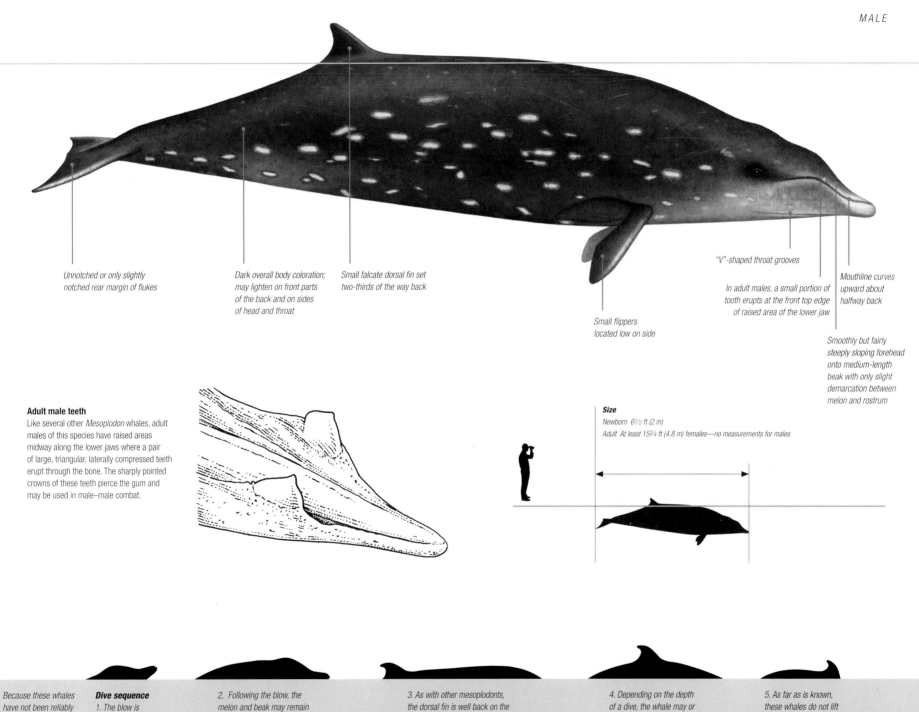

MALE

Unnotched or only slightly
notched rear margin of flukes

Dark overall body coloration;
may lighten on front parts
of the back and on sides
of head and throat

Small falcate dorsal fin set
two-thirds of the way back

"V"-shaped throat grooves

In adult males, a small portion of
tooth erupts at the front top edge
of raised area of the lower jaw

Mouthline curves
upward about
halfway back

Small flippers
located low on side

Smoothly but fairly
steeply sloping forehead
onto medium-length
beak with only slight
demarcation between
melon and rostrum

Adult male teeth
Like several other *Mesoplodon* whales, adult
males of this species have raised areas
midway along the lower jaws where a pair
of large, triangular, laterally compressed teeth
erupt through the bone. The sharply pointed
crowns of these teeth pierce the gum and
may be used in male–male combat.

Size
Newborn 6½ ft (2 m)
Adult At least 15¾ ft (4.8 m) females—no measurements for males

Because these whales
have not been reliably
identified at sea, this
sequence of images
is speculative.

Dive sequence
1. The blow is
likely invisible or
faint, depending
on the conditions.

2. Following the blow, the
melon and beak may remain
exposed above the surface
for a short time before the
dorsal fin appears.

3. As with other mesoplodonts,
the dorsal fin is well back on the
body and may not be seen until
the head has submerged.

4. Depending on the depth
of a dive, the whale may or
may not arch its back high
before submerging.

5. As far as is known,
these whales do not lift
their flukes as they slip
beneath the surface.

STRAP-TOOTHED WHALE

Family Ziphiidae

Species *Mesplodon layardii*

Other common names Layard's beaked whale

Taxonomy No subspecies recognized; nothing known about population structure

Similar species Could be confused with Andrews' and Gray's beaked whales, but the strap-toothed whale's long white beak, black melon, and gray blaze make it distinctive

Birth weight Unknown

Adult weight Probably up to at least 4,000 lb (1,800 kg)

Diet Squid

Group size 2–6 individuals

Major threats Possible threats include noise, especially from naval sonar; entanglement in fishing gear or debris; and ingestion of plastics

IUCN status Data Deficient

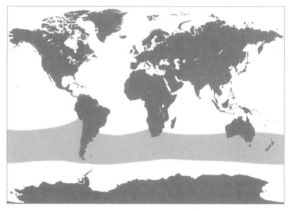

Range and habitat A southern hemisphere endemic whose normal distribution is generally described as being limited to cold-temperate waters south of 35°S and north of 60°S, possibly circumpolar. Strandings have occurred as far north as north-eastern Brazil and on the coast of Burma in the northern Indian Ocean at 16°N. At-sea sightings have all been in waters at least 6,600 ft (2,000 m) deep.

Identification checklist

- Small head with a moderately bulbous melon and fairly long beak; mouthline has only a slight smooth rise beginning about halfway back from the tip
- Erupted tooth ("strap tooth") in lower jaw of adult male, slanting back toward the melon at an angle of about 45 degrees, often coated with greenish-brown diatoms
- Small, falcate dorsal fin positioned far back on the body
- White beak and throat, large off-white to gray "blaze" beginning just behind the blowhole and sweeping back toward the dorsal fin
- White markings in genital area and on rear margins of flukes

Anatomy

This is the largest and one of the most strikingly colored of the *Mesoplodon* species. The black-and-white color pattern is bold and distinctive. It consists of a prominent black melon and "mask," a white beak and throat, a large off-white to grayish swath ("blaze") beginning just behind the blowhole and sweeping back toward the dorsal fin, and white markings in the genital area and on the rear margins of the flukes. The common name comes from the two erupted teeth ("strap teeth") in the lower jaws of adult males, which slant back toward the melon at an angle of about 45 degrees, wrapping over and pressing against the upper jaw, severely limiting the gape. These teeth are often coated with greenish-brown diatoms (algae), making them relatively inconspicuous.

Behavior

Strap-toothed whales occur in small groups ranging in size from two to six. Little is known about their behavior as they are rarely observed and identified at sea.

Food and foraging

The diet is dominated by various small oceanic squids. Interestingly, even though the fully erupted teeth of adult males limit their gape, they consume prey of similar size to that eaten by females and juveniles. Another interesting point is that these relatively large beaked whales eat fairly small prey—squid weighing less than 3½ oz (100 g)— similar to what is eaten by smaller cetaceans, such as spotted dolphins.

Life history

Virtually nothing is known about life history.

Conservation and management

There are no estimates of abundance for strap-toothed whales but based on the incidence of strandings in southern Africa, they may be fairly common in parts of their range. They have not been hunted regularly anywhere, nor are they known to be taken regularly as bycatch in fisheries. Like other beaked whales, they would be susceptible to bycatch in any part of their range where driftnet fisheries are conducted, and also to the harmful effects of underwater noise, such as naval sonar.

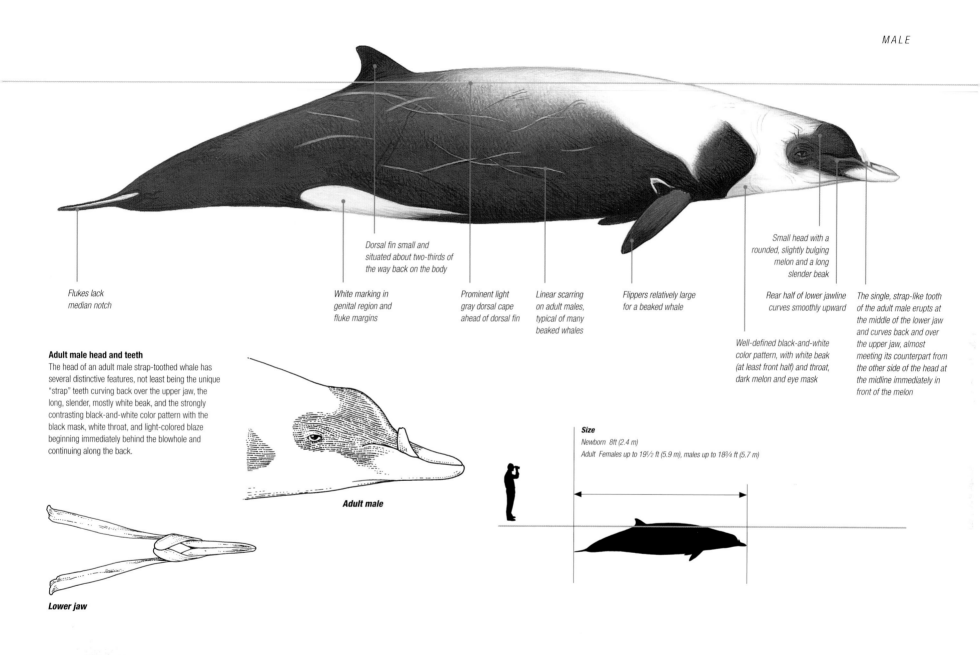

MALE

Dorsal fin small and situated about two-thirds of the way back on the body

Small head with a rounded, slightly bulging melon and a long slender beak

Flukes lack median notch

White marking in genital region and fluke margins

Prominent light gray dorsal cape ahead of dorsal fin

Linear scarring on adult males, typical of many beaked whales

Flippers relatively large for a beaked whale

Rear half of lower jawline curves smoothly upward

Well-defined black-and-white color pattern, with white beak (at least front half) and throat, dark melon and eye mask

The single, strap-like tooth of the adult male erupts at the middle of the lower jaw and curves back and over the upper jaw, almost meeting its counterpart from the other side of the head at the midline immediately in front of the melon

Adult male head and teeth
The head of an adult male strap-toothed whale has several distinctive features, not least being the unique "strap" teeth curving back over the upper jaw, the long, slender, mostly white beak, and the strongly contrasting black-and-white color pattern with the black mask, white throat, and light-colored blaze beginning immediately behind the blowhole and continuing along the back.

Adult male

Lower jaw

Size
Newborn 8ft (2.4 m)
Adult Females up to 19½ ft (5.9 m), males up to 18¾ ft (5.7 m)

Dive sequence
1. When a strap-toothed whale comes to the surface and the viewing conditions are good, the observer may have a chance at least to glimpse the long white beak and somewhat bulbous melon. When conditions are right, the blow may be visible but it is low and bushy.

2. As the animal continues its surfacing "roll," a portion of the beak may remain visible. If the animal is an adult male, the wrap-around "strap" teeth might be detected toward the back of the beak but since they are usually covered in greenish-brown diatoms, they can easily be overlooked. Stalked barnacles attached to the tusks might draw attention, appearing as limp strands of seaweed.

3. The rest of the surfacing will be much like that of other Mesoplodon species, with the small falcate dorsal fin positioned far back on the body.

4. As the back arches the small dorsal fin becomes more prominent.

5. Finally, as the animal dives below the surface, only the dorsal fin and the rear portion of the back are visible.

TRUE'S BEAKED WHALE

Family Ziphiidae

Species *Mesoplodon mirus*

Other common names None

Taxonomy The apparently separate distributions and possible morphological differences between True's beaked whales in the North Atlantic and southern hemisphere suggest that there are separate populations and maybe even subspecies

Similar species Most closely related to Sowerby's beaked whale, it can be difficult to distinguish True's beaked whales at sea unless an adult male is spotted—the exposed teeth at the tip of the lower jaw and the color pattern of the head make it distinctive

Birth weight Unknown

Adult weight At least 2,250 lb (1,020 kg) males; 3,100 lb (1,400 kg) females

Diet Squid and fish

Group size Small, probably fewer than 10 most of the time

Major threats Bycatch in driftnet and longline fisheries in areas where these overlap the species' distribution

IUCN status Data Deficient

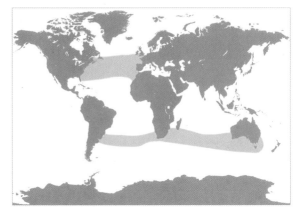

Range and habitat Formerly thought to be endemic to the North Atlantic, this species is now also known to occur regularly in the southern hemisphere, especially the South Atlantic. It is generally considered anti-tropical although there are records from subtropical areas in the Atlantic, such as the Bahamas, Florida, Canary Islands, and southeastern Brazil.

Identification checklist

- Rounded melon sloping onto a relatively short beak—sometimes described as dolphin-like
- Slightly erupted teeth (tusks) visible in males at tip of beak
- Small, triangular or falcate dorsal fin positioned well behind the mid-body
- Coloration pattern differs between North Atlantic and southern hemisphere populations but both forms are dark on the upper side and have a dark eye patch and white throat, with whiteness extending onto sides of head
- Southern form has a distinctive pale, grayish-white band beginning just ahead of the dorsal fin and spreading back to encompass the entire rear quarter of the body

Anatomy

The body is fairly robust and tapered at both ends. The rounded melon slopes smoothly onto the rostrum and the beak is relatively short with a mouthline that rises somewhat toward the back. The two teeth, which erupt slightly in adult males, are situated at the tip of the lower jaw. The dorsal fin is not large and varies in shape from triangular to falcate and hooked. The coloration seems to differ between North Atlantic and southern hemisphere animals. During an observation of a group off the east coast of the United States, the body appeared medium to brownish gray above and pale to white below, and the dark gray dorsal fin contrasted with the lighter gray body. Animals in the southern hemisphere exhibit a pale grayish-white band beginning in front of the dorsal fin and continuing back along the midline and expanding to encompass the rear quarter of the body.

Behavior

Few at-sea observations have been documented in detail and groups have been small, numbering only one to three individuals. The blow is inconspicuous under most sea conditions but when visible is low and columnar. These whales have been seen to breach repeatedly—up to 24 times at intervals of 20–60 seconds—falling back on their side or belly.

Food and foraging

Examinations of the stomach contents from stranded animals showed that both squid and fish are consumed. Observations in the North Atlantic have all been in water at least 3,300 ft (1,000 m) deep so they are likely capable of diving to considerable depths.

Life history

Based on data from stranded females, calves, and fetuses, it is estimated that the gestation period may be about 430 days, lactation lasts at least 300 days and the inter-birth interval is closer to 2 years than 1 year.

Conservation and management

The most significant threat is probably bycatch in driftnet and longline fisheries in areas where True's beaked whales occur. These whales, like other beaked whales, may also be affected by disturbance from human-caused noise, particularly seismic surveys and naval sonar.

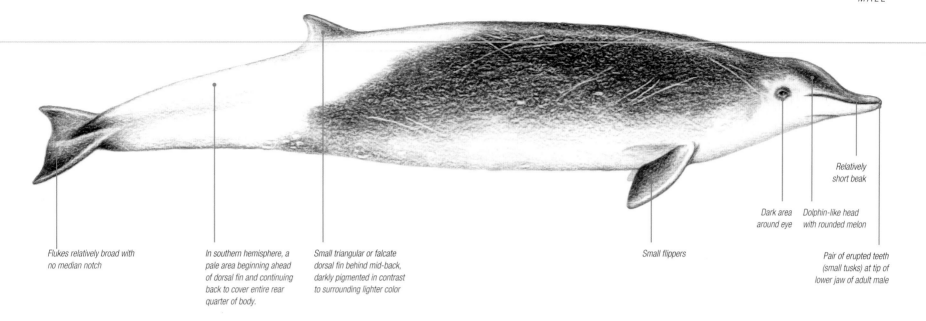

MALE

Flukes relatively broad with
no median notch

In southern hemisphere, a
pale area beginning ahead
of dorsal fin and continuing
back to cover entire rear
quarter of body.

Small triangular or falcate
dorsal fin behind mid-back,
darkly pigmented in contrast
to surrounding lighter color

Small flippers

Dark area
around eye

Dolphin-like head
with rounded melon

Relatively
short beak

Pair of erupted teeth
(small tusks) at tip of
lower jaw of adult male

Adult male head features
The dolphin-like head includes a smoothly sloping
transition from the rounded melon onto a relatively
short rostrum (beak). The two teeth at the tip of
the lower jaw erupt only in adult males. The
mouthline curves upward slightly toward the
rear—again, somewhat dolphin-like.

Size
Newborn 7¼ ft (2.2 m)
Adult 16½ ft (5 m) males; 16¾ ft (5.1 m) females

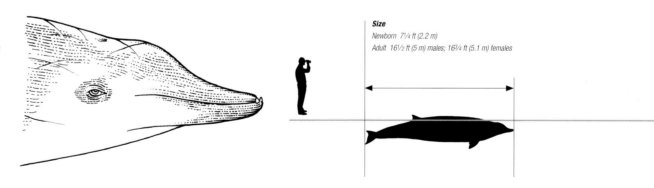

Dive sequence
*1. During a typical
surfacing, the rostrum
and melon clear the
surface first.*

*2. Much of the back follows while
the head is still visible to about the
level of the eye; the short, columnar
but very faint blow may be visible
as the animal appears to skim
along the surface.*

*3. The blow dissipates
quickly and is virtually
gone by the time the
dorsal fin appears.*

*4. The whale may then arch
somewhat as the head sinks
and only the back and dorsal
fin are apparent at the surface.*

*5. Finally, the body
sinks out of view.*

PERRIN'S BEAKED WHALE

Family Ziphiidae

Species *Mesoplodon perrini*

Other common names None

Taxonomy Described and named in 2002 after several stranded specimens in California initially identified as Hector's beaked whales were determined to represent a new species—it is most closely related to the pygmy beaked whale

Similar species Similar to Hector's beaked whale which, however, is known only from the southern hemisphere and therefore probably does not overlap in range with Perrin's

Birth weight Unknown

Adult weight Unknown

Diet Squid

Group size Unknown

Major threats Likely noise especially from naval sonar, gill-net entanglement, and entanglement in and ingestion of marine debris

IUCN status Data Deficient

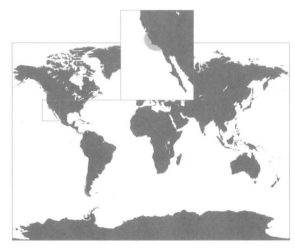

Range and habitat This species is known only from the eastern North Pacific Ocean, with strandings as far south as San Diego, California and as far north as Monterey, California. It is probably confined to offshore waters, at depths of more than 3,300 ft (1,000 m).

Identification checklist

- Small head with relatively short beak and slightly bulging melon
- Straight mouthline
- Triangular exposed portion of tooth (tusk) on each side near tip of lower jaw, in adult males
- Small to medium-sized, somewhat falcate dorsal fin behind the mid-back
- Coloration dark on upper side (including around eyes), light underneath, with white throat and lower jaw, a light gray zone on the side above and in front of the flipper, and a dark stripe running from the eye to just behind the blowhole
- Body flecked with cookie-cutter shark bites and scars, as well as long linear tooth-rake scars on adult males

Anatomy

The body is typical of the genus—compressed along the sides, especially in the relatively short tail region. The beak is shorter than that of most other *Mesoplodon* species, other than Hector's and pygmy beaked whales. The melon bulges slightly and the mouthline is straight. As in other beaked whales, dentition is limited to a single pair of laterally compressed teeth in the lower jaw, which erupt only in adult males. The teeth of Perrin's beaked whales are set just behind the tip of the jaw and the erupted portion takes the form roughly of an isosceles triangle, with the front edge being smoothly convex. These erupted teeth may be colonized by stalked barnacles.

Behavior

There is no information on group size or behavior, largely because this species has never been reliably identified at sea. Linear scars on the body of a stranded adult male suggest that as with most other beaked whales, the tusks of adult male Perrin's at least occasionally inflict wounds on other males, perhaps during aggressive interactions.

Food and foraging

The sparse information from stomach contents points to squid as a major contributor to the diet. Whether these whales also consume fish is unknown at present.

Life history

Virtually nothing is known about life history.

Conservation and management

Other than its apparently very limited range and possibly low numbers (no abundance estimate is available), there is no reason for conservation concern since this species has not been hunted or taken regularly as bycatch in fisheries. However, the potential threats of underwater noise from seismic surveys and naval sonar, entanglement in fishing gear set in deep waters within this whale's range, and the ingestion of plastic debris must always be kept in mind.

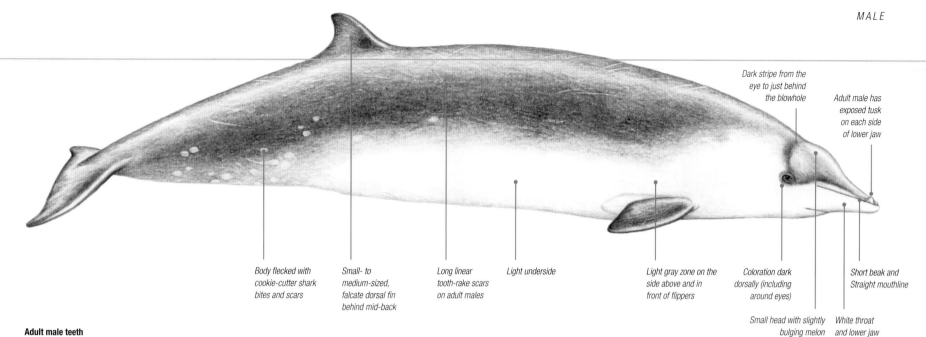

MALE

Dark stripe from the eye to just behind the blowhole

Adult male has exposed tusk on each side of lower jaw

Body flecked with cookie-cutter shark bites and scars

Small- to medium-sized, falcate dorsal fin behind mid-back

Long linear tooth-rake scars on adult males

Light underside

Light gray zone on the side above and in front of flippers

Coloration dark dorsally (including around eyes)

Short beak and Straight mouthline

Small head with slightly bulging melon

White throat and lower jaw

Adult male teeth

The two large, triangular, laterally compressed teeth of adult male Perrin's beaked whales are set very near to the tips of the lower jaws. Like those of other beaked whales, these teeth are exposed outside the closed mouth and are the presumed cause of the long linear scars seen on these whales' bodies.

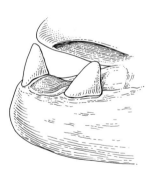

Side view

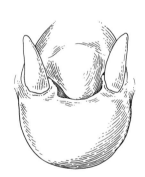

Front view

Size

Newborn Likely around 6½ ft (2 m)

Adult males 12¾ ft (3.9 m), females 14½ ft (4.4 m)

Dive sequence

1. Because these whales have not been seen and identified at sea, the profiles shown here are speculative. However, it is likely that the beak sometimes leads as the animal surfaces, quickly followed by the slightly bulging melon.

2. The back may then appear to roll at the surface.

3. The fairly small- to medium-sized dorsal fin then appears behind the mid-back.

4. It is unlikely that these animals spend much time at the surface or engage in aerial activity before diving for long periods.

PYGMY BEAKED WHALE

Family Ziphiidae

Species *Mesoplodon peruvianus*

Other common names Lesser beaked whale, Peruvian beaked whale

Taxonomy Most closely related to another poorly known species, Perrin's beaked whale

Similar species Similar to other beaked whales in its range but adults are smaller than those of all other beaked whale species

Birth weight Unknown

Adult weight Unknown

Diet Fish and probably squid

Group size Unknown but likely small

Major threats Gill-net entanglement is definitely a threat as a large proportion of the records of this species have involved animals that died in gill nets—they are likely also vulnerable to noise, especially from naval sonar

IUCN status Data Deficient

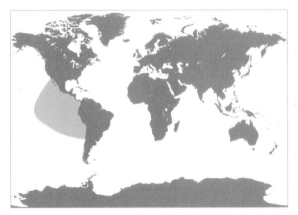

Range and habitat This species is known only from the Pacific Ocean and is probably endemic to that basin but with a fairly extensive range centered in the eastern tropical Pacific. Strandings or catches have been documented from northern Chile, Peru, Mexico, and southern California. It probably occurs mainly in deep offshore waters with depths of more than 3,300 ft (1,000 m).

Identification checklist

- Small head with slight bulge in melon
- Mouthline arches up toward the rear, with tusks of adult males in lower jaws
- Small to medium, somewhat falcate dorsal fin behind the mid-back
- Adult males have a broad white swath across the back
- Heavily scarred on sides of body and the peduncle
- Adult females lack color patterning; they are generally gray to brown with little scarring

Anatomy

This, the smallest, beaked whale has a body shape typical of the genus—compressed at the sides, especially in the short tail region. The beak is proportionally shorter than those of most other *Mesoplodon* species, other than Hector's and Perrin's beaked whales. The melon bulges slightly in front of the blowhole but slopes smoothly onto the short beak. The mouthline curves upward in both sexes but more strongly in adult males. As in other *Mesoplodon* species, dentition is limited to a single pair of laterally compressed teeth in the lower jaw, which erupt only in adult males. The erupted teeth, or tusks, are set in the raised area behind the middle of the lower jaw—they tilt forward and extend well above the top of the rostrum.

Behavior

There is little information on group size or behavior, largely because this species has rarely been identified at sea. Linear scars on adult males suggest that as with most other beaked whales, the tusks of adult male pygmy beaked whales sometimes inflict wounds on other males, perhaps during aggressive interactions.

Food and foraging

The sparse information available from stomach contents suggests that small deep- and mid-water fish are important to the diet in at least some areas. These whales probably also consume squid but this cannot be confirmed at present.

Life history

Virtually nothing is known about life history.

Conservation and management

Although there are no abundance estimates, the species may be fairly common in parts of its range including the southern Gulf of California and offshore Central America. The evidence of bycatch in Peru demonstrates this whale's vulnerability to entanglement in gill nets, which are used extensively in much of its range. Pygmy beaked whales are not known to have been hunted regularly anywhere. Although no mass strandings have been reported, these animals, like other beaked whales, are potentially threatened by underwater noise from seismic surveys and naval sonar.

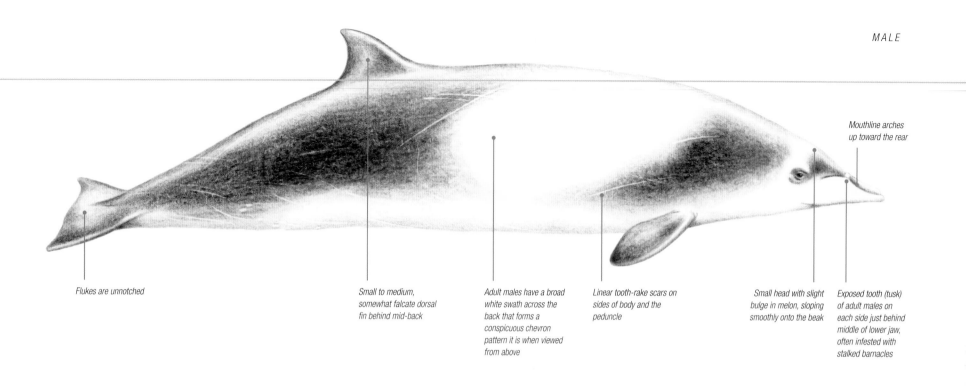

MALE

Mouthline arches up toward the rear

Flukes are unnotched

Small to medium, somewhat falcate dorsal fin behind mid-back

Adult males have a broad white swath across the back that forms a conspicuous chevron pattern it is when viewed from above

Linear tooth-rake scars on sides of body and the peduncle

Small head with slight bulge in melon, sloping smoothly onto the beak

Exposed tooth (tusk) of adult males on each side just behind middle of lower jaw, often infested with stalked barnacles

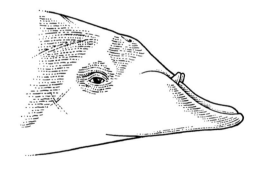

Adult male head features
The head of an adult male pygmy beaked whale shows the strongly upcurved lower jaw and how the tusks, which erupt behind mid-jaw and tilt forward, extend above the plane of the rostrum, possibly functioning as weapons in male–male combat, or perhaps only signaling maturity to potential mates.

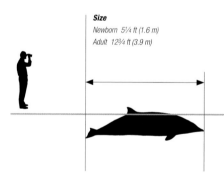

Size
Newborn 5¼ ft (1.6 m)
Adult 12¾ ft (3.9 m)

Dive sequence
1. Pygmy beaked whales are not inclined to show much of their features when surfacing. The blow is very faint, if visible at all, under most circumstances and only the sloping melon and a small part of the rostrum is likely to appear, briefly.

2. The head disappears from view and only the back is visible.

3. Soon followed by the appearance of the small- to medium-sized dorsal fin, which is slightly falcate or hooked.

4. The dorsal fin remains visible as the animal arches at the surface.

5. Finally, it sinks out of view, normally without lifting the flukes clear of the surface.

STEJNEGER'S BEAKED WHALE

Family Ziphiidae

Species *Mesoplodon stejnegeri*

Other common names Bering Sea beaked whale, sabre-toothed beaked whale

Taxonomy No subspecies have been proposed, nor is there any information to use as a basis for describing separate populations

Similar species In the Bering Sea region this is the only *Mesoplodon* species likely to be encountered—Baird's and Cuvier's beaked whales can occur throughout the range but the larger body and longer beak of Baird's and the shorter "goose" beak of Cuvier's, should make it possible to differentiate them

Birth weight Not known

Adult weight Not known

Diet Deep-water squids from at least two families, Gonatidae and Cranchiidae

Group size 3–4 and sometimes up to 15 individuals

Major threats Occasional reports of deliberate kills in Japan and entanglements in driftnets; the usual concerns about potential impacts of underwater noise and ingestion of marine debris also apply to this rarely seen species

IUCN status Data Deficient

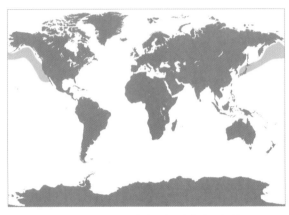

Range and habitat This species is endemic to the sub-Arctic and cold-temperate North Pacific and Bering Sea, with stranding records suggesting greatest abundance around the Aleutian Islands, in the deep waters of the southwestern Bering Sea, and in the Sea of Japan. Strandings and gill-net entanglements have occurred as far south as southern California in the east and Honshu (Japan) in the west.

Identification checklist

- Typical *Mesoplodon* features—small flippers and small, slightly falcate dorsal fin positioned behind the mid-back
- Beak appears to be nothing more than an extreme tapering of the body, with a smooth sloping transition from melon to rostrum
- Gently arched mouthline of adult males with a prominent, forward-tilted tusk at the front of the arch on each side
- Dark "skull cap" extending from rostrum down around eyes gives a "helmeted" appearance
- Except on lower jaw and throat, body of adult males is dark gray to black with heavy scarring, females and juveniles are dark on the upper side and light underneath with white lower jaw and throat

Anatomy

The most striking features of this *Mesoplodon* are found on the adult male. The melon is flat and slopes smoothly onto the rostrum. The lower jaws are arched, providing platforms for the sharp-tipped, forward-tilting teeth that erupt as tusks on adult males. As in Hubbs' beaked whale, these teeth can protrude above the rostrum when the mouth is closed. Body color is dark gray to black, with lighter tones on the front underside of the flippers. A dark "cap" extends from the rostrum around the eyes. Females often have an unusual pattern of white on the undersides of the flukes.

Behavior

Stejneger's beaked whales have been seen in close-knit groups of 3 or 4 and up to 15 individuals. A mass stranding of four females in the Aleutians suggests a degree of sex segregation. The extensive body scarring, together with the dentition and facial anatomy, suggest that adult males fight, possibly for social dominance or access to females.

Food and foraging

The diet has been studied from the stomach contents of stranded individuals. These whales appear to rely on deep-water squid. From the known squid prey species, it can be inferred that Stejneger's beaked whales make regular dives deeper than 660 ft (200 m).

Life history

Males can live for at least 36 years, judging by counts of tooth layers. Analyses of strandings data have pointed to the birth season being mainly spring. It has been suggested that these whales migrate, with peak occurrence in the southern reaches of their range in winter and spring. The presence of cookie-cutter shark bites on animals found in the Aleutian Islands suggests that they spend part of the year in warmer waters.

Conservation and management

Infrequent sightings suggest these whales are fairly rare. They have been found dead in driftnets. Disturbance from underwater noise from naval sonar and seismic surveys is also a threat. Mass strandings have occurred in the Aleutian Islands—beaked whales are not prone to such events except when exposed to naval sonar.

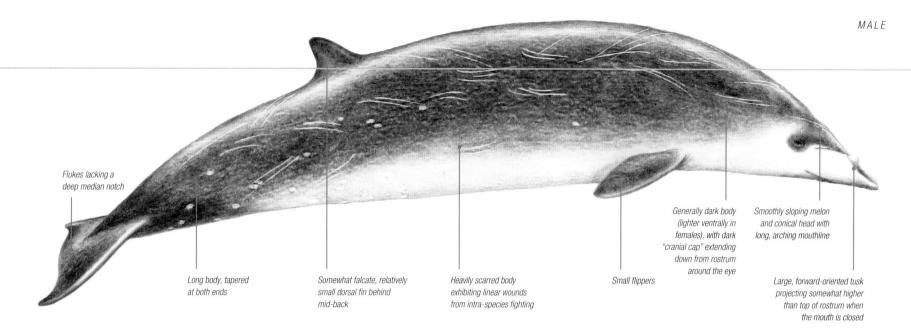

MALE

Flukes lacking a
deep median notch

Long body, tapered
at both ends

Somewhat falcate, relatively
small dorsal fin behind
mid-back

Heavily scarred body
exhibiting linear wounds
from intra-species fighting

Small flippers

Generally dark body
(lighter ventrally in
females), with dark
"cranial cap" extending
down from rostrum
around the eye

Smoothly sloping melon
and conical head with
long, arching mouthline

Large, forward-oriented tusk
projecting somewhat higher
than top of rostrum when
the mouth is closed

Adult male head features
The head of the adult male has a smoothly sloping
melon and conical head with a long, arched mouthline.
A large, forward-pointing tusk projects higher than the
top of rostrum when the mouth is closed. A dark "skull
cap" extends down from the rostrum around the eye.

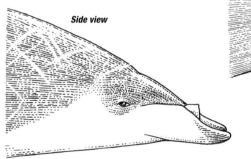

Side view

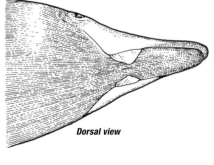

Dorsal view

Size
Newborn 7 ft (2.1 m)
Adult Up to 17¾ ft (5.4 m)

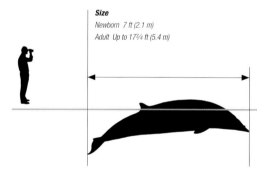

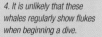

Surface behavior
The rarity of sightings at sea suggests that these
whales are not easily detected and identified and
therefore their profile at the surface may be
relatively inconspicuous— small groups, blow
hard to discern, little time spent and relatively
undemonstrative at the surface, vessel-shy etc.

Dive sequence
1. The tapered head, with its
gently arched lower jaw, as well
as the helmet-like cranial cap
may be seen first as the animal
breaks the surface. The rather
inconspicuous blow may
or may not be visible .

2. Next, the top of the
head and back appear.

3. As the head disappears, more
of the back and the small dorsal
fin appear.

4. It is unlikely that these
whales regularly show flukes
when beginning a dive.

SPADE-TOOTHED BEAKED WHALE

Family Ziphiidae

Species *Mesoplodon traversii*

Other common names None, formerly Bahamonde's beaked whale

Taxonomy One of the most recently described species of cetacean although the first specimen—assigned to a different genus—was found in 1874

Similar species Adults are similar in appearance to Gray's beaked whale or strap-toothed (Layard's) beaked whale

Birth weight Unknown

Adult weight Unknown

Diet Unknown but likely deep-water squid and fish, similar to most other *Mesoplodon* whales

Group size Unknown

Major threats None known

IUCN status Data Deficient

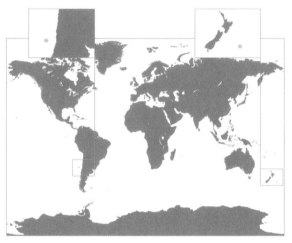

Range and habitat Thus far the species is documented to occur only in mid-latitudes of the South Pacific Ocean (Chile and New Zealand). Nothing definite is known about habitat but this is almost certainly a deep-water species.

Identification checklist

- Presumably low-angle surfacing profile typical of the genus; the relatively long beak might be partially visible at times
- Dorsal fin positioned well behind the mid-back, falcate, neither exceptionally large nor small
- Adult female coloration: white belly and flanks, with whiteness extending above the flippers and onto the sides of the face; dorsal surface and all appendages as well as the prominent melon and beak are dark gray or black, and there is a black eye patch.
- Juveniles and adult males may have a color pattern broadly similar to that of adult females, based on photographic evidence of a degraded juvenile carcass.

Anatomy

Only four records of occurrence exist, two based on skull fragments, one on a lower jaw with teeth (from an adult male), and the most recent on two animals that stranded together in New Zealand—one an adult female and the other a juvenile male. Photographs of the head profile of the relatively fresh adult female indicated a prominent melon and moderately long beak. The juvenile male's profile showed a much flatter melon sloping gently onto the rostrum. As in most other species in the genus, the teeth of the adult male erupt through the gum and may be visible outside the closed mouth.

Behavior

Because this species has never been observed and identified as such at sea, it is impossible to say anything certain about its behavior. Presumably it does not form large groups and occurs mainly in offshore waters.

Food and foraging

Nothing is known about food or foraging but it is reasonable to assume that, like the other beaked whales, spade-toothed beaked whales forage deep in the water column and subsist primarily on squid and fish.

Life history

Nothing is known about the life history of this rare and extremely poorly known species.

Conservation and management

For a rare and extremely poorly known species such as this one, it is impossible to characterize threats or to prescribe appropriate conservation actions.

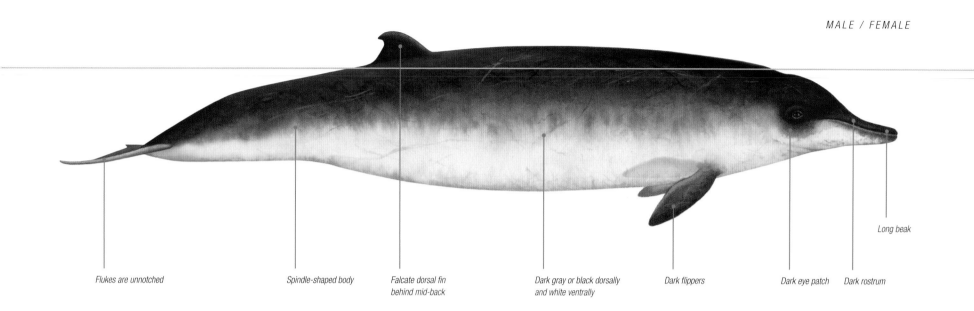

MALE / FEMALE

Flukes are unnotched

Spindle-shaped body

Falcate dorsal fin
behind mid-back

Dark gray or black dorsally
and white ventrally

Dark flippers

Dark eye patch

Dark rostrum

Long beak

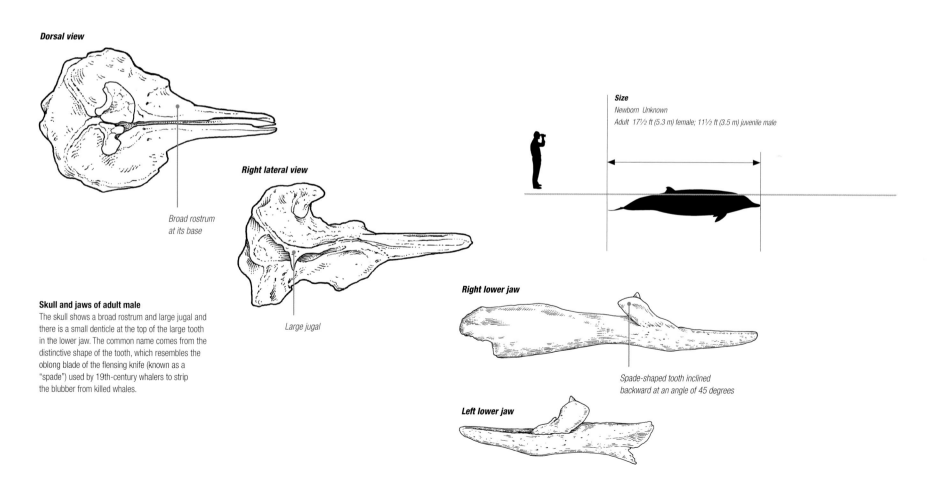

Dorsal view

Broad rostrum
at its base

Right lateral view

Large jugal

Size
Newborn Unknown
Adult 17½ ft (5.3 m) female; 11½ ft (3.5 m) juvenile male

Right lower jaw

Spade-shaped tooth inclined
backward at an angle of 45 degrees

Left lower jaw

Skull and jaws of adult male
The skull shows a broad rostrum and large jugal and
there is a small denticle at the top of the large tooth
in the lower jaw. The common name comes from the
distinctive shape of the tooth, which resembles the
oblong blade of the flensing knife (known as a
"spade") used by 19th-century whalers to strip
the blubber from killed whales.

SHEPHERD'S BEAKED WHALE

Family Ziphiidae

Species *Tasmacetus shepherdi*

Other common names Tasman beaked whale, Tasman whale

Taxonomy The unusual dentition, with a full complement of teeth in upper and lower jaws, makes Shepherd's beaked whale unique within the Ziphiidae and no subspecies are recognized

Similar species Could be confused with other large ziphiids that occur in its range, but in good sighting conditions it can be distinguished by its coloration pattern and features of the head (long dark beak and steeply rising, lighter-colored melon)

Birth weight Unknown

Adult weight Unknown

Diet Both squid and bottom fish

Group size Small, usually 3–6 individuals

Major threats None known with certainty although possible threats include noise, especially from naval sonar; entanglement in fishing gear or debris; and ingestion of marine debris

IUCN status Data Deficient

Range and habitat This species is widely distributed in the southern hemisphere between 30°S and 55°S, in cold-temperate waters. It has been observed in waters from a few hundred to many thousands of feet in depth. Most strandings have been in New Zealand, while live observations have taken place around Tristan da Cunha and Gough Island, and off northern New Zealand and southern Australia.

Identification checklist

- Body robust with external features typical of ziphiids: a pair of throat grooves that converge toward the front, small flippers set low on the sides, flukes without a central notch, and a falcate (dolphin-like) dorsal fin set well behind the mid-back
- Fairly long, well-defined beak with a rather pointed tip
- Melon prominent and bluff, more like that of bottlenose whales than *Mesoplodon* species
- Two large teeth erupt (but are barely visible) at tip of lower jaw of adult males, and both sexes have rows of smaller erupted teeth along upper and lower jaws
- Coloration consists of a dark beak and dark eye patches, a pale melon, a whitish to light gray shoulder, and flank patches continuous with the white belly

Anatomy

Until the 1980s, Shepherd's beaked whales were known only from strandings, with no photographs or examinations of live individuals or fresh stranded specimens to provide a basis for describing their appearance. The distinctive coloration pattern is now well known and there appears to be little difference in this patterning between males and females or between old and young. The species is unique among the ziphiids (beaked whales) in having a full set of functional teeth lining the upper and lower jaws (17–21 pairs above, 22–28 pairs below) as well as the usual pair of larger teeth at the tip of the lower jaw that erupt only in adult males (but are not very conspicuous). The forehead is steep and the beak well defined, apparently becoming longer with age. Linear scars on the bodies of adult males are probably the result of aggressive interactions among them.

Behavior

On the few occasions when the species has been seen, the blow was either visible (in sightings from the air) or invisible to only slightly visible (sightings from shipboard). Groups have tended to be small with three to six animals. They display no aerial behavior, such as breaching or fluking. The beak may be lifted clear of the water when surfacing, at least some of the time.

Food and foraging

Information on diet comes from examinations of the stomach contents of only two stranded individuals, one of which had eaten mostly bottom-living fish and the other, squid. Nothing is known about diving behavior but given its primarily offshore distribution, this species is likely an adept diver.

Life history

Virtually nothing is known.

Conservation and management

Shepherd's beaked whales have never been exploited on a significant scale anywhere in their range. They may be affected, as are other beaked whales, by disturbance from noise associated with offshore oil and gas development and military activities, as well as by ingestion of marine debris.

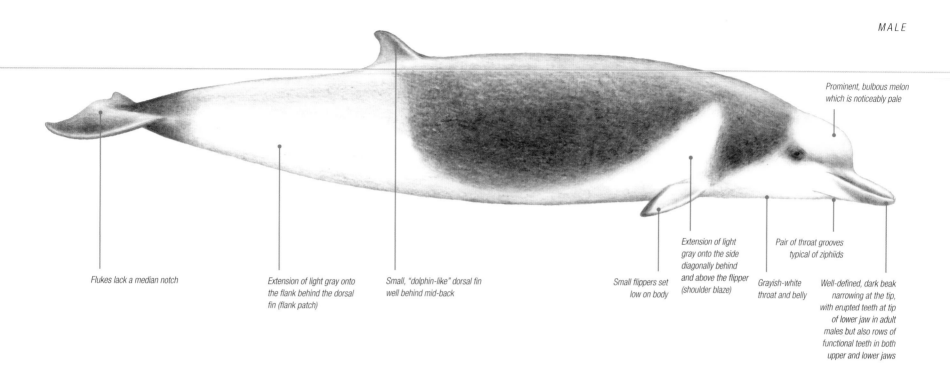

MALE

Prominent, bulbous melon
which is noticeably pale

Flukes lack a median notch

Extension of light gray onto
the flank behind the dorsal
fin (flank patch)

Small, "dolphin-like" dorsal fin
well behind mid-back

Small flippers set
low on body

Extension of light
gray onto the side
diagonally behind
and above the flipper
(shoulder blaze)

Grayish-white
throat and belly

Pair of throat grooves
typical of ziphiids

Well-defined, dark beak
narrowing at the tip,
with erupted teeth at tip
of lower jaw in adult
males but also rows of
functional teeth in both
upper and lower jaws

Adult male head features

The adult male (as well as the female) has a fairly
long, well-defined beak and a bulbous melon
(forehead). Color elements include a dark beak and
area around the eye, a pale melon, and a white
throat that extends up to the level of the eye. The
two erupted teeth at the tip of the lower jaw (only
in adult males) are present but inconspicuous.

Size
Newborn Probably around 10 ft (3 m)
Adult 21¾–22½ ft (6.6–6.8 m)

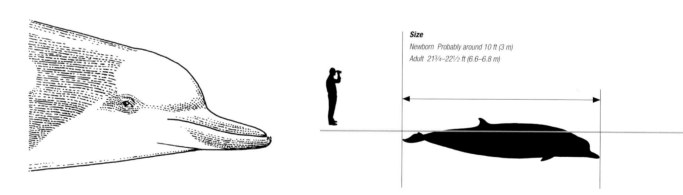

Dive sequence

1. The beak may come out
of the water at an angle of around
40 degrees just ahead of the
prominent, relatively steep melon as
the animal surfaces. The blow may
be barely visible but generally is not
seen during shipboard sightings.

2. The beak and melon
are visible as the roll at
the surface begins.

3. The small dolphin-like
dorsal fin, set well behind
the midpoint of the back,
rises out of the water.

4. The back arches,
making the dorsal fin more
prominent, while the melon
dips below the surface.

5. As the animal arches for a final
dive after a series of shallow dives
near the sea surface, it is unlikely
to raise its flukes. Observations
suggest that these whales generally
do not exhibit aerial behavior—
however this is based on little
evidence as sightings are rare.

CUVIER'S BEAKED WHALE

Family Ziphiidae

Species *Ziphius cavirostris*

Other common names Goose-beaked whale

Taxonomy The only species of this genus; no subspecies recognized; distinct populations likely to exist but not yet characterized

Similar species Easily confused with other similar-sized ziphiids, particularly female and juvenile *Mesoplodon* species—the relatively short beak helps distinguish Cuvier's from other beaked whales

Birth weight 550–660 lb (250–300 kg)

Adult weight 4,850–6,400 lb (2,200–2,900 kg)

Diet Mainly deep-water oceanic squid, also some fish and shrimp

Group size Small, generally 3–4 individuals, occasionally up to 10

Major threats Noise especially from naval sonar, entanglement in gill nets and debris, ingestion of marine debris, such as plastic bags

IUCN status Least Concern globally but Mediterranean subpopulation Vulnerable

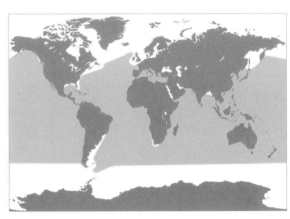

Range and habitat This species is nearly cosmopolitan in cool-temperate to tropical waters. It is primarily found in continental or island slope areas at depths of 3,300 ft (1,000 m) and more.

Identification checklist

- Typical features of many ziphiids —small falcate (curved) dorsal fin about two-thirds back, small flippers that fit into "pockets" on the body wall, the lack of a median notch between proportionally large flukes, and "V"-shaped throat grooves
- Relatively short beak and distinctively upcurved mouthline—head profile is reminiscent of a goose beak
- Single pair of conical forward-slanting teeth at tip of the lower jaw, erupted only in adult males, exposed when mouth is closed
- Dark gray (adult males) to reddish brown (adult females), white head of adult males, numerous cookie-cutter shark scars, and many tooth rakes on adult males

Anatomy

A robust ziphiid with a noticeably short, peculiarly goose-like beak. The species name *cavirostris* refers to a well-developed basin, or cavity, in the top of the skull immediately in front of the nasal bones. Adult males, like adult male *Mesoplodon* whales, have a dense rostrum that could serve in sound production or male–male combat.

Behavior

They tend to travel in small groups averaging 3–4, with occasional sightings of 10 or more as well as of solitary individuals. The members of a group tend to dive at the same time for an average of about an hour, sometimes longer. Groups, especially those consisting of adult males, can be stable over years and stay in relatively small areas.

Food and foraging

Although often characterized as squid specialists, Cuvier's beaked whales appear to be opportunistic, preying on a wide variety of open-ocean, mid- to deep-water organisms. As they descend on a deep dive, whales start clicking continuously at a rate of about two clicks per second when they reach a depth of 1,300 –1,640 ft (400–500 m), searching for prey. Upon detecting a prey item, the click repetition rate increases until it becomes a buzz. The whales then cease clicking as they ascend to the surface and remain silent until the next deep dive.

Life history

Almost nothing is known about the life history of this species. Sexual maturity is thought to be reached at an average body length of 19 ft (5.8 m) in females and 18 ft (5.5 m) in males.

Conservation and management

Probably the world's most abundant ziphiid, given its broad geographical distribution and the number of strandings as well as at-sea observations. The threat of greatest concern is intense mid-frequency naval sonar, which has been responsible for numerous mass strandings of Cuvier's and other beaked whales. Required use of acoustic pingers in a driftnet fishery off California apparently was effective in eliminating beaked whale bycatch.

MALE

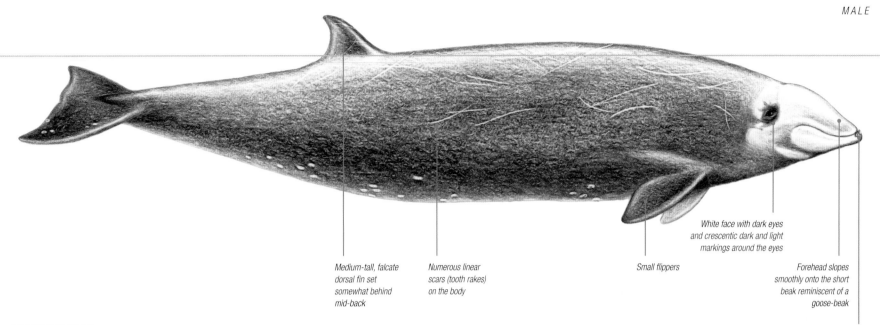

Medium-tall, falcate
dorsal fin set
somewhat behind
mid-back

Numerous linear
scars (tooth rakes)
on the body

White face with dark eyes
and crescentic dark and light
markings around the eyes

Small flippers

Forehead slopes
smoothly onto the short
beak reminiscent of a
goose-beak

Small pair of erupted
teeth at tip of the lower
jaw (often obscured by
stalked barnacles)

Adult head appearance
The adult male is readily distinguished by the white front part of
the head (often extending across the upper surface of the neck)
as well as the two forward-tilting teeth at the tip of the lower jaw.
Adult females have no erupted teeth and a more muted but still
complex pigmentation pattern including the dark areas and light
swirl markings around the eyes.

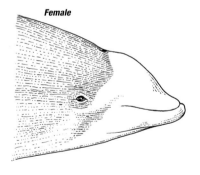

Female

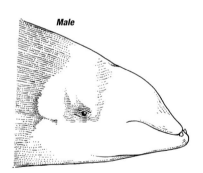

Male

Size
Newborn 9 ft (2.7 m)
Adult average 20 ft (6.1 m) but range up to 23 ft (6.9 m)

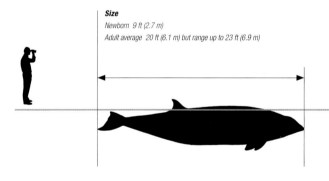

Dive sequence
1. The short beak and head appears
first above the surface. The low,
bushy blow quickly follows.

2. The often heavily scarred back and
relatively tall, falcate dorsal fin emerge
after the head has submerged.

3. These whales do not lift the flukes
clear of the surface when diving, and
rarely breach. Generally they can only
be well observed in calm seas.

River dolphin spyhopping (right)

Spyhopping is a behavior exhibited by many whale and dolphin species that allows them to view the world above the water's surface. When a river dolphin spyhops, its slender beak is easily observed.

TOOTHED WHALES
River Dolphins

River dolphins consist of four species of small-bodied cetaceans distributed between four families. Most river dolphin species occupy freshwater rivers of South America and Asia, although the Franciscana dolphin inhabits saltwater estuaries and nearshore habitats. These dolphins range in size from approximately 6½–10 ft (2–3 m) in length and are characterized by elongated and narrow beaks. Despite similarities in their anatomy and habitats, the river dolphin families are not closely related.

River dolphin characteristics

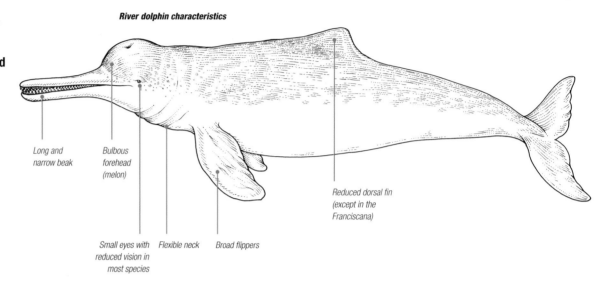

Long and narrow beak

Bulbous forehead (melon)

Small eyes with reduced vision in most species

Flexible neck

Broad flippers

Reduced dorsal fin (except in the Franciscana)

- The skin of river dolphins can span a wide spectrum of colors, including gray, black, brown, yellow, white, and pink. Skin coloration often changes as an individual ages.

- River dolphins tend to live in muddy waters with low visibility. As a result, many of these dolphins have small eyes and poor vision. The Ganges River dolphin lacks lenses and the eyes are likely to be used as light-sensing organs. All species utilize echolocation for navigation, and the boto possesses stiff whiskers, or vibrissae, on its snout to help guide itself through murky water.

- The diet of river dolphins largely consists of fish, although some individuals are reported to eat crustaceans, mollusks, and even turtles.

- Unlike in most toothed whales, the vertebrae in the necks of river dolphins are unfused, giving the neck flexibility. The boto in particular uses its flexible neck to help it navigate through flooded forests.

- River-dwelling species tend to be solitary or congregate in small and loose groups. Those that inhabit estuaries and coastal waters, such as the Fransciscana, can congregate in groups with one to two dozen individuals, but large pods that characterize some oceanic dolphins are not observed in river dolphin communities.

- All species of river dolphin are protected, and the Ganges River dolphin is classed as an endangered species. The baiji or Chinese river dolphin, is considered to be functionally extinct, meaning that the wild population is no longer viable. There have been no confirmed sightings of the species since 2004. Other river dolphin species are either vulnerable to extinction or too poorly studied for an accurate conservation assessment to be made.

Skull side view

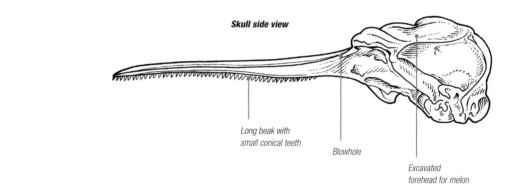

Long beak with small conical teeth

Blowhole

Excavated forehead for melon

Skin coloration

River dolphins vary in skin coloration and patterns. The baiji (Lipotes vexillifer) is counter-shaded with a light gray or white underside and gray upper side. The Ganges River dolphin (Platanista minor) typically is uniformly brown but can be gray-blue to gray. The Amazon River dolphin (Inia geoffrensis) ranges from gray to carnation pink that may become blotchy with age.

BAIJI

Family Lipotidae

Species *Lipotes vexillifer*

Other common names Yangtze river dolphin, Chinese river dolphin

Taxonomy Closely related to the Amazon River dolphin and Franciscana

Similar species The Yangtze finless porpoise inhabits the same range as the baiji

Birth weight About 13 lb (6 kg)

Adult weight 275 lb (125 kg) males; 525 lb (238 kg) females

Diet Any available species of freshwater fish, including silver carp, common carp, and grass carp

Group size Usually 2–6 animals, sometimes up to 16 individuals

Major threats Habitat degradation, river traffic, fishing gear, reduction of fishery resources, and water pollution

IUCN status Declared Functionally Extinct in 2007

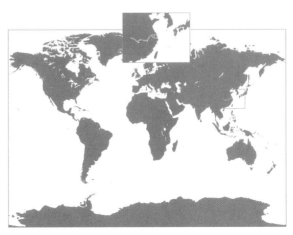

Range and habitat This species inhabited the mainstream of the middle and lower reaches of the Yangtze River; individuals might occasionally enter some tributary lakes during intense flooding. Some individuals were seen in the Fuchun River, immediately south of the Yangtze in 1955.

Identification checklist

- Very long, narrow, and slightly upturned beak
- Oval blowhole
- Very small eyes
- Low triangular dorsal fin located about two-thirds of body length from snout
- Broad rounded flippers

Anatomy

The baiji is countershaded with light bluish gray on the upper side and pale gray to white on the underside. The long, narrow beak seen in adults is slightly blunter in juveniles. The eyes are very small and they are set high up on the sides of the head.
There are 62–68 teeth in the upper jaw and 64–72 in the lower jaw. The tooth enamel is ornamented with irregular vertical ridges.

Behavior

Baiji were generally found in eddy countercurrents below meanders and channel convergences. They lived in small groups. In the 1980s, the most common group size was 2–6 animals; the largest group observed was about 16 individuals. The largest recorded range of a recognizable baiji was more than 120 miles (200 km) from the initial sighting location.

Food and foraging

The baiji was an opportunistic feeder, taking any available species of freshwater fish, the only selection criterion being size and they selected fish that were not too large for them to swallow.

Life history

The sexes had roughly the same growth rate until they were about four years old, which was the age of sexual maturity for males. Following sexual maturity, males grew more slowly than females. Females attained sexual maturity at about six years and they continued to grow until around the age of eight.

Conservation and management

Despite all the efforts made to conserve the baiji since the early 1980s, the population declined drastically in less than two decades as rapid industrialization along the Yangtze valley has led to massive habitat degradation.

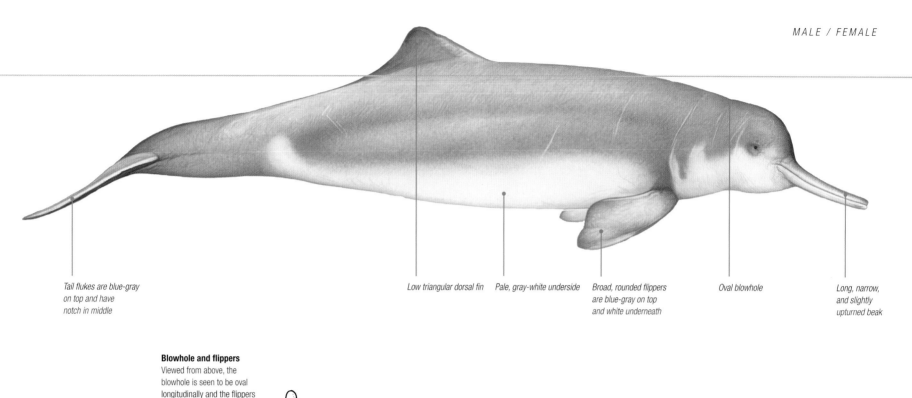

Tail flukes are blue-gray on top and have notch in middle

Low triangular dorsal fin

Pale, gray-white underside

Broad, rounded flippers are blue-gray on top and white underneath

Oval blowhole

Long, narrow, and slightly upturned beak

Blowhole and flippers
Viewed from above, the blowhole is seen to be oval longitudinally and the flippers are broad and rounded.

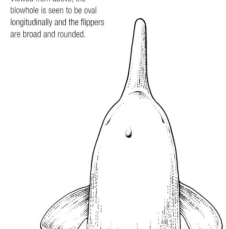

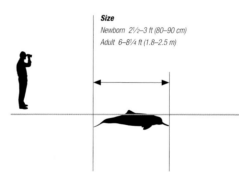

Size
Newborn 2½–3 ft (80–90 cm)
Adult 6–8¼ ft (1.8–2.5 m)

Blow
The baiji usually surfaces without causing white water and breathes in a smooth manner. The duration of the blow is 0.2 to 0.6 seconds.

Dive sequence
This species has a sequence of several short breathing intervals (10–30 seconds) alternating with a longer one, the longest being up to 200 seconds.

FRANCISCANA

Family Pontoporiidae

Species *Pontoporia blainvillei*

Other common names La Plata dolphin, Franciscana dolphin

Taxonomy Closest relatives are the Amazon River dolphins of the genus *Inia*, living in rivers of northern South America

Similar species Burmeister's porpoise and tucuxi

Birth weight 11–13 lb (5–6 kg)

Adult weight 44–88 lb (20–40 kg), with females averaging larger, 71 lb (32 kg) to males 57 lb (26 kg)

Diet Wide variety of small fish, mollusks, squid, and shrimp

Group size Normally 2–3, up to 30

Major threats Gill-net entanglement and drowning in trawl nets

IUCN status Vulnerable

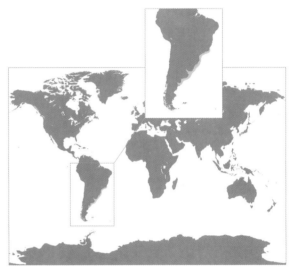

Range and habitat This species is found along coasts and in estuaries of the western South Atlantic, from southeast Brazil to northern Argentina—mostly in waters up to 130 ft (40 m) deep.

Identification checklist

- Very small dolphin, which is uniformly brownish
- Slightly falcate, rounded dorsal fin
- Flexible neck
- Broad flippers with ridges that have the distinct appearance of fingers
- Shy, with quiet surfacing

Anatomy

The Franciscana resembles other river dolphins. It has a very long, slim beak in adulthood, broad flippers, and a small, rounded dorsal fin. The relative length of the beak increases with age, and holds a large number of small, sharp teeth (50-62 pairs in upper and lower jaws), more than in most other cetacean species.

Behavior

Franciscanas live in turbid waters, in small groups, with little display when they surface. Leaping is rare and bow-riding unknown. The species probably does not disperse from its natal area, has a small home range, and lives in small matrilineal groups.

Food and foraging

Mostly forages opportunistically on the seabed, taking a variety of small prey fish and squid, and occasionally mollusks and shrimps—the latter especially in young animals. Apparent cooperative feeding behavior has been reported, but is not thought to be usual.

Life history

Franciscanas reach sexual maturity at a very young age (two to five years), producing a calf every year or two, and dying young compared to other dolphins. Most do not reach 20 years of age. The combination of the small body size of males compared to females, relatively small testes, the lack of fighting scars on the skin—in contrast to the Amazon River dolphin, for example—and the small group size points to an unusual, perhaps unique, social system in this dolphin. Mate guarding and paternal assistance in rearing the young have been hypothesized to explain these observations, but conclusive evidence is lacking. Breeding is very seasonal in the southern part of the range, but much less so further north.

Conservation and management

There is good evidence that the level of fishery mortality in this species exceeds the birth rate, and so populations are declining. Fishery deaths occur throughout the species' range in Brazil, Uruguay, and Argentina.

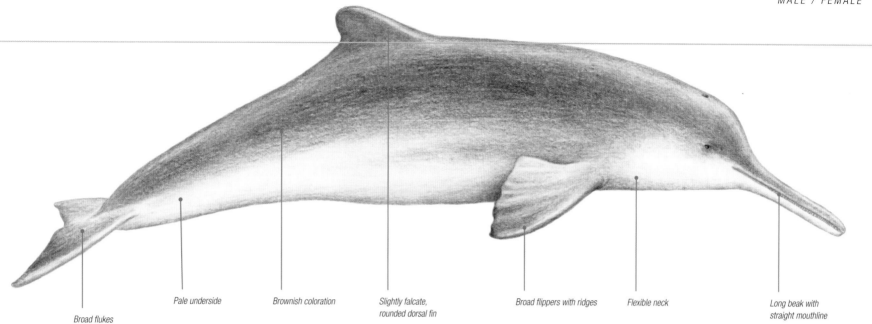

Broad flukes

Pale underside

Brownish coloration

Slightly falcate, rounded dorsal fin

Broad flippers with ridges

Flexible neck

Long beak with straight mouthline

Flippers with "fingers"
The flipper of the Franciscana is unlike that of any other dolphin, with ridges which correspond to the "finger" bones within the flipper. The unique shape of this species owes much to its origins in rivers, but today it occurs only in the sea.

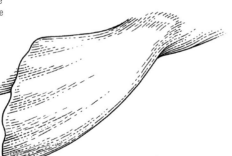

Size
Newborn 2¼–2½ ft (70–80 cm)
Adult 4¼–5 ft (1.3–1.5 m) female; 3¾–4¼ ft (1.1–1.3 m) male

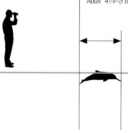

Dive sequence
1. The beak and head break the surface first.

2. Next the back and dorsal fin briefly appear.

3. The dolphin quickly submerges below surface.

AMAZON RIVER DOLPHIN

Family Iniidae

Species *Inia geoffrensis*

Other common names Boto, pink dolphin

Taxonomy Although two other populations previously recognized as subspecies have been proposed as distinct species, *I. boliviensis*, and *I. araguaiaensis*, some doubts remain about the recognition of these isolated populations as separate species

Similar species Franciscana and the Asian river dolphins

Birth weight 22–29 lb (10–13 kg)

Adult weight 220–456 lb (100–207 kg) with the larger males averaging 340 lb (154 kg) and females averaging 220 lb (100 kg)

Diet Fish, rarely crabs and turtles

Group size Typically 1–5, but up to 40

Major threats Hunting for use as fish bait, net entanglement

IUCN status Data Deficient

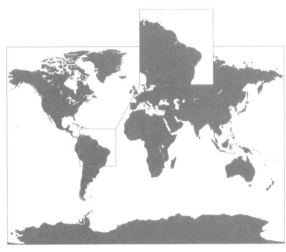

Range and habitat This species is confined to freshwater habitats, such as rivers, lakes, channels, and flooded forest throughout most of the Amazon and Orinoco basins. Seasonal movements occur but no migrations.

Identification checklist

- Body color is dark gray (calves) to bright pink (adult males)
- Long, robust, narrow beak
- Distinctive bulbous melon
- Long, low dorsal fin or ridge
- Body and neck flexible

Anatomy

The flexible body, broad flippers, and unfused neck vertebrae separate these river dolphins from their more familiar marine counterparts—dolphins within the family Delphinidae. This freshwater species is known for the pink coloration of adult males.

Behavior

This species is well adapted to life in shallow waters of the Amazon basin, happily swimming in depths of less than 6½ ft (2 m), and in the tangled vegetation of the flooded forest. Group size is typically small, though aggregations of up to 40 may occur in favored locations. Living in waters with poor visibility, Amazon River dolphins primarily use echolocation to forage. Males are much larger than females, and probably use their size to compete for sexual access. Uniquely among cetaceans, males use objects such as rocks and branches in socio-sexual display.

Food and foraging

Amazon River dolphins are unique in having two types of teeth—conical at the front for grasping, and cusped at the back for crushing. This adaptation allows them to eat armored catfish and even, on occasion, turtles. A broad variety of fish dominates the diet, varying seasonally as water levels rise and fall, with commensurate impacts on the aquatic ecology.

Life history

Births occur year round, peaking in the low-water season. Calves remain with their mothers for at least two years. Simultaneous lactation and pregnancy is quite common, so mature females spend most of their lives nursing their most recent calf. Females give birth to their first calf at between 7 and 10 years of age.

Conservation and management

Amazon River dolphins live their entire lives close to human populations, which rely on the same waterways for food and transportation. The deliberate hunting of them for use as fish bait has had disastrous consequences over the past two decades. This is in addition to frequent accidental entanglement in gill nets. The IUCN status reflects the lack of knowledge about Amazon River dolphins, and masks what is almost certainly a continuing decline in numbers throughout their geographic range.

MALE / FEMALE

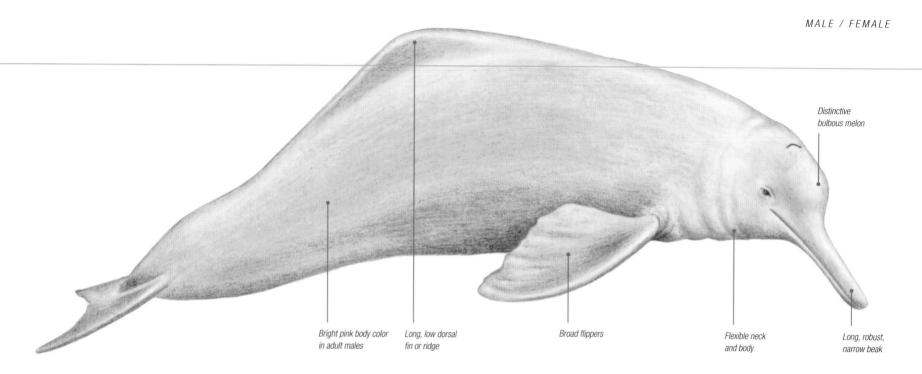

Distinctive
bulbous melon

Bright pink body color
in adult males

Long, low dorsal
fin or ridge

Broad flippers

Flexible neck
and body

Long, robust,
narrow beak

Teeth in lower jaw

These dolphins have 31–36 small, conical teeth in both
the upper and lower jaws. The posterior teeth in the
lower jaw are especially broad at the base and enhance
the ability of this species to crush hard-bodied prey
such as armored fish.

Side view

Overhead view

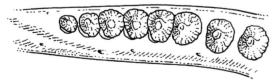

Size

Newborn 2½–3 ft (80–90 cm)
Adult 6–8½ ft (1.8–2.6 m)

Dive sequence

1. The beak and forehead
break the surface of the
water first.

2. Next, the back and dorsal
fin break the surface.

3. Then the back arches briefly
before the dolphin disappears
below the surface.

4. The flukes remain below
the surface when diving.

GANGES RIVER DOLPHIN

Family Platanistidae

Species *Platanista gangetica*

Other common names Susu, blind river dolphin, South Asian dolphin

Taxonomy Two subspecies recognized *P.p. gangetica* (Ganges River dolphin) and *P.p. minor* (Indus River dolphin)

Similar species None

Birth weight 9–11 lb (4–5 kg)

Adult weight 154–198 lb (70–90 kg)

Diet Freshwater fish (such as gobies) and invertebrates (such as prawns)

Group size Solitary or in pairs—rarely form groups of 6–10

Major threats Hydroelectric power and irrigation projects, hunting, and fisheries bycatch

IUCN status Endangered

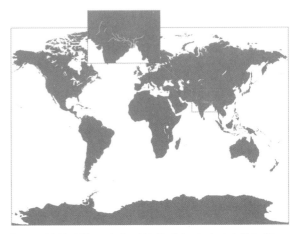

Range and habitat This species is restricted to shallow, freshwater habitats of the Ganges-Brahmaputra-Meghna and Karnaphuli-Sangu river systems of Nepal, India, and Bangladesh. It is generally found in waters with salinity levels below 10 ppt (parts per thousand).

Identification checklist

- Long, narrow, prominent beak
- Single slit instead of crescent-shaped blowhole
- Triangular dorsal fin located two-thirds down the back
- Broad flippers

Anatomy

This is a stocky dolphin with a creased, flexible neck. Females are slightly larger than males. They have a dark brown appearance with a slightly darker upper surface and pinkish underside. Tiny, pinhole eyes are located just above the corners of the upturned mouth. They are poorly developed and lack crystalline lenses, hence the nickname "blind dolphin." This species has prominent external ears and the melon is rounded with a longitudinal ridge. The dolphins have 26–39 upper teeth and 26–35 lower teeth. This species is commonly called "susu" supposedly in reference to the noise it makes when it breathes.

Behavior

These dolphins exhibit short surfacing patterns and can stay underwater between 30–90 seconds and longer. They are not acrobatic but frequently swim on their sides (especially in captivity) with their flipper trailing the muddy bottom. They rely on sound to navigate the muddy, turbid waters in which they are typically found, emitting high-frequency echolocation clicks to detect prey at short range.

Food and foraging

Mainly bottom feeders, they take clams, prawns, gobies, catfish, and carp. To maximize foraging opportunities during the day, they frequently aggregate at river confluence and countercurrent pools. They show peak feeding activity during the morning and afternoon hours.

Life history

They reach sexual maturity at body lengths of more than 5¾ ft (1.7 m) and at around 10 years of age. Calving can occur year round but peaks from December to January and March through May. Very little is known about their reproductive behavior. Females give birth to a single calf. Gestation can last between 9 and 11 months, and average lifespan is about 30 years.

Conservation and management

They are recognized as the National Aquatic Animal of India and are protected legally. Population abundance is 2,500–3,000 animals. Habitat degradation resulting from construction of dams and barrages, poaching for bait, fisheries bycatch, pollution, and irrigation projects are major threats affecting population viability.

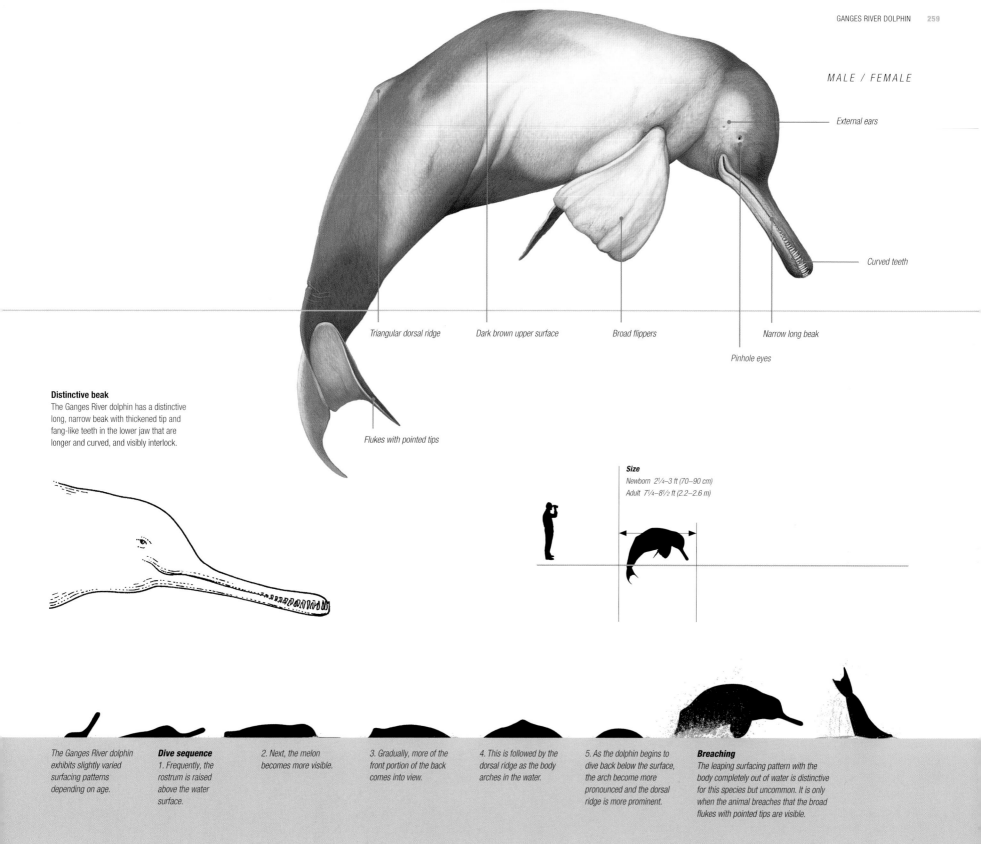

MALE / FEMALE

External ears

Curved teeth

Triangular dorsal ridge

Dark brown upper surface

Broad flippers

Narrow long beak

Pinhole eyes

Flukes with pointed tips

Distinctive beak
The Ganges River dolphin has a distinctive long, narrow beak with thickened tip and fang-like teeth in the lower jaw that are longer and curved, and visibly interlock.

Size
Newborn 2¼–3 ft (70–90 cm)
Adult 7¼–8½ ft (2.2–2.6 m)

The Ganges River dolphin exhibits slightly varied surfacing patterns depending on age.

Dive sequence
1. Frequently, the rostrum is raised above the water surface.

2. Next, the melon becomes more visible.

3. Gradually, more of the front portion of the back comes into view.

4. This is followed by the dorsal ridge as the body arches in the water.

5. As the dolphin begins to dive back below the surface, the arch become more pronounced and the dorsal ridge is more prominent.

Breaching
The leaping surfacing pattern with the body completely out of water is distinctive for this species but uncommon. It is only when the animal breaches that the broad flukes with pointed tips are visible.

TOOTHED WHALES
Porpoises

Porpoises are shy cetaceans that typically frequent shallow, coastal waters. Seven species of porpoise are recognized across three genera (*Neophocaena*, *Phocoena*, and *Phocoenoides*). The family includes the critically endangered vaquita (*Phocoena sinus*), which occupies the smallest geographic area of any marine mammal species, under 900 square miles (2,300 square km). Porpoises may be confused with small-bodied oceanic dolphins (Delphinidae), but porpoises lack the distinctive beaks that characterize most dolphin species.

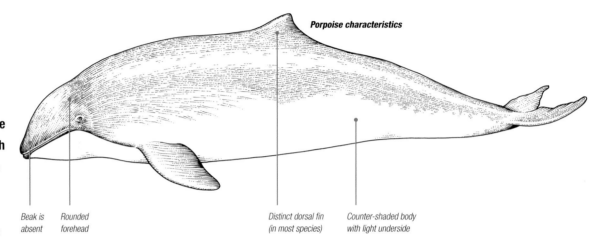

Harbor porpoise underwater (right)
Porpoises often can be confused with oceanic dolphins in the wild. However, the snouts of porpoises, such as this harbor porpoise (Phocoena phocoena), are rounded in profile and lack prominent beaks.

Porpoise characteristics

Beak is absent

Rounded forehead

Distinct dorsal fin (in most species)

Counter-shaded body with light underside

- In general, porpoises are small cetaceans with adult body lengths of 4¼–7½ ft (1.3–2.3 m).
- Porpoises tend to have lighter undersides, but some species such as the Dall's porpoise (*Phocoenoides dalli*) have distinctive markings. Burmeister's porpoise (*Phocoena spinipinnis*) has very dark gray skin that turns to black shortly after death.
- Although most porpoises occupy shallow waters, the Dall's porpoise is known to dive quite deeply, to more than 300 ft (90 m).
- Most porpoises are exclusively marine, but there is a distinct population of the narrow-ridged finless porpoise (*Neophocaena asiaeorientalis*) that occupies freshwaters of the Yangtze River in China.
- The diet of porpoises largely consists of fish and cephalopods (squid), although some individuals are reported to eat crustaceans, including krill.
- Porpoises are not as gregarious as oceanic dolphins and typically live in small groups. They do not often approach boats or perform complex acrobatics that are associated with some dolphin species.

- The leading edges of the flippers and dorsal fins of the harbor and Burmeister's porpoises possess tubercles (small bumps). The dorsal fin of the Dall's porpoise is triangular in shape and the fin of the spectacled porpoise is rounded. As their name implies, the finless porpoises (genus *Neophocaena*) lack dorsal fins and instead possess a row of dorsal protuberances.
- Although the most common porpoises, namely the harbor and Dall's porpoises (*Phocoena phocoena* and *Phocoenoides dalli*), have been assigned a conservation status of Least Concern, the finless porpoises (*Neophocaena* spp.) are categorized as Vulnerable, and the vaquita (*Phocoena sinus*) is Critically Endangered. Gill nets used in fishing pose the greatest threat to porpoises.
- The teeth of porpoises are compressed and spatulate, as opposed to the conical teeth of oceanic dolphins (Delphinidae).

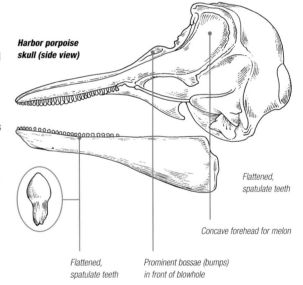

Harbor porpoise skull (side view)

Flattened, spatulate teeth

Concave forehead for melon

Flattened, spatulate teeth

Prominent bossae (bumps) in front of blowhole

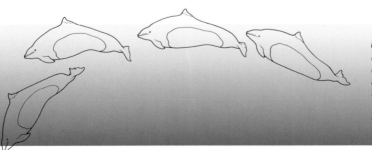

Underwater dive
Porpoises do not typically leave the water while swimming, but rather roll forward as they surface. The flukes typically remain below the surface when diving, but the dorsal fin (if present) is often observed. There is little disturbance of the water when most species dive, although the Dall's porpoise may produce a characteristic splash. The blow of a porpoise is rarely seen.

NARROW-RIDGED FINLESS PORPOISE

Family Phocoenidae

Species *Neophocaena asiaeorientalis*

Other common names Finless porpoise

Taxonomy Recently recognized as a separate species with two subspecies Yangtze finless porpoise
(*N. a. asiaeorientalis*) and the East Asian finless porpoise (*N. a. sunameri*)

Similar species Indo-Pacific finless porpoise

Birth weight 11–22 lb (5–10 kg)

Adult weight 88–154 lb (40–70 kg)

Diet Small fish, squid, crustacean

Group size 1–5, occasionally 20

Major threats Bycatch, habitat loss, and degradation

IUCN Status Threatened

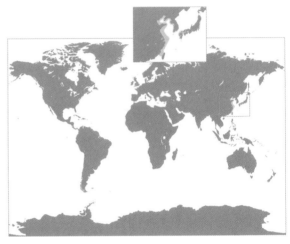

Range and habitat This species inhabits coastal waters of eastern China through
the Korean Peninsula to Japan. There is also a population inhabiting the Yangtze
River in China.

Identification checklist

- No dorsal fin
- No beak
- Narrow dorsal ridge

Anatomy

The body of the narrow-ridged finless porpoise is relatively slender
and flexible. As the name suggests, it lacks a dorsal fin and has
a narrow ridge that runs along the middle of the back. The shape
of the ridge varies slightly among different populations. Small,
horny bumps or tubercles are scattered along the ridge. The
function of tubercles is unknown but they seem to be used in
contact behavior between animals. The forehead is round,
lacking a beak, and 15–20 pairs of spade-shaped teeth are
in the upper and lower jaws.

Behavior

These porpoises rarely show aerial behavior and usually swim
quietly. Only the back is shown at the surface. However, on
occasion they actively porpoise to chase fish or mate.

Food and foraging

Finless porpoises feed on various small prey from benthic
shrimp and fish to schooling fish and cephalopods.

Life history

Narrow-ridged finless porpoises live for about 20 years and attain
sexual maturity at around 4–6 years old for both sexes. Females
give birth every 2 years after 11 months gestation. Calving peaks
in spring for most populations but in winter for the population in
the Ariake Sound, Japan. This difference may be related to
different prey availability for nursing mothers and weaned
calves. Mothers nurse their calf for about six to seven months
after birth. These porpoises are usually found in a loose group
of just a few animals.

Conservation and management

Due to their shallow-water habitat, narrow-ridged finless porpoises
are vulnerable to habitat loss and degradation, bycatch, water
pollution, and boat traffic. In the Seto Inland Sea, one of the major
habitats of this species in Japan, the abundance estimated in
1999–2000 was only 30–40 percent of that in the late 1970s.
Annual decline of five to seven percent was also estimated for
the population in the Yangtze River, China.

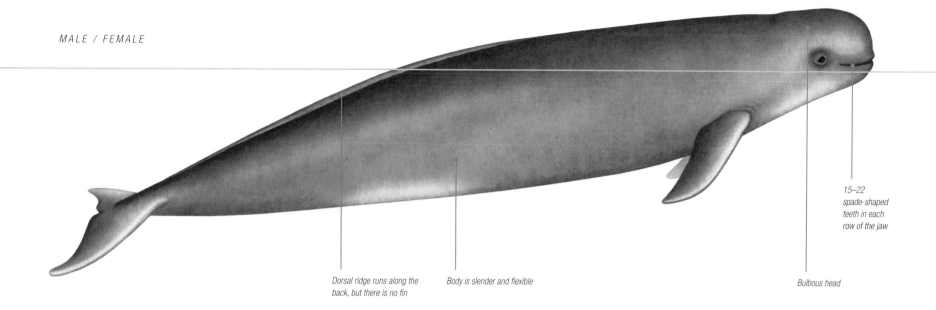

MALE / FEMALE

15–22
spade-shaped
teeth in each
row of the jaw

Dorsal ridge runs along the
back, but there is no fin

Body is slender and flexible

Bulbous head

Tubercles and dorsal ridge
The narrow-ridged finless porpoise has a narrow dorsal ridge
with small tubercles. In the Indo-Pacific finless porpoise, the
area scattered with tubercles is much wider and almost flat
or slightly concave.

Narrow-ridged finless

Indo-Pacific finless

Size
Newborn 2–2½ ft (60–80 cm)
Adult 5¼–6¼ ft (1.6–1.9 m)

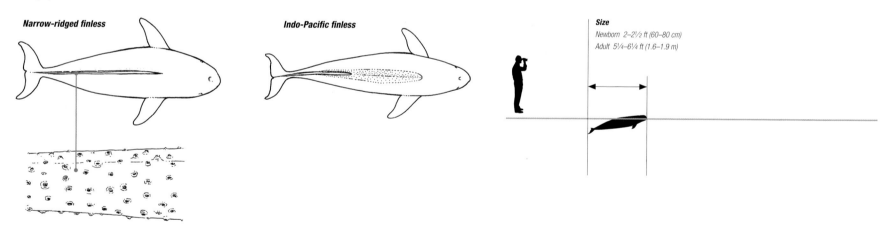

Dive sequence
1. First, the head breaks
the surface quietly,
a blow is rarely seen.

2. Next the rounded
back follows.

3. Flukes do not
appear on the surface.

4. Finless porpoises
usually disturb the
surface very little.

INDO-PACIFIC FINLESS PORPOISE

Family Phocoenidae

Species *Neophocaena phocaenoides*

Other common names None

Taxonomy Recently recognized as a species separate from the narrow-ridged finless porpoise *N. asiaeorientalis*

Similar species Almost indistinguishable from the narrow-ridged species at sea without a good view of the dorsal ridge

Birth weight 11–22 lb (5–10 kg)

Adult weight 88–154 lb (40–70 kg)

Diet Small fish, cephalopods, crustaceans

Group size 1–5, occasionally up to 20

Major threats Accidental mortality in fishing gear (bycatch) especially gill nets, vessel strikes, habitat loss and degradation

IUCN status Vulnerable

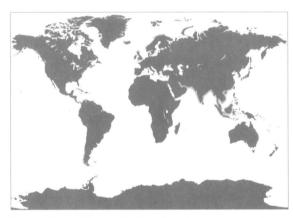

Range and habitat This species inhabits shallow tropical and subtropical waters of the Indian Ocean and Pacific Ocean, in a coastal band from Iran and the Arabian Gulf eastward throughout much of Southeast Asia and northward to at least the Taiwan Strait. They are frequently present in tidal creeks, estuaries, and inter-island channels. Although most of their habitat is shallow—less than 160 ft (50 m) deep—and near shore, they have been observed 30–149 miles (50–240 km) from shore in the Yellow and East China seas.

Identification checklist

- Rounded head with no beak
- No dorsal fin but low ridge with the apex far back near tail
- Wide patch of tubercles along the back, narrowing posteriorly
- Tubercled area along the back in front of the ridge appears flat or gently concave
- Generally dark to medium gray body, often with white or light areas on lips and throat

Anatomy

The body is relatively slender and flexible, with a rounded (bulbous) head, no beak, and a peduncle that narrows dramatically toward the flukes. There are 15–22 pairs of small, spade-shaped teeth in the upper and lower jaws. As the name implies, Indo-Pacific finless porpoises lack a dorsal fin. Instead they have a dorsal ridge that is much farther back on the body—around 75–90 percent of the way back. Small horny prominences, called tubercles, are present on a flattened or slightly concave area of the back anterior to the dorsal ridge. The function of these tubercles is unknown.

Behavior

Finless porpoises are not notably gregarious and usually occur in small groups of two to five. Larger aggregations may form in food-rich areas. These animals do not bow-ride or exhibit aerial behavior and tend to be hard to approach although they sometimes follow a fast-moving boat and play in its stern wake. When chasing prey or socializing, they sometimes porpoise almost clear of the surface but normally they roll quietly and show their back only briefly at the surface.

Food and foraging

Finless porpoises are considered opportunistic predators, consuming a large variety of small organisms including bottom-living fish and crustaceans, schooling fish, and cephalopod mollusks, such as squid, cuttlefish, and octopus. They are not deep divers, usually remaining submerged for less than a minute, and never for longer than a few minutes.

Life history

These porpoises likely live for about 20 years and attain sexual maturity at 4–6 years. Females give birth every 2 years after 11 months of gestation. The calving season is prolonged, spanning the months of June to March. Calves are nursed for about 6–7 months.

Conservation and management

Because of their coastal, shallow-water distribution in mild latitudes, finless porpoises are often in close proximity to human activities and therefore are affected by the loss and degradation of habitat, pollution, and vessel traffic. Entanglement in fishing gear, especially gill nets, is probably the greatest threat although vessel strikes and noise disturbance are also potential concerns.

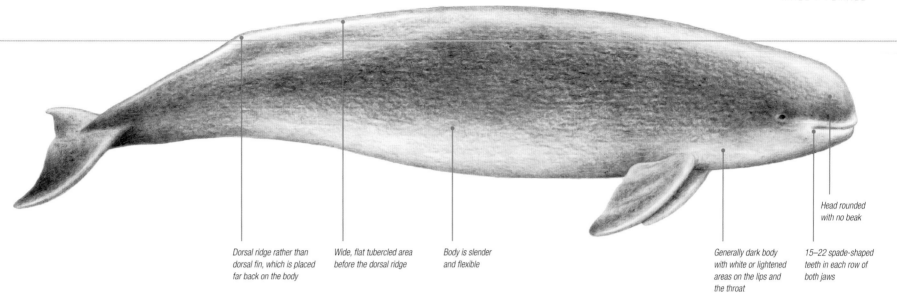

MALE / FEMALE

Dorsal ridge rather than
dorsal fin, which is placed
far back on the body

Wide, flat tubercled area
before the dorsal ridge

Body is slender
and flexible

Generally dark body
with white or lightened
areas on the lips and
the throat

15–22 spade-shaped
teeth in each row of
both jaws

Head rounded
with no beak

Tubercles and dorsal ridge
The two finless porpoise species are superficially very similar.
They can be distinguished primarily by differences in the
shape and extent of the tubercled area and dorsal ridge. The
Indo-Pacific species has a flatter and much wider tubercled
area in front of the relatively short dorsal ridge, which is
positioned far back, ending just before the tail stock.

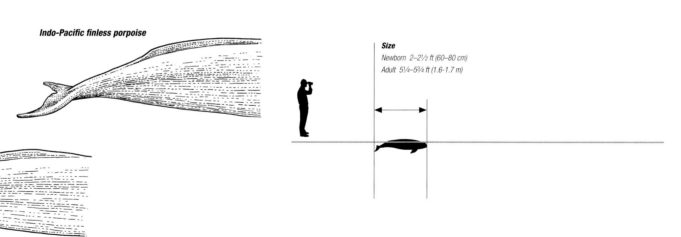

Indo-Pacific finless porpoise

Narrow-ridged finless porpoise

Size
Newborn 2–2½ ft (60–80 cm)
Adult 5¼–5¾ ft (1.6-1.7 m)

Finless porpoises are among the
most cryptic cetaceans and are
very difficult to detect and observe
in any but ideal sea conditions.

Dive sequence
1. The head generally breaks
the surface quietly and the
blow is usually not visible.

2. The back comes into view
as the animal rolls smoothly
forward, causing little surface
disturbance.

3. The apex of the dorsal ridge
becomes especially pronounced, far
back on the body, as the animal
arches to begin a terminal dive.

4. The flukes normally
do not appear above
the surface.

SPECTACLED PORPOISE

Family Phocoenidae

Species *Phocoena dioptrica*

Other common names No other English names

Taxonomy There are no recognized subspecies and this is sometimes considered as a separate genus under the name *Australophocaena dioptrica*

Similar species May be confused with similar-sized delphinids of the genus *Cephalorhynchus* and, in South American waters, also with Burmeister's porpoise

Birth weight Not known, probably between 22 and 33 lb (10 and 15 kg)

Adult weight Probably up to 265 lb (120 kg)

Diet Fish (mainly anchovies), krill, small squid, and stomatopods

Group size The few sightings reported concern small groups, probably mother-calf pairs, and groups of 3, the maximum group size reported is 10

Major threats Incidental catches, ocean pollution, noise pollution, and global warming

IUCN status Data Deficient

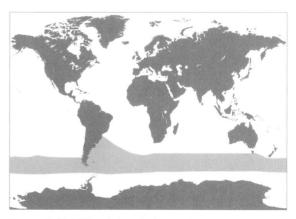

Range and habitat This species' range is circumpolar in temperate and sub-Antarctic and Antarctic waters of the Southern Ocean. The few known observations show that this species occurs in oceanic as well as shelf waters.

Identification checklist

- Black back and upper flanks with white lower flanks and belly
- Stocky, robust body shape
- Triangular or rounded dorsal fin in mid-position; sexual dimorphism exhibited
- Spectacle-like markings around the eye
- Small head, no demarcated beak
- Small, rounded flippers

Anatomy

The head is typically porpoise-like: small and rounded with little to no beak developed. The lips are black and there are 17–23 pairs of teeth in the upper and lower jaws. The black-and-white coloration is striking. There is a dark mouth-to-flipper stripe, which is sometimes rather faint. The flippers, fluke, and fluke peduncle vary in color from whitish to all-black. The dorsal fin exhibits convex leading and trailing edges. Males have large, rounded dorsal fins.

Behavior

Although rarely seen, observations indicate that these animals are elusive. Spectacled porpoises are capable of rapid swimming and occasionally approach boats.

Food and foraging

The main prey of the spectacled porpoise are apparently pelagic schooling fish such as anchovies, but since very few stomachs have been examined the present understanding of its diet may be rather incomplete. Observed school sizes are less than a handful indicating a pairwise or even single individual foraging. But knowledge on foraging strategies within this species is rather poor.

Life history

Mating and calving occur mainly in the spring and midsummer. The gestation period is estimated to last 11 months. Sexual maturity is reached at approximately 4½ ft (1.3 m) in length and at an age of around two years in females and 4¾ ft (1.4 m) and four years in males. Males grow somewhat larger than females. The lifespan is unknown. Gestation lasts 11 months and lactation 6–15 months.

Conservation and management

Data is deficient on this species' status. Incidental catches are known to occur.

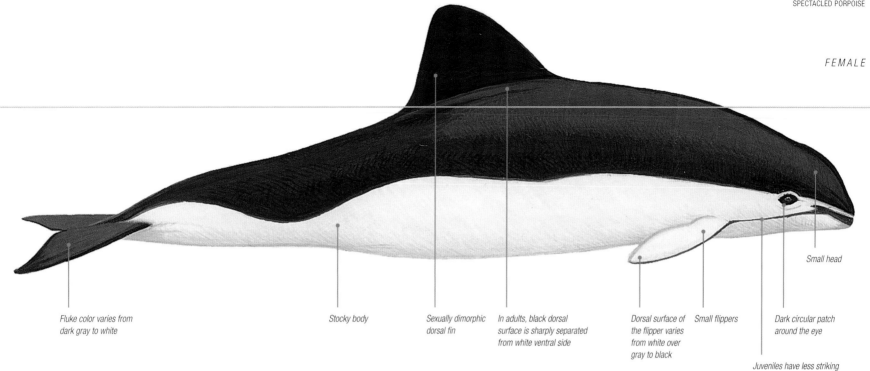

FEMALE

Fluke color varies from
dark gray to white

Stocky body

Sexually dimorphic
dorsal fin

In adults, black dorsal
surface is sharply separated
from white ventral side

Dorsal surface of
the flipper varies
from white over
gray to black

Small flippers

Dark circular patch
around the eye

Small head

Juveniles have less striking
coloration, usually dark gray
on the dorsal side and light
gray ventrally, and a flipper
stripe from the gape to
the flipper

Sexual dimorphism
Compared to females, adult males possess
a much larger, more rounded dorsal fin.

Male

Female

Size
Newborn 3–3¾ ft (90–110 cm)
Adult 4¾–7 ft (1.4–2.1 m) female; 5–7½ ft (1.5–2.3 m) male

Dive sequence
*Very few observations have been
made of this species but diving
behavior seems to be similar to
that of the other true porpoises.*

*1. There is no visible blow.
Initially, the head and anterior
dorsal part of the body appears.*

*2. Next, the triangular
dorsal fin appears.*

*3. The dorsal fin and
back are clearly visible,
while the porpoise rolls
at the surface.*

*4. This sequence happens in fast
succession and is usually repeated
several times before the porpoise
initiates another longer dive.*

HARBOR PORPOISE

Family Phocoenidae

Species *Phocoena phocoena*

Other common names Common porpoise, puffing pig

Taxonomy Four subspecies are recognized: *P. p. phocoena* from the North Atlantic, *P. p. relicta* from the Black Sea and eastern Mediterranean, *P. p. vomerina* from the eastern North Pacific, and an unnamed fourth subspecies from the western North Pacific

Similar species May be confused with Dall's porpoise along the North Pacific distributional range of the species

Birth weight 11–22 lb (5–10 kg)

Adult weight 99–220 lb (45–100 kg) females; 77–165 lb (35–75 kg) males

Diet A wide array of smaller fish and occasionally squid

Group size 1–10, usually 1–2

Major threats Incidental catches, ocean pollution, noise pollution

IUCN status Least Concern, but some local populations are severely threatened

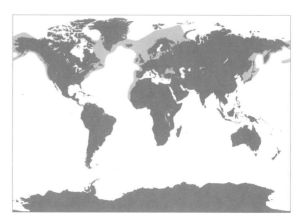

Range and habitat This species is found in temperate and subarctic waters of the North Atlantic and North Pacific. It occurs primarily in coastal waters, but also over continental shelves. Several occurrences offshore as well as in fresh water have been observed.

Identification checklist

- Back and tail flukes are dark gray, almost black
- Chin and belly are light gray, almost white
- Short, stocky body shape
- Low, triangular dorsal fin
- No demarcated beak
- Relatively small, rounded flippers

Anatomy

The darker coloration of the back becomes variable at the sides, where it merges with the lighter underside. Most individuals have a dark patch around the eye and a stripe connecting the eye and the flipper. Females grow somewhat larger than males. The harbor porpoise is one of the smallest cetacean species, although size varies considerably throughout its range, with larger animals occurring off California and Northwest Africa. As the other true porpoises, harbor porpoises have spade-shaped teeth—20–29 pairs of teeth in the upper and lower jaws.

Behavior

Usually calm at the surface, although bursts of speed associated with feeding are common. Harbor porpoises adapt their behavior to local conditions. In some areas they are considered shy, in others they may jump out of the water or approach vessels. They often rest at the surface with the blowhole as the most elevated part of the body (logging). Although usually seen alone or in small groups, harbor porpoises may occur in larger feeding aggregations. They may dive deeper than 660 ft (200 m) and dives of more than five minutes have been recorded.

Food and foraging

The diet mainly consists of small shoaling fish. They feed on a wide range of species including herring, sand eel, gobies, and codfish such as whiting and cod. Foraging often takes place close to the sea bottom. Social feeding also occurs.

Life history

Mating and giving birth occur in midsummer. The gestation period lasts 10–11 months. Sexual maturity is reached at lengths of 4¼–4¾ ft (1.3–1.4 m) in males and 4¾–5 ft (1.4–1.5 m) in females, and between two and five years in both sexes. Females are able to give birth in successive years, being pregnant and lactating at the same time, but may also rest between births. Maximum lifespan is more than 20 years, but most porpoises do not live beyond 12 years.

Conservation and management

Throughout most of its range, incidental catches in set fishing gear is a serious threat for harbor porpoises. Noise and other disturbance as well as pollution also impact the species.

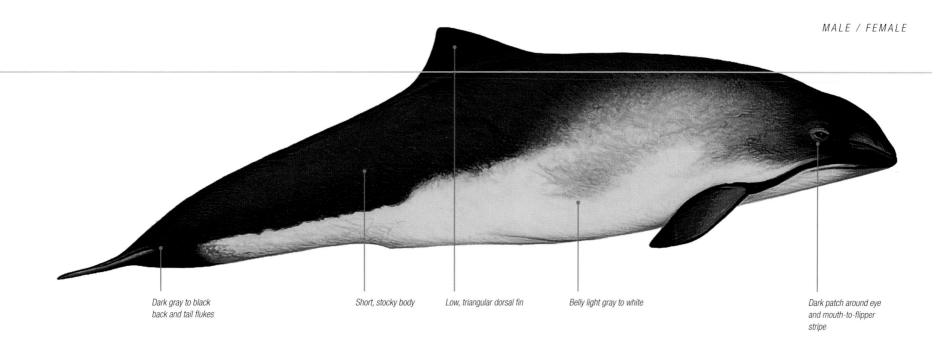

MALE / FEMALE

Dark gray to black
back and tail flukes

Short, stocky body

Low, triangular dorsal fin

Belly light gray to white

Dark patch around eye
and mouth-to-flipper
stripe

Dorsal fin variation
The dorsal fin exhibits some variation throughout the entire range
of the species. Some individuals have dorsal fins of a triangular
shape, others approach the dolphin-like falcate fin shape.

Falcate type

Intermediate type

Triangular type

Size
Newborn 2⅛–2½ ft (65–80 cm)
Adult 4¾–6½ ft (1.4–2 m) females; 4½–6 ft (1.3–1.8 m) male

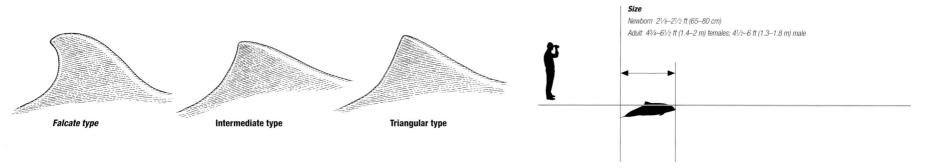

Dive sequence
*1. There is no visible blow.
Initially, the head and front
dorsal part of the body appears.*

*2. This is followed by
the triangular dorsal fin.*

*3. The dorsal fin and back
are clearly visible, while the
porpoise rolls at the surface.*

*4. This sequence happens in
fast succession and is usually
repeated several times before
the porpoise initiates another
longer dive.*

Breaching
*The harbor porpoise
rarely jumps out of
the water.*

VAQUITA

Family Phocoenidae

Species *Phocoena sinus*

Other common names Gulf of California harbor porpoise

Taxonomy Most closely related to *P. spinipinnis*

Similar species None

Birth weight 16½–22 lb (7.5–10 kg)

Adult weight 121 lb (55 kg)

Diet More than 20 species of mainly fish and squid

Group size Average 2 animals, maximum 10

Major threats Bycatch in gill and entangling nets

IUCN status Critically Endangered

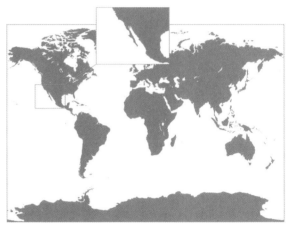

Range and habitat This species occurs in the northern and Upper Gulf of California, in mainly shallow waters at less than 100 ft (30 m) depth.

Identification checklist

- Dorsal fin appears tall in proportion to body size
- Black patches around the eyes and mouth
- Pectoral fins are proportionally larger than in other porpoises
- Very brief surfacing times

Anatomy

The most conspicuous characteristic is the black patches around the eyes and mouth. The pigmentation pattern is dark gray on the upper side, light gray at the sides, with a white underside. The flukes, fins, and dorsal fin are relatively larger than in other porpoises. The dorsal fin has an average height of 6¾ in (17 cm).

Behavior

Very little is known about the behavior of this species. They do not leap or splash and are shy, generally avoiding vessels. Nothing is known of their mating behavior or system, although the large size of testes suggests sperm competition among males.

Food and foraging

Vaquitas are opportunistic predators that feed on a wide variety of small-bodied, shallow-water benthic and pelagic fish, and squid. Many prey species, such as croakers and toadfish, produce sound and it is thought vaquitas may use passive listening to locate prey. None of the main prey of the vaquita are commercially important species, although some are taken as bycatch in trawl fisheries.

Life history

Vaquitas tend to aggregate in loose groups. These loose aggregations have a short duration and shift locations in short periods of time. Vaquita are seasonal breeders with most births occurring around March. The gestation period is probably 10–11 months. The oldest specimen is a 21-year-old female. Age at sexual maturity has been difficult to estimate due to the absence of juvenile samples. However all females less than three years were immature and all females older than six years were mature. Most female harbor porpoises mature at age three and give birth every other year.

Conservation and management

The main risk to vaquita survival is bycatch in gill and entangling nets. The vaquita is recognized as the most endangered marine mammal species in the world. The only management measure to prevent its extinction is to eliminate bycatch throughout its habitat in the Upper and Northern Gulf of California. It is estimated that the current population size is fewer than 100 individuals.

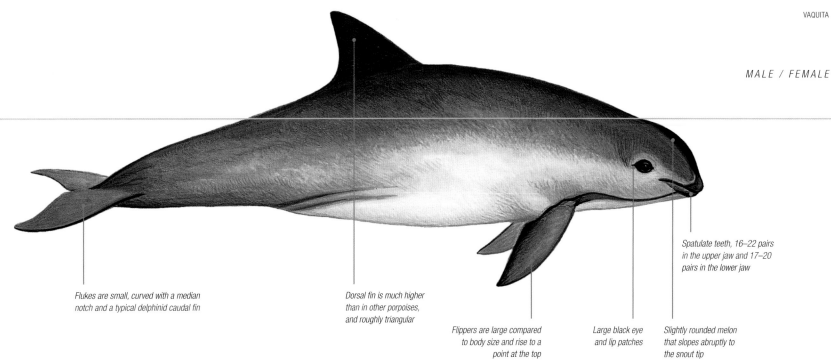

Spatulate teeth, 16–22 pairs in the upper jaw and 17–20 pairs in the lower jaw

Flukes are small, curved with a median notch and a typical delphinid caudal fin

Dorsal fin is much higher than in other porpoises, and roughly triangular

Flippers are large compared to body size and rise to a point at the top

Large black eye and lip patches

Slightly rounded melon that slopes abruptly to the snout tip

Dorsal fin comparison
The vaquita's dorsal fin is taller and more falcate in proportion to the body length than in other true porpoise species. It has an average height of 6¾ in (17 cm).

Burmeister's porpoise

Harbor porpoise

Vaquita

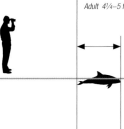

Size
Newborn 2¼–2½ ft (69–80 cm)
Adult 4¼–5 ft (1.3–1.5 m)

Blow
This is a small shy animal that is not easy to spot at sea. The blow of such a small cetacean is not visible at a distance.

Dive sequence
1. On most occasions there is hardly any time to observe the rostrum, although there are a few photographs that clearly show the black patches around the eyes and snout. The dive begins with the emergence of the back and the very prominent dorsal fin.

2. The arching movement is continuous and too fast, lasting only around one second.

3. On the final diving movement only the tip of dorsal fin is visible. More often than not the peduncle is not shown.

Fluking
This species does not show flukes.

Breaching
This is not a breaching species. At most, and very rarely, it shows the front half of body.

BURMEISTER'S PORPOISE

Family Phocoenidae

Species *Phocoena spinipinnis*

Other common names No other English names

Taxonomy No subspecies are presently recognized

Similar species May be confused with Commerson's or Chilean dolphins or the spectacled porpoise

Birth weight 9–15 lb (4–7 kg)

Adult weight 230 lb (105 kg) maximum (range not known)

Diet Species of anchovy and hake, and a wide array of smaller fish (both pelagic and bottom-dwelling) and occasionally squid and crustaceans

Group size 2–6, sometimes up to 70 animals in aggregation

Major threats Incidental catches, ocean pollution, and noise pollution

IUCN status Least Concern

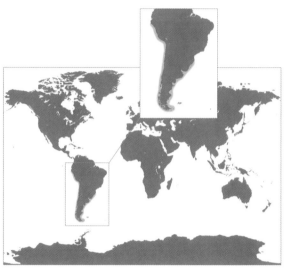

Range and habitat The Burmeister's porpoise is found along the coasts of South America. Due to the cold Humboldt Current, the range extends further north on the Pacific seaboard.

Identification checklist

- Almost entirely black
- Rather prominent tubercles on leading edge of dorsal fin
- Short, stocky body shape
- Low and dorsal fin, placed far back, tilted backward, with a much longer leading than trailing edge
- No demarcated beak
- Relatively large, rounded flippers

Anatomy

The head is blunt and beakless, and typically porpoise-like. The flippers appear relatively small. The coloration exhibits varying shades of brown and dark gray. Around the eye there is a conspicuous dark eye spot. Teeth are spade-shaped as in other true porpoises, with between 14 and 23 pairs of teeth in both the upper and lower jaws.

Behavior

This species is rather elusive and difficult to detect even under calm weather conditions. No general migration patterns have been detected but movements are probably connected to shifts in prey abundance and may be onshore–offshore or north–south. They are usually calm at the surface, although bursts of speed associated with feeding are common. Burmeister's porpoises apparently only rarely jump out of the water or approach vessels.

Food and foraging

The diet mainly consists of small shoaling fish from pelagic as well as bottom habitats. They feed on a wide range of species.

Life history

Mating and calving occur in midsummer. Birth size ranges between 2¾ and 3 ft (85–90 cm). Sexual maturity is reached at about 5¼ ft (1.6 m) in males and 5 ft (1.55 m) in females. The age at maturity for both sexes is not known but may be similar to other phocoenids. Gestation lasts 11–12 months, meaning that females may not be able to give birth in successive years. The maximum lifespan is not known.

Conservation and management

Throughout most of its range, incidental catch in set fishing gear is a serious threat for the porpoises. Noise and other disturbance as well as pollution also impact the species.

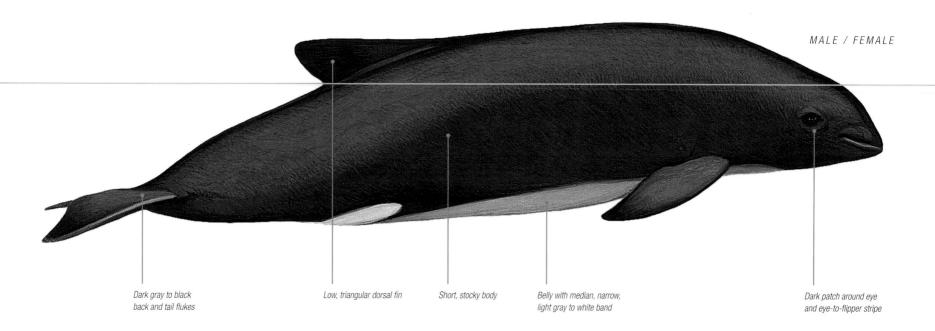

MALE / FEMALE

Dark gray to black
back and tail flukes

Low, triangular dorsal fin

Short, stocky body

Belly with median, narrow,
light gray to white band

Dark patch around eye
and eye-to-flipper stripe

Dorsal fin tubercles
All porpoise species, except Dall's porpoise, have epidermal
tubercles in the form of small raised protuberances along the
leading edge of the dorsal fin—they are developed shortly after
birth. The function of the tubercles remains unknown, although
study suggests that they may have hydrodynamic importance.

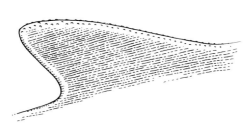

Size
Newborn 2¾ ft (85 cm)
Adult 4¾–6½ ft (1.4–2 m) female; 4½–6 ft (1.35–1.8 m) male
Maximum adult length is 6½ ft (2 m)

Dive sequence
1. There is no visible
blow. Initially, the head
and front upper part of
the body appears.

2. The triangular dorsal
fin follows.

3. The dorsal fin and back
become clearly visible, while the
porpoise rolls at the surface.

4. This sequence happens in rapid
succession and is usually repeated
several times before the porpoise
initiates another longer dive.

Breaching
Burmeister's porpoises
very rarely jump out
of the water.

DALL'S PORPOISE

Family Phocoenidae

Species *Phocoenoides dalli*

Other common names True's porpoise (for the *Truei* type of the species)

Taxonomy Two subspecies recognized: *P. d. dalli* *Dalli*-type Dall's porpoise and *P. d. truei* *Truei*-type porpoise

Similar species Along the coasts of the North Pacific the species may be confused with the harbor porpoise

Birth weight 28–42 lb (13–19 kg)

Adult weight 154–353 lb (70–160 kg) female; 176–440 lb (80–200 kg) male

Diet Schooling fishes, such as mackerel, herring, anchovy; mesopelagic fishes, such as deep-sea smelt and lanternfish, and squid

Group size Usually 1–10, larger groups occur

Major threats Incidental catches, hunting

IUCN Status Least Concern

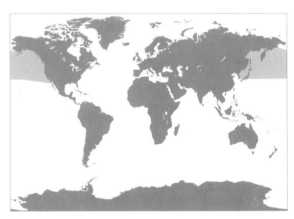

Range and habitat This species is found in temperate and subarctic waters of the North Pacific and neighboring seas. It occurs in oceanic waters and deeper inshore waters.

Identification checklist

- Back, front of body, and head are dark gray, almost black
- Back of lower flanks and belly are white
- White frosting develops on dorsal fin and flukes with age
- Short and extremely stocky body shape
- Low, triangular dorsal fin, tilted forward in adult males
- Small head, no demarcated beak
- Small, rounded flippers

Anatomy

There are two distinct forms of coloration in which the front extent of the white areas of the flanks and belly differ. In one type the white area begins immediately before the flippers, while in the other it is farther back, well behind the flippers. Males grow larger than females and in adult males the leading edge of the dorsal fin is tilted forward, there is an exaggerated postanal hump, a deepened tail stock, and the trailing edge of the flukes is convex. As in the other true porpoises, Dall's porpoise has spade-shaped teeth. The teeth are very small, and the number highly variable, but usually between 21–28 pairs in the upper and lower jaws. With up to 98 vertebrae, Dall's porpoise has more vertebrae than any other cetacean—an adaptation to fast, dynamic swimming

Behavior

Usually very fast-swimming and active at the surface, Dall's porpoise often create a "V"-shaped splash swimming at the surface, called rooster-tailing. When moving more slowly, they behave much like other porpoises. This species may bow-ride on fast-moving vessels, but rarely leaps out of the water. Although usually seen in smaller groups, Dall's porpoise may occur in larger feeding aggregations.

Food and foraging

The diet mainly consists of small pelagic schooling fish such as mackerel, herring, and anchovies as well as squid. They may dive deeper than 1,640 ft (500 m) to reach their prey.

Life history

Mating and giving birth occur in the summer. The gestation period lasts 10–12 months. Sexual maturity is reached at lengths of 6–6½ ft (1.8–2 m) in males and 5¾–6¼ ft (1.7–1.9 m) in females. The age at which sexual maturity is attained is between three-and-a-half and eight years in males, and four and seven years in females. Females are able to give birth in successive years, being pregnant and lactating at the same time, but may also rest between births. Maximum lifespan is more than 20 years, but most individuals do not live beyond 12 years.

Conservation and management

Incidental catches in fishing gear is a serious threat for Dall's porpoise. Pollution, noise, and habitat alteration also impact the species.

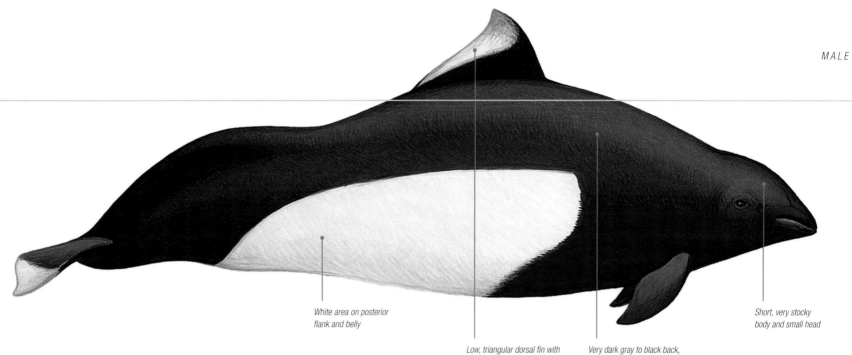

MALE

White area on posterior
flank and belly

Low, triangular dorsal fin with
white fringe in adults, tilted
forward in adult males

Very dark gray to black back,
head, and anterior body

Short, very stocky
body and small head

Color morphs

Dall's porpoise occurs in two color
morphs: one where the white flank/belly
patch extends forward to the level of the
dorsal fin (*dalli*-type) and one where the
patch extends to the level of the flippers
(*truei*-type). The *dalli*-type occurs
throughout the range of the species, the
truei-type occurs along the east coast of
Japan and the Kurile Islands and in the
Okhotsk Sea.

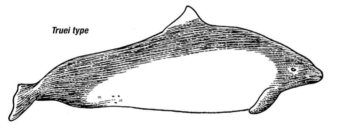

Truei type

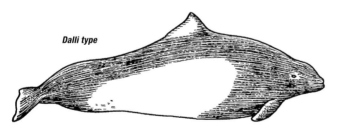

Dalli type

Size
Newborn 3–3¾ ft (90–110 cm)
Adult 5¾–7¼ ft (1.7–2.2 m) female; 6–8 ft (1.8–2.4 m) male

Dive sequence
*1. There is no visible blow.
During fast swimming, the
back and dorsal fin of the
porpoise appear with a
splash of water.*

*2. During slow
swimming, the head
and the triangular
dorsal fin appear first*

*3. The dorsal fin and
back are clearly visible,
while the porpoise rolls
at the surface.*

*4. This sequence happens in
rapid succession and is
usually repeated several times
before the porpoise initiates
another longer dive.*

Breaching
*Dall's porpoise rarely
jumps out of the water.*

APPENDICES

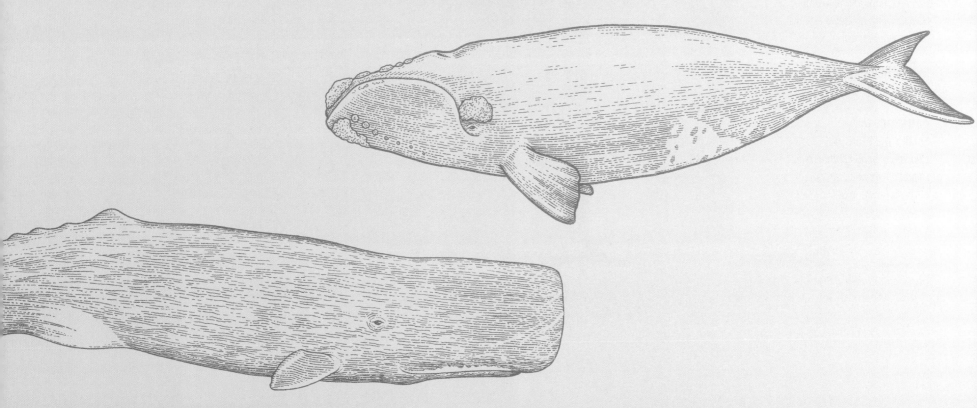

CLASSIFICATION OF CETACEANS

MYSTICETI

FAMILY BALAENIDAE

Balaena mysticetus—bowhead whale
Eubalaena australis—southern right whale
Eubalaena glacialis—North Atlantic right whale
Eubalaena japonica—North Pacific right whale

FAMILY ESCHRICHTIIDAE

Eschrichtius robustus—gray whale

FAMILY BALAENOPTERIDAE

Balaenoptera acutorostrata—common minke whale
Balaenoptera bonaerensis—Antarctic minke whale
Balaenoptera borealis—sei whale
Balaenoptera edeni—Bryde's whale
Balaenoptera musculus—blue whale
Balaenoptera omurai—Omura's whale
Balaenoptera physalus—fin whale
Megaptera novaeangliae—humpback whale

FAMILY NEOBALAENIDAE

Caperea marginata—pygmy right whale

ODONTOCETI

FAMILY DELPHINIDAE

Cephalorhynchus commersonii—Commerson's dolphin
Cephalorhynchus eutropia—Chilean dolphin
Cephalorhynchus heavisidii—Heaviside's dolphin
Cephalorhynchus hectori—Hector's dolphin
Delphinus capensis—long-beaked common dolphin
Delphinus delphis—short-beaked common dolphin
Feresa attenuata—pygmy killer whale
Globicephala macrorhynchus—short-finned pilot whale
Globicephala melas—long-finned pilot whale
Grampus griseus—Risso's dolphin
Lagenodelphis hosei—Fraser's dolphin
Lagenorhynchus acutus—Atlantic white-sided dolphin
Lagenorhynchus albirostris—white-beaked dolphin
Lagenorhynchus australis—Peale's dolphin

Lagenorhynchus cruciger—hourglass dolphin
Lagenorhynchus obliquidens—Pacific white-sided dolphin
Lagenorhynchus obscurus—dusky dolphin
Lissodelphis borealis—northern right whale dolphin
Lissodelphis peronii—southern right whale dolphin
Orcaella brevirostris—Irrawaddy dolphin
Oracella heinsohni—Australian snubfin dolphin
Orcinus orca—killer whale
Peponocephala electra—melon-headed whale
Pseudorca crassidens—false killer whale
Sotalia fluviatilis—tucuxi
Sotalia guianensis—Guiana dolphin
Sousa chinensis—Indo-Pacific humpback dolphin
Sousa plumbea—Indian humpback dolphin
Sousa sahulensis—Australian humback dolphin
Sousa teuszii—Atlantic humpback dolphin
Stenella attenuata—pantropical spotted dolphin
Stenella clymene—clymene dolphin
Stenella coeruleoalba—striped dolphin
Stenella frontalis—Atlantic spotted dolphin
Stenella longirostris—spinner dolphin
Steno bredanensis—rough-toothed dolphin
Tursiops aduncus—Indo-Pacific bottlenose dolphin
Tursiops truncatus—common bottlenose dolphin

FAMILY PHYSETERIDAE

Kogia breviceps—pygmy sperm whale
Kogia simus—dwarf sperm whale
Physeter macrocephalus—sperm whale

FAMILY ZIPHIIDAE

Berardius arnuxii—Arnoux's beaked whale
Berardius bairdii—Baird's beaked whale
Hyperoodon ampullatus—Northern bottlenose whale
Hyperoodon planifrons—southern bottlenose whale
Indopacetus pacificus—Longman's beaked whale
Mesoplodon bidens—Sowerby's beaked whale
Mesoplodon bowdoini—Andrews' beaked whale
Mesoplodon carlhubbsi—Hubbs' beaked whale

Mesoplodon densirostris—Blainville's beaked whale
Mesoplodon europaeus—Gervais' beaked whale
Mesoplodon hotaula—Deraniyagala's beaked whale
Mesoplodon ginkgodensis—ginkgo-toothed beaked whale
Mesoplodon grayi—Gray's beaked whale
Mesoplodon hectori—Hector's beaked whale
Mesoplodon layardii—strap-toothed whale
Mesoplodon mirus—True's beaked whale
Mesoplodon perrini—Perrin's beaked whale
Mesoplodon peruvianus—pygmy beaked whale
Mesoplodon stejnegeri—Stejneger's beaked whale
Mesoplodon traversii—Spade toothed whale
Tasmacetus shepherdi—Shepherd's beaked whale
Ziphius cavirostris—Cuvier's beaked whale

FAMILY PLATANISTIDAE

Platanista gangetica—Ganges River dolphin

FAMILY INIIDAE

Inia geoffrensis—Amazon River dolphin

FAMILY LIPOTIDAE

Lipotes vexillifer—baiji

FAMILY PONTOPORIIDAE

Pontoporia blainvillei—franciscana

FAMILY MONODONTIDAE

Delphinapterus leucas—white whale or beluga
Monodon monoceros—narwhal

FAMILY PHOCOENIDAE

Neophocaena asiaeorientalis—narrow-ridged finless porpoise
Neophocaena phocaenoides—Indo-Pacific finless porpoise
Phocoena dioptrica—spectacled porpoise
Phocoena phocoena—harbor porpoise
Phocoena sinus—vaquita
Phocoena spinipinnis—Burmeister's porpoise
Phocoenoides dalli—Dall's porpoise

GLOSSARY

Allomaternal care
A female taking care of offspring other than her own.

Anterior
Located near the front of the body.

Anthropogenic
Something that is caused by humans and may negatively affect whales. For example, anthropogenic noise comes from commercial ship traffic, seismic and oil exploration, acoustic research, and military sonar.

Balaenid
A family of baleen whales that demonstrates common descent and includes right whales and the bowhead whale.

Baleen
Plates of keratin (epithelial tissue, which is the same material that makes up the hair, claws, and

fingernails of mammals) that hang from the roof of the mouth of mysticete whales; also called whalebone.

Baleen whale
One of two major groups (suborders) of whales that lack teeth as adults and possess baleen. *See also* Mysticete.

Beak
The forward-projecting jaws of a whale, and a characteristic feature of some whales, for example, beaked whales.

Benthic
Living at or near the sea bottom.

Blaze
A streak-like marking of an animal; usually light colored and set against a dark background, for exampe. the white and ocher blazes on the flanks of the Atlantic white-sided dolphin).

Blow
Exhaled air mixed with water and oil released by whales from the blowhole; it can appear low and bushy or tall and columnar, depending on the species, and is a useful field-identification feature.

Blowhole
Nostril opening at the top of the head of whales that functions in respiration; odontocetes have one and mysticetes possess two blowholes.

Blubber
A thick layer of fat in the deepest layer of the skin of whales. Blubber provides insulation and fat (energy) storage as well as smoothing out the body shape that assists the animal in swimming.

Bossae
Prominent paired bumps near the blowhole.

Bow-riding
An energy-saving behavior of some toothed whales (e.g. dolphins) in which they position themselves immediately in front of a ship or large whale to ride the wave.

Breaching
A behavior that involves leaping completely out of the water. Many reasons have been suggested for breaching including signaling, dominance, or warning other whales of danger.

Bycatch
That portion of a catch or harvest that includes non-targeted animals including dolphins, taken incidentally through fishing.

Calf
A juvenile cetacean.

Callosities
Hard, thickened, raised skin patches on the head of right whales that are often inhabited by whale lice and barnacles. The callosity patterns are unique to individuals, which makes them a useful identification feature.

Cape
Dark region on the back of a toothed whale that is located in front of the dorsal fin and often extends onto the sides of the animal. For example, a dark gray dorsal cape is seen in the Clymene dolphin.

Caudal
Refers to the fluke or posterior portion of the body.

Cephalopod
A type of mollusk (invertebrate animal) that possesses a large head, eyes, and tentacles around the mouth; examples are squid and octopus. Cephalopods are the primary prey of some whales such as beaked whales.

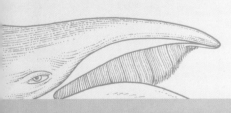

Cetacean
Marine mammals that include whales, dolphins, and porpoises.

Circumpolar
Range of distribution of whales that includes the high latitudes of the northern or southern hemispheres. For example, the narwhal and beluga have a circumpolar distribution in Arctic and subarctic waters.

Clicks
Typically a broadband frequency of sound (30–150 kHz) of short duration that is used by toothed whales in navigation and foraging. A narrow-band frequency of sound is characteristic of a whistle that is used for individual recognition by some dolphins.

Countershaded
A coloration pattern in which the upper (dorsal) surface of the body is dark and the underside (ventral) surface is lighter in color (for example, common bottlenose dolphin and rough-toothed dolphin). This is a type of camouflage since this coloration eliminates shadows that might result from the top-lit area of the ocean and make the whale conspicuous to predators.

Cow
A female whale.

Demersal
Living near the bottom of the sea.

Dorsal
Referring to the upper surface of the back.

Dorsal fin
Located on the back (dorsal surface) of the body of most whales, it provides a control surface to stabilize the animal.

Drift nets
A type of fishing where nets hang vertically in the water without being anchored at the bottom. Non-targeted animals including dolphins get trapped in the nets or entangled in discarded nets.

Echinoderms
A lineage of invertebrate animals; examples include sea stars and sea urchins.

Echolocation
The production of high-frequency sounds and their reception by reflected echoes, used by odontocetes to navigate and locate prey.

Electroreceptor sensitivity
The biological ability to perceive natural electrical stimuli suggested for the function of vibrissae that occur on the rostrum of the Guiana dolphin.

Estrus
The period of sexual receptivity in females.

Euryphagous
Being a generalist; being able to eat a variety of prey.

Falcate
Refers to a dorsal fin that is curved backward or sickle-shaped. For example, a falcate dorsal fin is characteristic of fin whales.

Flipper
Front limbs of cetaceans that are variously shaped; for example, long and narrow in the humpback whale for rapid locomotion, and short and paddle-shaped in right whales, which aids in maneuvering and slow turns.

Flipper-slapping
A behavior that involves striking the surface of the water with the flippers; it likely functions in non-verbal communication.

Flukes
The horizontally flattened tail of whales. The shape of flukes varies among whales although most provide high lift. The underside of flukes develop distinctive scars and markings with time, which enables identification in the field.

Gape
In the mouth of cetaceans, the distance between open jaws.

Gestation
The period of time between fertilization and birth.

Gill nets
A type of fishing using vertical nets with meshes sized so that a fish that puts its head through is entangled by the gills when it tries to withdraw. Non-targeted species including dolphins can be trapped in the nets as bycatch or entangled in discarded nets.

Hybridization
An animal resulting from a cross between animals of different species. One of the few examples in cetaceans is the Clymene dolphin, which is a hybrid formed by the crossing of spinner and striped dolphins.

GLOSSARY

IUCN status
Based on recognition of threatened and endangered animals (the Red List updated annually) by a global conservation organization, International Union for the Conservation of Nature.

Juvenile
Young whales.

Krill
Shrimp-like crustaceans that are the primary food for many baleen whales.

Lactation
The production of milk by females to nurse the young.

Lateral
On the side of the body or vertical plane of a structure.

Lek
Mating strategy or courtship display in which males display to attract females; male humpback whales singing on the breeding grounds has been referred to as a floating lek.

Lobtailing
Behavior that involves forceful slapping of the flukes against the water; also referred to as tail-slapping. Lobtailing likely also functions in non-verbal communication.

Logging
Behavior in which animals lie still at the surface.

Longline
A type of fishing that uses a long line with baited hooks. Non-targeted species including dolphins can be caught as bycatch or entangled in discarded long lines.

Mass stranding
Three or more individuals of the same species that intentionally swim or are unintentionally trapped ashore by waves or receding tides. In some cases mass strandings of whales, for example, beaked whales in the Bahamas in 2000, have been linked to military sonar.

Melon
Bulbous forehead of toothed whales containing fats used to focus echolocation sounds.

Mesopelagic
Refers to an intermediate ocean depth ranging from 660 ft (200 m) to 3,330 ft (1,000 m) in depth.

Mysticete
One of two major groups of whales including baleen whales that possess baleen such as fin and humpback whales.

Odontocete
One of two major groups of whales named for their possession of teeth; toothed whales include dolphins and killer whales.

Pectoral flippers
Pair of fore-flippers in whales. Flipper shape reflects swimming behavior. For example, the narrow, elongate flippers of humpback whales facilitate fast swimming, while the broad flippers of bowhead and right whales aid in slow turns.

Peduncle
Refers to the muscular portion of the tail of a whale, located between the dorsal fin and the tail.

Pelagic
Referring to the open ocean.

Philopatry
Tendency of an individual to stay in its home area.

Photo-identification
Technique that involves the collection and use of photographs of diagnostic features of marine mammals for identification purposes.

Pod
Social group of whales. Most whale pods are family groups. For example, killer whale pods comprise related females and their offspring.

Porpoise
Family of small odontocetes that are characterized by an indistinct beak, robust body, and spade-shaped teeth.

Porpoising
Behavior that involves low leaps made by some cetaceans, notably porpoises at the surface of the water. Porpoising functions to conserve energy by minimizing drag on the body when swimming.

Posterior
Located near the back of an animal.

Purse seine

Fishing using long deep nets that are set around fish to form a circular wall, then gathered at the bottom and drawn up to form a "purse." In the 1960s large numbers of pantropical spotted, spinner, and common dolphins, which swim with yellowfin tuna, were trapped and killed in the nets.

Resident

Refers to an ecotype of killer whale with a distinctive morphology, genetics, behavior and ecology, e.g. resident killer whales differ from transients in feeding exclusively on fish (especially salmon and trout). Resident killer whales live in waters along the west coast of North America.

Rorqual

From the Norwegian word for furrow whale, rorqual refers to the throat grooves of balaenopterid whales (Balaenopteridae), which allow the mouth to expand during engulfment feeding; they include blue, sei, Bryde's, Omura's, minke, fin, and humpback whales.

Rostrum

Refers to the beak or snout of whale; includes the upper jaw.

Schools

Structured social groups observed in toothed whales characterized by long-term associations of individuals. Examples are various delphinid species.

Sonar

High-frequency sound system used by toothed whales to echolocate.

Spyhopping

Behavior in which a whale raises its head above the water, likely to permit a view of potential prey or other whales.

Tail stock

See Peduncle.

Toothed whale

One of two major groups (suborders) of whales that possess teeth. *See also* Odontocete

Transient

Refers to an ecotype of killer whale that moves from one place to another; typically refers to killer whales that range along the west coast of North America. This ecotype differs from residents in morphology, genetics, and ecology. For example, transient killer whales feed almost exclusively on marine mammals.

Umbilicus

Refers to the scar for attachment of the umbilical cord to the fetus.

Ventral

Positioned on the whale's underside.

Vibrissae

Whiskers on the head of some whales that likely serve a sensory role.

Wake-riding

Refers to whales swimming in the wake of a boat.

Whale lice

A type of crustacean found attached to the skin of some whales, especially mysticetes, and that feeds on the upper layer of whale skin. Whale lice are not actually related to the lice of land mammals.

Zooplankton

Refers to animal plankton that are found in the upper portions of the ocean.

RESOURCES

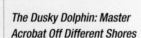

BOOKS

Biology of Marine Mammals
Edited by John E. Reynolds III
and Sentiel A. Rommel
(Smithsonian Books, 1999)

The Bowhead Whale
Edited by John J. Burns, J. Jerome
Montague, and Cleveland J. Cowles
(Special Publication Number 2. Society
for Marine Mammalogy, 1993)

**Cetacean Societies: Field
Studies of Dolphins and Whales**
Edited by Janet Mann,
Richard C. Connor, Peter L. Tyack,
and Hal Whitehead
(University of Chicago Press, 2000)

**A Complete Guide
to Antarctic Wildlife**
Hadoram Shirihai
(A & C Black, 2007)

**Conservation and Management
of Marine Mammals**
Edited by John R. Twiss
and Randall R. Reeves
(Smithsonian Institution Press, 1999)

**The Cultural Lives of Whales
and Dolphins**
Hal Whitehead and Luke Rendell
(University of Chicago Press, 2014)

**The Dolphin and Whale
Career Guide**
Thomas B. Glen (Omega Publishing
Company, Chicago, 1997)

**Dolphins Down Under: Understanding
the New Zealand Dolphin**
Liz Slooten and Steve Dawson
(Otago University Press, 2013)

**The Dusky Dolphin: Master
Acrobat Off Different Shores**
Edited by Bernd Würsig
and Melany Würsig
(Academic Press, 2009)

Encyclopedia of Marine Mammals
Edited by William F. Perrin, Bernd
Würsig, & J.G.M. Thewissen
(Academic Press, 2nd edition, 2009)

**Handbook of Marine Mammals,
Volume 3: The Sirenians
and Baleen Whales**
Edited by Sam H. Ridgway
and Sir Richard Harrison
(Academic Press, 1985)

**Handbook of the Mammals of
the World, Vol 4: Sea Mammals**
Edited by Don E. Wilson
and Russell A. Mittermeier
(Lynx Edicions, 2014)

**Marine Mammals: Evolutionary
Biology**
Annalisa Berta, James L. Sumich,
and Kit M. Kovacs
(Academic Press, 3rd edition, 2015)

**Marine Mammals of the World:
A Comprehensive Guide to their
Identification**
Thomas A. Jefferson, Marc A. Webber,
and Robert L. Pitman
(Academic Press, 2nd edition, 2015)

**Marine Mammal Research:
Conservation Beyond Crisis**
Edited by John E. Reynolds, William
F. Perrin, Randall R. Reeves, Suzanne
Montgomery and Tim J. Ragen
(Johns Hopkins University Press, 2005)

**National Audubon Society Guide
to Marine Mammals of the World**
Randall R. Reeves, Brent S. Stewart,
Phillip J. Clapham, and James
A. Powell (Knopf, 2002)

**Return to the Sea: The Life
and Evolutionary Times
of Marine Mammals**
Annalisa Berta
(University of California Press, 2012)

**Starting Your Career
as a Marine Mammal Trainer**
Terry S. Samansky
(DolphinTrainer.com, 2002)

**The Urban Whale: North Atlantic
Right Whales at the Crossroads**
Edited by Scott D. Kraus
and Rosalind M. Rolland
(Harvard University Press, 2007)

**The Walking Whales: From Land
to Water in Eight Million Years**
J.G.M. 'Hans' Thewissen
(University of California Press, 2014)

Whales, Dolphins, and Porpoises
Mark Carwardine
(Dorling Kindersley, 2002)

**Whales and Dolphins of the
Southern African Subregion**
Peter B. Best
(Cambridge University Press, 2009)

**Whales, Whaling, and Ocean
Ecosystems**
Edited by James A. Estes, Douglas
P. DeMaster, Daniel F. Doak, Terrie
M. Williams, and Robert L. Brownell
(University of California Press, 2007)

USEFUL WEB SITES

American Cetacean Society
acsonline.org
The mission of the American Cetacean
Society is to protect whales, dolphins,
porpoises, and their habitats through
public education, research grants, and
conservation actions.

Cascadia Research Collective
www.cascadiaresearch.org
Research on endangered marine
mammals. This site contains useful
fact sheets and information.

**Cetacean Ecology, Behavior,
and Evolution Lab (CEBEL)**
www.cebel.com.au
A multi-disciplinary group that works
at the interface of animal behavior,
population ecology, and evolutionary
biology to understand the structure,
dynamics, history, and trajectories of
cetacean populations.

European Cetacean Society (ECS)
www.EuropeanCetaceanSociety.eu
Professional biologists and others
interested in whales and dolphins.

Hawaiian Odontocetes
www.cascadiaresearch.org/Hawaii/
species.htm
Research and observations
for tropical odontocetes.

**IUCN Red List of Threatened
Species**
www.iucnredlist.org/
Provides taxonomic, conservation
status, and distribution information on
plants, fungi, and animals.

**Murdoch University's Cetacean
Research Unit (MUCRU)**
www.mucru.org
An academic research group focusing
on the conservation and management
of cetaceans and dugongs. The group
strives to design and conduct rigorous
applied and empirical research that
supports industry and government in
meeting their environmental, regulatory,
and statutory responsibilities.

The Society for Marine Mammalogy
www.marinemammalscience.org
Marine Mammal Society's Committee
on Taxonomy—the authoritative list is
updated annually.

NOTES ON CONTRIBUTORS

EDITOR

Annalisa Berta has been Professor of Biology at San Diego State University, California for more than 30 years specializing in the anatomy and evolutionary biology of marine mammals. Past President of the Society of Vertebrate Paleontology and co-Senior Editor of the *Journal of Vertebrate Paleontology*, Berta has authored/co-authored numerous scientific articles and several books for the specialist and non-scientist, including *Return to the Sea: The Life and Evolutionary Times of Marine Mammals* (University of California Press, 2012) and *Marine Mammals: Evolutionary Biology, 3rd Ed.* (Academic Press, 2015).

Contribution: Introduction pp. 6–7, Phylogeny & Evolution pp.10–15, Anatomy & Physiology pp.16–19, How to Use the Species Directory pp. 64–5, Glossary pp. 278–81.

CONTRIBUTORS

Alex Aguilar is Professor and Director of the Institute of Biodiversity Research of the University of Barcelona, Spain. His research focuses primarily on population biology and conservation of marine mammals, although he has also conducted studies on the ecotoxicology, anatomy, and physiology of these animals, as well as on the history of whaling.

Contribution (with Morgana Vighi): Fin Whale pp. 100–101.

Renee Albertson completed her Ph.D. in 2014 at Oregon State University (OSU) where her research focused on the population structure and phylogeography of rough-toothed dolphins. She currently teaches several courses at OSU and continues to study the population structure of cetaceans, including humpback whales.

Contribution: Rough-Toothed Dolphin pp. 178–9.

Sarah G. Allen has studied marine birds and mammals for more than 30 years, mostly in California and for two seasons in Antarctica. Currently, she is the Ocean and Coastal Resources Program lead, for the Pacific West Region of the U.S. National Park Service. She received her B.S., M.S., and Ph.D. from the University of California, Berkeley. She co-authored the *Field guide to the Marine Mammals of the Pacific Coast: Baja, California, Oregon, Washington, British Columbia* (UC Press, California Natural History Guide Series).

Contribution: Rorqual Whales & Gray Whale introduction pp. 80–81, Common Minke Whale pp. 86–7, Sei Whale pp.90–91, Blue Whale pp. 94–5, Omura's Whale pp. 98–9.

Masao Amano is Professor of Marine Mammalogy at Nagasaki University, Japan. He is currently interested in the behavior and social structure of sperm whales and bottlenose dolphins, and in the biology of finless porpoises around the Nagasaki area.

Contribution: Narrow-Ridged Finless Porpoise pp. 262–3.

Ana Rita Amaral has a Ph.D. in Evolutionary Biology and is currently a Postdoctoral Research Fellow at the Center for Ecology, Evolution, and Environmental Changes (University of Lisbon, Portugal) and at the American Museum of Natural History (USA). She has been pursuing research on the evolutionary history and molecular ecology of the Delphininae.

Contribution: Long-Beaked Common Dolphin pp. 116–17, Short-Beaked Common Dolphin pp. 118–19, Clymene Dolphin pp. 170–71.

Jessica Aschettino completed her Masters degree in 2010 at Hawaii Pacific University where she researched the population size and structure of melon-headed whales in Hawaii. She is a Reseach Associate with Cascadia Research Collective and has extensive field experience with tropical odontocetes. She currently works as a marine scientist and project manager performing research and monitoring in the Atlantic and Pacific Oceans.

Contribution: Pygmy Killer Whale pp. 120–21, Short-Finned Pilot Whale pp. 122–3, Long-Finned Pilot Whale pp. 124–5, Melon-Headed Whale pp. 152–3, False Killer Whale pp. 154–5, Pantropical Spotted Dolphin pp. 168–9, Spinner Dolphin pp. 176–7.

Lars Bejder is Professor at Murdoch University, Australia, and an Adjunct Associate Professor at Duke University, USA. His research interests include analysis and development of quantitative methods to evaluate complex animal social structures, evaluation of the impacts of human activity on cetaceans (coastal development, tourism, habitat degradation), and fundamental biology and ecology, including assessing abundance and habitat use of marine wildlife. He works closely with wildlife management agencies to optimize the conservation and management outcomes of his research.

Contribution (with Krista Nicholson): Indo-Pacific Bottlenose Dolphin pp. 180–81.

Susan Chivers is a Research Scientist at NOAA's Southwest Fisheries Science Center in La Jolla, California. For more than 30 years, Susan has studied cetacean populations in the eastern North Pacific, specializing in life history studies of small dolphin species.

Contribution: Life History pp. 32–9.

Frank Cipriano is Director of the GTAC—a core molecular research facility at San Francisco State University, California. His research interests include the systematics, behavioral ecology, and conservation of cetaceans, development of tools for molecular forensics and species identification, and the factors influencing evolution, diversification, and speciation in marine ecosystems.

Contribution: Atlantic White-Sided Dolphin pp. 130–31.

Rochelle Constantine is the Director of the Joint Graduate School in Coastal and Marine Science at the University of Auckland, New Zealand. She runs the Marine Mammal Ecology Group and takes a multi-disciplinary, collaborative research approach to understanding the ecology of whales and dolphins ranging from the South Pacific Islands to Antarctica.

Contribution: Bryde's Whale pp. 92–3.

Steve Dawson received his Ph.D. in 1990 for his work on Hector's dolphin. His work, and that of his graduate students, focuses on the conservation biology, ecology, and acoustic behavior of dolphins and whales. He and his partner Professor Elisabeth Slooten lead long-term studies of Hector's dolphin (30 years), sperm whales (25 years), and Fiordland bottlenose dolphins (25 years). Dawson is a Professor in the Marine Science Department, University of Otago, New Zealand.

Contribution: Commerson's Dolphin pp. 108–109, Chilean Dolphin pp. 110–11,

NOTES ON CONTRIBUTORS

Heaviside's Dolphin pp. 112–13, Hector's Dolphin pp. 114–15.

Natalia A. Dellabianca is a biologist, Assistant Researcher at the Centro Austral de Investigaciones Científicas (CADIC-CONICET), Ushuaia, Tierra del Fuego, Argentina, and a member of the research group of the AMMA project. Her main research interests include the effects of climate variability on marine mammals of southern South America and Antarctica.

Contribution (with Natalie Goodall): Peale's Dolphin pp. 134–5, Hourglass Dolphin pp. 136–7.

Louella Dolar is Adjunct Professor at the Institute for Environmental and Marine Sciences (IEMS), Silliman University, Philippines. Her research interests include marine mammal ecology and natural history, marine mammal–fishery interactions, fish biology, and fishery assessments.

Contribution: Fraser's Dolphin pp. 128–9.

Eric Ekdale is a Research Scientist and Adjunct Instructor in the Department of Biology at San Diego State University, California. His research interests include the evolution of sensory and feeding systems in baleen whales, as well as the evolutionary relationships among marine mammal species.

Contribution: Identification Keys pp. 48–55, Right Whales introduction pp. 66–7, Pygmy Right Whale pp. 78–9, Oceanic Dolphins introduction pp.

106–107, Narwhal & Beluga introduction pp. 194–5, Narwhal pp. 196–9, River Dolphins introduction pp. 250–51, Porpoises introduction pp. 260–61.

Ari S. Friedlaender is an Associate Professor at Oregon State University and his research focuses on using tag technology to study the foraging ecology of marine mammals around the world. He is particularly interested in the feeding behavior of baleen whales, specifically in Antarctica.

Contribution: Food & Foraging pp. 26–31, Range pp. 40–41, Habitat pp. 42–3.

Anders Galatius is a researcher at Aarhus University, Denmark and received his Ph.D. from the University of Copenhagen 2009. He works on the morphology, evolution, and ecology of dolphins, porpoises, and seals. He is a member of several expert groups on marine mammal ecology.

Contribution (with Carl Kinze): White-Beaked Dolphin pp. 132–3, Spectacled Porpoise pp. 266–7, Harbor Porpoise pp. 268–9, Burmeister's Porpoise pp. 272–73, Dall's Porpoise pp. 274–5.

R. Natalie P. Goodall is a botanist, zoologist, and founder and director of the Museo Acatushún de Aves y Mamíferos Marinos Australes (AMMA) at Estancia Harberton, Tierra del Fuego, Argentina, which houses a large collection of marine mammals

and birds of southernmost South America. Her museum trains 8–10 university-level interns per month during the warmer months.

Contribution (with Natalia Dellabianca): Peale's Dolphin pp. 134–5, Hourglass Dolphin pp. 136–7.

Armando Jaramillo-Legorreta is a researcher for the Marine Mammals Research and Conservation Group of the National Institute of Ecology in charge of the study of habitat use and acoustic monitoring of the vaquita population. He is the current President of the Mexican Society of Marine Mammals.

Contribution: (with Lorenzo Rojas-Bracho): Vaquita pp. 270–1.

Robert D. Kenney is mostly retired from the research faculty at the University of Rhode Island's Graduate School of Oceanography, where he spent more than three decades studying North Atlantic right whales and other charismatic marine megafauna.

Contribution: Southern Right Whale pp. 68–9, North Atlantic Right Whale pp. 70–3, North Pacific Right Whale pp. 74–5, Bowhead Whale pp. 76–7.

Carl Chr. Kinze is a cetacean specialist who holds a Ph.D. and works in the fields of taxonomy, nomenclature, history of cetology, and zoogeography. He resides in Copenhagen, Denmark.

Contribution (with Anders Galatius): White-Beaked Dolphin pp. 132–3,

Spectacled Porpoise pp. 266–7, Harbor Porpoise pp. 268–9, Burmeister's Porpoise pp. 272–3, Dall's Porpoise pages 274–5.

Tony Martin is Professor of Animal Conservation at the University of Dundee, U.K. Prior to this, he worked on aquatic mammals and seabirds at the Sea Mammal Research Unit and British Antarctic Survey, with emphasis on polar and equatorial regions.

Contribution (with Vera da Silva): Tucuxi pp. 156–7, Guiana Dolphin pp. 158–9, Franciscana pp. 254–5, Amazon River Dolphin pp. 256–7.

Krista Nicholson is affiliated with Murdoch University where she earned her M.Sc. studying bottlenose dolphins of Shark Bay, Western Australia. Her main research interests are the improvement of methods to investigate population demographics for better conservation and management of cetacean populations.

Contribution (with Lars Bejder): Indo-Pacific Bottlenose Dolphin pp. 180–81.

Greg O'Corry-Crowe is a molecular and behavioral ecologist who combines genetic analysis with field techniques to study marine mammals including the beluga whale. He is particularly interested in investigating the effects of ecosystem and climate change on marine apex predators and in applying both scientific research and local

knowledge to effective co-management and conservation.

Contribution: Beluga pp. 200–203.

Guido J. Parra is a Colombian-born biologist who currently leads the Cetacean Ecology, Behavior, and Evolution Lab (CEBEL) within the School of Biological Sciences, Flinders University, South Australia. He has broad research interests in population ecology, behavioral ecology, and conservation biology. His research seeks to understand marine mammal ecology, behavior, and evolution to address pressing conservation issues.

Contribution: Australian Snubfin Dolphin pp. 148–9, Indo-Pacific Humpback Dolphin pp. 160–61, Indian Humpback Dolphin pp. 162–3, Australian Humpback Dolphin pp. 164–5, Atlantic Humpback Dolphin pp. 166–7.

Heidi Pearson is an Assistant Professor of Marine Biology at the University of Alaska Southeast. Her research focuses on the behavior and ecology of marine mammals with an emphasis on dusky dolphins, humpback whales, and sea otters. Her particular interests include the evolution of social structure and the ecosystem services provided by marine mammals.

Contribution: Behavior pp. 20–5, Conservation & Management pp. 44–5, Surface Behaviors pp. 56–9, How & Where to Watch pp. 60–61, Common Bottlenose Dolphin pp. 182–3.

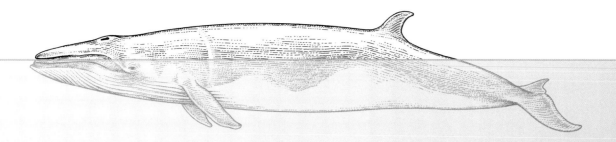

Robert Pitman is a marine ecologist at Southwest Fisheries Science Center, La Jolla, California, and during the last 40 years he has conducted marine mammal surveys throughout the world. Currently his work focuses on the status and ecology of killer whales in Australia and Antarctica.

Contribution: Killer Whale pp. 150–51.

Randall Reeves is a consultant based in Hudson, Quebec, Canada, who has served since 1996 as chairman of the IUCN Cetacean Specialist Group. His main areas of interest and expertise are marine mammal biology and conservation. He currently also serves as chair of the Committee of Scientific Advisers of the U.S. Marine Mammal Commission and the IUCN Western Gray Whale Advisory Panel.

Contribution: Northern Right Whale Dolphin pp. 142–3, Southern Right Whale Dolphin pp. 144–5, Striped Dolphin pp. 172–3, Atlantic Spotted Dolphin pp. 174-5, Beaked Whales introduction pp. 204–205, Arnoux's Beaked Whale pp. 206–207, Baird's Beaked Whale pp. 208–209, Northern Bottlenose Whale pp. 210–211, Southern Bottlenose Whale pp. 212–13, Longman's Beaked Whale pp. 214–15, Sowerby's Beaked Whale pp. 216–17, Andrews' Beaked Whale pp. 218–19, Hubbs' Beaked Whale pp. 220–21, Blainville's Beaked Whale pp. 222–3, Gervais' Beaked Whale pp. 224–5, Ginkgo-Toothed Beaked Whale pp. 226–7, Gray's Beaked Whale pp. 228–9, Hector's Beaked Whale pp.

230–31, Deraniyagala's Beaked Whale pp. 232-3, Strap-Toothed Whale pp. 234–5, True's Beaked Whale pp. 236–7, Perrin's Beaked Whale pp. 238–9, Pygmy Beaked Whale pp. 240–41, Stejneger's Beaked Whale pp. 242–3, Spade-Toothed Beaked Whale pp. 244–5, Shepherd's Beaked Whale pp. 246–7, Cuvier's Beaked Whale pp. 248–9, Indo-Pacific Finless Porpoise pp. 264–5.

Lorenzo Rojas-Bracho coordinates the Marine Mammal Research and Conservation in Mexico's INECC. He chairs the international recovery team of the vaquita (CIRVA) and has chaired different international committees, workshops, and working groups related to the management of cetaceans. He is a member of the CMS Scientific Council's Aquatic Mammals Working Group and the IUCN's Cetacean Specialist Group and The Red List Authority.

Contribution: Gray Whale pp. 82–5 (with Jorge Urbán Ramírez), Vaquita pp. 270–71 (with Armando Jaramillo-Legorreta).

Vera M. F. da Silva is a researcher at the National Institute for Amazonian Research in Brazil and Head of the Aquatic Mammals Lab and a member of the CSG/IUCN. She is an active field biologist and coordinator of the Projeto Boto, a long-term project for the conservation of river dolphins in Brazil.

Contribution (with Tony Martin): Tucuxi pp. 156–7, Guiana Dolphin pp. 158–9,

Franciscana pp. 254–5, Amazon River Dolphin pp. 256–7.

Fred Sharpe received his Ph.D. in Behavioral Ecology at Simon Fraser University, Canada and has been studying humpback whales in Southeast Alaska since 1987. Fred is a Level Four large whale disentangler and is a recipient of the Fairfield Award for Innovative Marine Mammal Research and the Society for Marine Mammalogy's Science Communication Award.

Contribution: Humpback Whale pp. 102–105.

Brian D. Smith has directed the Wildlife Conservation Society's Asian Cetacean Program since 2001 and currently serves as the Asia Coordinator of the IUCN Species Survival Commission Cetacean Specialist Group and a member of the Society for Marine Mammalogy Conservation Committee.

Contribution: Irrawaddy Dolphin pp. 146–7.

Mridula Srinivasan is a Marine Ecologist and Co-Founder and Director, Marine Mammal Program at Terra Marine Research Institute (TeMI), India, a non-profit organization. TeMI is a non-profit organization based in Karwar, India, dedicated to studying and conserving India's marine and terrestrial biodiversity through community-invested solutions.

Contribution: Risso's Dolphin pp.

126–7, Pacific White-Sided Dolphin pp. 138–9, Dusky Dolphin pp. 140–41, Ganges River Dolphin pp. 258–9.

Jorge Urbán Ramírez is a member of the Scientific Committee of the International Whaling Commission (IWC) and the Cetacean Specialist Group of the International Union for Conservation of Nature (IUCN). Since 1988 Jorge has been the Coordinator of the Marine Mammal Research Program of the Universidad Autónoma de Baja California Sur (UABCS) at La Paz, México.

Contribution (with Lorenzo Rojas-Bracho): Gray Whale pp. 82–5.

Morgana Vighi is currently completing her Ph.D. at the University of Barcelona, Spain, investigating the population structure of North Atlantic fin whale and South Atlantic right whale and the effects of whaling exploitation on this last population. She received a M.Sc. in Marine Biology and she has been working since 2007 with cetaceans, regularly taking part in Mediterranean field expeditions and collaborating with different research projects focusing on various aspects of cetacean biology.

Contribution (with Alex Aguilar): Fin Whale pp. 100–101.

Kristi West is an Associate Professor of Biology at Hawaii Pacific University where she directs the university's marine mammal stranding program. Her research interests include determining

threats to cetacean population health in Hawaii and the greater Pacific.

Contribution: Sperm Whales introduction pp. 184–5, Sperm Whale pp. 186–9, Pygmy Sperm Whale pp. 190–91, Dwarf Sperm Whale pp. 192–3.

Alexandre Zerbini is a cetacean biologist with expertise in population abundance and assessment methods and in satellite tagging technology. He holds a Joint Associate research position with Cascadia Research Collective and the National Marine Mammal Laboratory, NOAA Fisheries in the U.S. He is also the Scientific Director of Instituto Aqualie, a non-profit organization in his home country of Brazil.

Contribution: Antarctic Minke Whale pp. 88–9.

Kaiya Zhou is a Professor at Nanjing Normal University and studied the baiji from the mid-1950s until it became extinct. His primary research interests are the conservation of dolphins and porpoises in Chinese waters.

Contribution: Baiji pp. 252–3.

INDEX

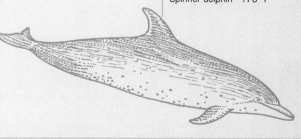

ACKNOWLEDGMENTS

ANNALISA BERTA

I want to thank the 37 whale biologists I had the pleasure of working with over the last two years for providing accurate species accounts and fascinating highlights of whale biology. Thanks, too to the artists for their detailed illustrations and the photographers for selecting stunning images that convey the visual beauty of these magnificent mammals of the sea. Robert Boessenecker is acknowledged for his advice on the life restorations of extinct whales. Finally, I am grateful to the entire production team at Ivy Press: Kate Shanahan, Commissioning Editor; Caroline Earle, Senior Project Editor and copy-editor Ruth O'Rourke-Jones for their expert guidance, organization and editorial skills.

PICTURE CREDITS

The publisher would like to thank the following individuals and organizations for their kind permission to reproduce the images in this book. Every effort has been made to acknowledge the pictures; however, we apologize if there are any unintentional omissions.

Alamy: 73t: /© Naturfoto-Online.

Corbis: 72t: /©Barrett & MacKay/All Canada Photos.

Ari S. Friedlaender: 26, 28r, 31, 40, 43tl.

Getty Images: 24 /Barcroft / Alexander Sofonov.

J. C. Lanaway: 61b, 202r.

Sergio Martinez/PRIMMA-UABCS: 84t.

Nature Picture Library: 195: /Doug Allan; 67: /Franco Banfi; 18tr, 34, 34–35, 36, 37bl, 37br, 44tr, 48t, 69t (small repeat, 54), 71t (small repeat, 54, 60), 75t (small repeat 54), 75t (small repeat 11, 54) 77t (small repeat 11, 54), 77b, 79t (small repeat 11, 53), 83t (small repeat 11, 54), 83c, 85b, 87t (small repeat 53), 89b, 91t (small repeat 55), 91b, 93t (thumbnail repeat 55, 60), 93b, 95t (thumbnail page 55), 97b, 101t (thumbnail repeat 55), 101b, 103t (small repeat 55, 61), 105b, 109b, 117t (small repeat 10, 50), 117b, 119t (small repeat page 50), 119b, 121t (small repeat 51), 123t (small repeat 53), 125t (small repeat 53), 127t (small repeat 51),129t (small repeat page 51), 129b, 131t (small repeat 51), 131b, 133t (small repeat 51, 61), 133b, 135t (small repeat 51), 137t (small repeat 51), 139t (small repeat 51), 141t (small repeat 51), 143t (small repeat 50), 143b, 145t (small repeat 50), 145b, 147t (small repeat 51), 147b, 151t (small repeat 53, 60), 153t (small repeat 51), 153b, 155t (small repeat 53), 155b, 161t (small repeat 50), 167t (small repeat 50), 167b, 169t (small repeat 50), 169b, 171t (small repeat 50), 171b, 173t (small repeat 50), 175t (small repeat 50), 175b, 177t (small repeat 50), 177b, 179t (small repeat 50), 179b, 181t (small repeat 50), 183t (small repeat 50, 61), 183b, 187t (small repeat 10, 54), 187c, 189b, 191t (small repeat 51), 191b, 193t (small repeat 10, 51), 193b, 197t (small repeat 10, 53), 201t (small repeat 53), 203b, 207t (small repeat 52), 209t (small repeat 52), 211t (small repeat 10, 52), 213t (small page 52), 215t (small repeat 52), 219t (small repeat 52), 221t (small repeat 52), 235t (small repeat 52), 253t (small repeat 10, 50), 259t (small repeat 10, 50), 267t (small repeat 51), 267b, 269t (small repeat 10, 51), 271t (small repeat 51), 271b, 273t (small repeat 51), 273b, 275t (small repeat 51), 275b: /Martin Camm (WAC); 44l, 49tc, 96, 251: /Mark Carwardine; 107, 185: /Brandon Cole; 59r: /Armin Maywald; 21, 58bl, 81, 105t, 189t /Doug Perrine; 57l, 205: /Todd Pusser; 188: /Luis Quinta; 64, 69b, 72–73b, 75b, 79b, 99t (small repeat 11, 53), 99b,111b, 113b, 115b, 121b, 123b, 125b, 127b, 135b, 137b, 139b, 141b, 149t (small repeat 51), 149b, 151b, 157b, 159t (small repeat 50), 159b, 161b, 163t (small repeat 50), 163b, 165t (small repeat 50), 165b, 173b, 181b, 199b, 207b, 209b, 211b, 213b, 215b, 217b, 221b, 223b, 225b, 227b, 229b, 231b, 233t (small repeat 52), 233b, 253b, 237b, 239b, 241b, 243b, 245t (thumbnail repeat 52), 247b, 249b, 253b, 255b, 257b, 259b, 263t (small page 51), 263b, 265b, 269b: /Rebecca Robinson; 59l: /Gabriel Rojo; 95bl, 97t, 104, 198, 199t, 202bl, 203t: /Doc White; 261: /Solvin Zankl.

NOAA, NMFS, Southwest Fisheries Science Center: 33br; 38; 39r: /W. Perryman; 32–33, 33tl: /D. Weller.

Dan Olsen, North Gulf Oceanic Society: 25bl.

Richard Palmer: 11tl, 12 (fossil whale restorations after Carl Buell), 14, 15 (after Robert Boessenecker), 16–17, 18cl, 19, 20, 22, 27, 28l, 29, 30, 35r, 45, 56, 58r.

Christopher Pearson: 25tl.

Heidi Pearson: 23, 25tr, 25br, 57r, 58tl.

Sandra Pond: 4–5, 18bl, 18br, 48b, 49r, 66 (all), 69cl, 71b, 72br, 75cl, 77cl, 79cl, 80 (all), 83bl, 87cl, 91cl, 93cl, 95br, 99cl, 101cl, 103bl, 103br, 106 (all), 109t (small repeat 51), 109cl, 111t (small repeat 51), 111cl, 113t (small repeat 51), 113cl, 115t (small repeat 51), 115cl, 117cl, 119cl, 121cl, 123cl, 125cl, 127cl, 129cl, 131cl, 133cl, 135cl, 137cl, 139cl, 141cl, 143cl, 145cl, 147cl, 149cl, 151cl, 153cl, 155cl, 157t (small repeat 50), 157cl, 159cl, 161cl, 163cl, 165cl, 167cl, 169cl, 171cl, 173cl, 175cl, 177cl, 179cl, 181cl, 183cl, 184 (all), 187b, 191cl, 193cl, 194 (all), 197bl, 197br, 201bl, 204 (all), 207cl, 209cl, 211cl, 213cl, 215cl, 217t (small repeat 52), 217cl, 219cl, 219br, 221cl, 223t (small repeat 52), 223cl, 225t (small repeat 52), 225cl, 227t (small repeat 52), 227cl, 229t (small repeat 52), 229cl, 231t (small repeat 52), 231cl, 233cl, 235cl, 237t (small repeat 52), 237cl, 239t (small repeat 52), 239cl, 241t (small repeat 52), 241cl, 243t (small repeat 52), 243cl, 245cl, 245br, 247t (small repeat 52), 247cl, 249t (small repeat 52), 249cl, 250 (all), 253cl, 255t (small repeat 10, 50), 255cl, 257t (small repeat 10, 50, 60), 257cl, 259cl, 260 (all), 263cl, 265t (small repeat 51), 265cl, 267cl, 269cl, 271cl, 273cl, 275cl, 276–287.

Nick Rowland: 13 (modified from Marx and Uhen, 2010), 37t (modified from Berta et al., 2015), 39l, 41, 60–61 (map).

Shutterstock: 43tc: /James Michael Dorsey; 43tr: /guentermanaus; 46–47: /Alberto Loyo; 2: /Tom Middleton; 43br: /Jonathan Nafzger; 8–9, 43bl: /pierre_j; 42: /Kristina Vackova; 7: /Chris G. Walker; 62–63: /Paul S. Wolf; 6: /Igor Zh.

Jorge Urbán/PRIMMA-UABCS: 84b, 85t.